Springer Series in Advanced Manufacturing

Series Editor

Professor D.T. Pham
Intelligent Systems Laboratory
WDA Centre of Enterprise in Manufacturing Engineering
University of Wales Cardiff
PO Box 688
Newport Road
Cardiff
CF2 3ET
UK

Other titles in this series

Assembly Line Design
B. Rekiek and A. Delchambre

Advances in Design
H.A. ElMaraghy and W.H. ElMaraghy (Eds.)

Effective Resource Management in Manufacturing Systems:
Optimization Algorithms in Production Planning
M. Caramia and P. Dell'Olmo

Condition Monitoring and Control for Intelligent Manufacturing
L. Wang and R.X. Gao (Eds.)

Optimal Production Planning for PCB Assembly
W. Ho and P. Ji

Trends in Supply Chain Design and Management: Technologies and Methodologies
H. Jung, F.F. Chen and B. Jeong (Eds.)

Process Planning and Scheduling for Distributed Manufacturing
L. Wang and W. Shen (Eds.)

Collaborative Product Design and Manufacturing Methodologies and Applications
W.D. Li, S.K. Ong, A.Y.C. Nee and C. McMahon (Eds.)

Decision Making in the Manufacturing Environment
R. Venkata Rao

Frontiers in Computing Technologies for Manufacturing Applications
Y. Shimizu, Z. Zhang and R. Batres

Reverse Engineering: An Industrial Perspective
V. Raja and K.J. Fernandes (Eds.)

Automated Nanohandling by Microrobots
S. Fatikow

A Distributed Coordination Approach to Reconfigurable Process Control
N.N. Chokshi and D.C. McFarlane

ERP Systems and Organisational Change
B. Grabot, A. Mayère and I. Bazet (Eds.)

ANEMONA
V. Botti and A. Giret

Theory and Design of CNC Systems
S.H. Suh, S.-K. Kang (et al.)

Kai Cheng
Editor

Machining Dynamics

Fundamentals, Applications and Practices

 Springer

Kai Cheng, BEng, MSc, PhD, FIET, MIMechE
Advanced Manufacturing & Enterprise
Engineering (AMEE) Department
School of Engineering and Design
Brunel University
Middlesex UB8 3PH
UK

ISBN 978-1-84628-367-3   ·   e-ISBN 978-1-84628-368-0

DOI 10.1007/978-1-84628-368-0

Springer Series in Advanced Manufacturing ISSN 1860-5168

British Library Cataloguing in Publication Data
Machining dynamics : fundamentals, applications and
  practices. - (Springer series in advanced manufacturing)
  1. Machine-tools - Dynamics 2. Machining
  I. Cheng, Kai
  621.9'02
ISBN-13: 9781846283673

Library of Congress Control Number: 2008923551

© 2009 Springer-Verlag London Limited

Apart from any fair dealing for the purposes of research or private study, or criticism or review, as permitted under the Copyright, Designs and Patents Act 1988, this publication may only be reproduced, stored or transmitted, in any form or by any means, with the prior permission in writing of the publishers, or in the case of reprographic reproduction in accordance with the terms of licences issued by the Copyright Licensing Agency. Enquiries concerning reproduction outside those terms should be sent to the publishers.

The use of registered names, trademarks, etc. in this publication does not imply, even in the absence of a specific statement, that such names are exempt from the relevant laws and regulations and therefore free for general use.

The publisher makes no representation, express or implied, with regard to the accuracy of the information contained in this book and cannot accept any legal responsibility or liability for any errors or omissions that may be made.

Cover design: eStudio Calamar S.L., Girona, Spain

Printed on acid-free paper

9 8 7 6 5 4 3 2 1

springer.com

Books are to be returned on or before
the last date below.

Preface

Machining dynamics plays an essential role in the performance of machine tools and machining processes, which directly affects the material removal rate, and workpiece surface quality as well as dimensional and form accuracy. However, despite its obvious technical and economic importance and tremendous progress in machining technology during the last few decades, machining dynamics still remains as one of the least understood manufacturing science topics. In industrial practices, machining parameters are still chosen primarily through empirical testing and the experience of machine operators and programmers. This approach is costly, and while databases have been developed from large numbers of empirical tests, these databases lose relevance as new tools, machines and workpiece materials are developed and applied. Furthermore, a better understanding of machining dynamics is becoming increasingly important for engaging in ultraprecision and micro manufacturing because of the machining accuracy, scale and complexity involved. Therefore, it is essential to systematically research the machining dynamics within the material removal and surface generation processes and machine operations with particular respect to the quantitative effects from machine tools, tooling, process variables and workpiece materials.

The advances in computational modelling, sensors, diagnostic equipment and analysis tools, surface metrology, and manufacturing science during the past decade have enabled academia and engineers to research the machining dynamics from a new dimension and therefore to have the potential for great industrial benefits, for instance, including:

- Analysis of the material removal dynamics, particularly the effects of cutting speeds and tooling geometry on the stress and temperature conditions at the tool-workpiece interface and thus the surface integrity and functionality.
- Multi-body dynamic analysis of the machine tool structure including the dynamic properties of interfaces between components such as spindles, slideways and drive systems, etc.

- Design of machine tool structures for dynamic repeatability, which is important in predictive control of the machine dynamic performance.
- Dynamic modelling of the machine systems (machine and machining processes) and on line/real time identification of the system modal parameters.
- Development of analytical solutions for the stability of complex contours machining and nonlinear models of interrupted machining.
- Development of novel algorithms (integrated with existing CAD/CAM/CAE tools) for compensation control of machining errors at real time.
- Ultraprecision and micromachining of various engineering materials with predictability, producibility and productivity.
- Modelling, simulation, control and optimization of precision machined surfaces including their surface texture, topography, integrity and functionality generation and formation.

This book aims to provide the state of the art of research and engineering practice in machining dynamics which is becoming increasingly important in modern manufacturing engineering. The book is concerned with machining dynamics in a comprehensive systematic manner and utilizing it proactively in manufacturing practice.

The advances in precision/ultraprecision machining, high speed machining, micro manufacturing, and computational modelling and analysis tools that have led to machining dynamics in the new context are the subject of the first chapter. The machine-tool-workpiece loop stiffness can place deterministic effects on the machining system's performance. Scientific understanding and comprehension of fundamentals of the loop and its dynamic behaviour in the process is central to the progress of this technology. Basic concepts and theory of machining instability and dynamics associated with the loop are therefore formulated in Chapter 2. Further advancements in the technology can be aided through a generalized theoretical understanding, scientific diagnostics and experimental analysis of machining dynamics as presented in Chapters 3 and 4. Following up those, a series of investigations are discussed on dynamics in tooling design, various machining processes, and design of precision machines. First, tooling design, tool wear and tool life are presented in Chapter 5. Machining dynamics in turning, milling and grinding processes are then studied in Chapters 6, 7 and 8, respectively. With the inexorable transition from conventional and precision machining, to ultraprecision and micro/nano machining, micro machining dynamics are starting to attract attention. Chapter 9 is devoted to the dynamics in ultraprecision machining using a single point diamond tool and the associated impact on nano-surface generation. Chapter 10 provides a dynamics-driven approach to precision machines design and thorough discussions on its implementation and application perspectives.

Owing to the diverse character of the subject, a single notation for the book has been difficult to achieve. For ease of working, therefore, a list of principal symbols and their meanings is included in the appropriate chapters as needed.

The diversity of the subject of machining dynamics has required that specialists in each of its main fields should prepare the chapters of this book. The comprehensive interest in the subject is evident, with 16 authors coming from 12 academic and industrial institutions. I am grateful to them all, for the benefit of their advice and expertise, and their patience in supplying with me their specialist chapters, and in many cases for lengthy subsequent dialogues.

This book can be used as a textbook for a final year elective subject on manufacturing engineering, or as an introductory subject on machining technology at the postgraduate level. It can also be used as a textbook for teaching advanced manufacturing technology in general. The book can also serve as a useful reference for manufacturing engineers, production supervisors, and planning and application engineers, as well as industrial engineers.

At Brunel University, I am indebted to my colleagues Dr Dehong Huo, Ms Sara Sun, Khalid Nor, Lei Zhou and Dr Rhys Morgan for their assistance in checking many of the details of the chapters. At the publisher, Springer-Verlag London Ltd, I have been appreciative of the support from Simon Rees, Anthony Doyle, Cornelia Kresser and Nicolas Wilson, as the book has developed from its draft outline form through various stages of its production.

Finally and most importantly, my greatest thanks have to be reserved for my wife, Lucy Lu, and Mike Cheng for their steadfast support and interest throughout the preparation of the book.

Brunel University *Kai Cheng*
West London, UK

Contents

List of Contributors xvii

1 Introduction 1
   1.1 Scope of the Subject 1
   1.2 Scientific and Technological Challenges and Needs 2
   1.3 Emerging Trends 4
   References 6

2 Basic Concepts and Theory 7
   2.1 Introduction 7
   2.2 Loop Stiffness within the Machine-tool-workpiece System ... 7
      2.2.1 Machine-tool-workpiece Loop Concept 7
      2.2.2 Static Loop Stiffness 8
      2.2.3 Dynamic Loop Stiffness and Deformation 9
   2.3 Vibrations in the Machine-tool System 10
      2.3.1 Free Vibrations in the Machine-tool System 10
      2.3.2 Forced Vibrations 13
   2.4 Chatter Occurring in the Machine Tool System 15
      2.4.1 Definition 15
      2.4.2 Types of Chatters 16
      2.4.3 The Suppression of Chatters 16
   2.5 Machining Instability and Control 17
      2.5.1 The Conception of Machining Instability 17

    2.5.2 The Classification of Machining Instability 19

Acknowledgements 19

References 19

3 Dynamic Analysis and Control 21

  3.1 Machine Tool Structural Deformations 21

    3.1.1 Machining Process Forces 22

    3.1.2 The Deformations of Machine Tool Structures and Workpieces 30

    3.1.3 The Control and Minimization of Form Errors 39

  3.2. Machine Tool Dynamics 43

    3.2.1 Experimental Methods 43

    3.2.2 The Analytical Modelling of Machine Tool Dynamics 47

  3.3. The Dynamic Cutting Process 54

    3.3.1. Mechanic of Dynamic Cutting 55

    3.3.2. The Dynamic Chip Thickness and Cutting Forces 59

  3.4. Stability of Cutting Process 63

    3.4.1 Stability of Turning 64

    3.4.2. The Stability of the Milling Process 68

    3.4.3. Maximizing Chatter Free Material Removal Rate in Milling 74

    3.4.4. Chatter Suppression-Variable Pitch End Mills 79

  3.5. Conclusions 82

  References 83

4 Dynamics Diagnostics: Methods, Equipment and Analysis Tools 85

  4.1 Introduction 85

  4.2 Theory 86

    4.2.1 An Example 88

    4.2.2 The Substructure Analysis 90

  4.3 Experimental Equipment 92

    4.3.1 The Signal Processing 92

    4.3.2 Excitation Techniques 93

    4.3.3 The Measurement Equipment 93

    4.3.4 Novel Approaches 94

    4.3.5 In-process Sensors 96

    4.3.6 Dynamometers 96

    4.3.7 The Current Monitoring 97

    4.3.8 The Audio Measurement... 97

    4.3.9 Capacitance Probes 97

    4.3.10 Telemetry and Slip Rings... 98

    4.3.11 Fibre-optic Bragg Grating Sensors....................................... 98

4.4 Chatter Detection Techniques 98

    4.4.1 The Topography... 100

    4.4.2 The Frequency Domain.. 100

    4.4.3 Time Domain 105

    4.4.4 Wavelet Transforms 109

    4.4.5 Soft Computing 110

    4.4.6 The Information Theory 111

4.5 Summary and Conclusions 111

Acknowledgements 112

References 112

5 Tool Design, Tool Wear and Tool Life 117

5.1 Tool Design 118

    5.1.1 The Tool-workpiece Replication Model 118

    5.1.2 Tool Design Principles... 120

    5.1.3 The Tool Design for New Machining Technologies............... 123

5.2 Tool Materials 124

    5.2.1 High Speed Steel... 124

    5.2.2 Cemented Carbide... 124

    5.2.3 Cermet... 125

    5.2.4 Ceramics 125

    5.2.5 Diamond.. 126

    5.2.6 Cubic Boron Nitride.. 127

5.3 High-performance Coated Tools 127

    5.3.1 Tool Coating Methods 128

    5.3.2 The Cutting Performance of PVD Coated Tools 129

    5.3.3 The Cutting Performance of CVD Coated Tools 132

5.3.4 Recoating of Worn Tools 133
5.4 Tool Wear 133
    5.4.1 Tool Wear Classification 134
    5.4.2 Tool Wear Evolution 136
    5.4.3 The Material-dependence of Wear 138
    5.4.4 The Wear of Diamond Tools 139
5.5 Tool Life 142
    5.5.1 The Definition of Tool Life 142
    5.5.2 Taylor's Tool Life Model 142
    5.5.3 The Extended Taylor's Model 144
    5.5.4 Tool Life and Machining Dynamics 145
References 148

6 Machining Dynamics in Turning Processes 151
6.1 Introduction 151
6.2 Principles 151
    6.2.1 The Turning Process 153
6.3 Methodology and Tools for the Dynamic Analysis and Control 154
6.4 Implementation Perspectives 155
6.5 Applications 156
    6.5.1 The Rigidity of the Machine Tool, the Tool Fixture
    and the Work Material 156
    6.5.2 The Influence of the Input Parameters 162
6.6 Conclusions 164
References 164

7 Machining Dynamics in Milling Processes 167
7.1 Introduction 167
    7.1.1 Forced Vibration 167
    7.1.2 Self-excited Vibration 168
    7.1.3 The Scope of This Chapter 169
    7.1.4 Nomenclature in This Chapter 170
7.2 The Dynamic Cutting Force Model for Peripheral Milling 171
    7.2.1 Oblique Cutting 172

- 7.2.2 The Geometric Model of a Helical End Mill 173
- 7.2.3 Differential Tangential and Normal Cutting Forces 174
- 7.2.4 Undeformed Chip Thickness 175
- 7.2.5 Differential Cutting Forces in X and Y Directions 178
- 7.2.6 Total Cutting Forces in X and Y Directions 180
- 7.2.7 The Calibration of the Cutting Force Coefficients 181
- 7.2.8 A Case Study: Verification 186
- 7.3 A Dynamic Cutting Force Model for Ball-end Milling 186
  - 7.3.1 A Geometric Model of a Ball-end Mill 186
  - 7.3.2 Dynamic Cutting Force Modelling 188
  - 7.3.3 The Experimental Calibration of the Cutting Force Coefficients 194
  - 7.3.4 A Case Study: Verification 198
- 7.4 A Machining Dynamics Model 200
  - 7.4.1 A Modularisation of the Cutting Force 200
  - 7.4.2 Machining Dynamics Modelling 203
  - 7.4.3 The Surface Generation Model 205
  - 7.4.4 Simulation Model 207
- 7.5 The Modal Analysis of the Machining System 207
  - 7.5.1 The Mathematical Principle of Experimental Modal Analysis 208
  - 7.5.2 A Case Study 209
- 7.6 The Application of the Machining Dynamics Model 213
  - 7.6.1 The Machining Setup 213
  - 7.6.2 Case 1: Cut 13 214
  - 7.6.3 Case 2: Cut 14 219
- 7.7 The System Identification of Machining Processes 224
  - 7.7.1 The System Identification 225
  - 7.7.2 The Machining System and the Machining Process 226
  - 7.7.3 A Case Study 227
  - 7.7.4 Summary 231
- References 231

8 Machining Dynamics in Grinding Processes 233
- 8.1 Introduction 233

8.2 The Kinematics and the Mechanics of Grinding 236
    8.2.1 The Geometry of Undeformed Grinding Chips 236
8.3 The Generation of the Workpiece Surface in Grinding 242
8.4 The Kinematics of a Grinding Cycle 248
8.5 Applications of Grinding Kinematics and Mechanics 253
8.6 Summary 259
References 261

9 Materials–induced Vibration in Single Point Diamond Turning 263

9.1 Introduction 263
9.2 A Model-based Simulation of the Nano-surface Generation 264
    9.2.1 A Prediction of the Periodic Fluctuation of Micro-cutting Forces 265
    9.2.2 Characterization of the Dynamic Cutting System 269
    9.2.3 A Surface Topography Model for the Prediction
    of Nano-surface Generation 271
    9.2.4 Prediction of the Effect of Tool Interference 275
    9.2.5 Prediction of the Effect of Material Anisotropy 277
9.3 Conclusions 278
Acknowledgements 279
References 279

10 Design of Precision Machines 283

10.1 Introduction 283
10.2 Principles 284
    10.2.1 Machine Tool Constitutions 284
    10.2.2 Machine Tool Loops and the Dynamics of Machine Tools 288
    10.2.3 Stiffness, Mass and Damping 290
10.3 Methodology 293
    10.3.1 Design Processes of the Precision Machine 293
    10.3.2 Modelling and Simulation 295
10.4 Implementation 298
    10.4.1 Static Analysis 298
    10.4.2 Dynamic Analysis 298
    10.4.3 A General Modelling and Analysis Process Using FEA 300

10.5 Applications..303
    10.5.1 Design Case Study 1: A Piezo-actuator
    Based Fast Tool Servo System 303
    10.5.2 Design Case Study 2: A 5-axis Micro-milling/
    grinding Machine Tool 313
    10.5.3 Design Case Study 3: A Precision Grinding Machine Tool............317
Acknowledgements 320
References 320
Index..323

List of Contributors

Erhan Budak
Faculty of Engineering and Natural Sciences,
Sabanci University,
Tuzla, Istanbul 34956,
Turkey

C.F. Cheung
Ultra-Precision Machining Centre,
Department of Industrial and Systems Engineering,
The Hong Kong Polytechnic University,
Hung Hom, Kowloon, Hong Kong

Kai Cheng
Advanced Manufacturing and Enterprise Engineering (AMEE),
School of Engineering and Design,
Brunel University,
Uxbridge, Middlesex UB8 3PH, UK

Xun Chen
School of Mechanical, Materials and Manufacturing Engineering,
The University of Nottingham,
Nottingham NG7 2RD, UK

J. Paulo Davim
Department of Mechanical Engineering,
University of Aveiro,
Campus Santiago,
3810-193 Aveiro,
Portugal

E. O. Ezugwu
Machining Research Centre,
Department of Engineering Systems,
London South Bank University,
London SE1 0AA, UK

Dehong Huo
Advanced Manufacturing and Enterprise Engineering (AMEE),
School of Engineering and Design,
Brunel University,
Uxbridge, Middlesex UB8 3PH, UK

J. Landre Jr.
Manufacturing Research Centre,
Mechanical and Mechatronics Engineering,
Pontifical Catholic University of Minas Gerais,
PUC Minas, Av. Dom José Gaspar, 500, Belo Horizonte, MG, Brazil

Xiongwei Liu
School of Aerospace, Automotive and Design Engineering,
University of Hertfordshire,
Hatfiled AL10 9AB, UK

W.B. Lee
Ultra-Precision Machining Centre,
Department of Industrial and Systems Engineering,
The Hong Kong Polytechnic University,
Hung Hom, Kowloon, Hong Kong

Yoshihiko Murakami
R&D Centre,
OSG Corporation,
Honnogahara 1-15, Toyokawa,
Aich Prefecture 442-8544, Japan

Neil D Sims
Advanc ed Manufacturing Research
Centre with Boeing,
Department of Mechanical
Engineering,
The University of Sheffield,
Sheffield S1 3JD, UK

W. F. Sales
Manufacturing Research Centre,
Mechanical and Mechatronics
Engineering,
Pontifical Catholic University of
Minas Gerais,
Av. Dom José Gaspar, 500, Belo
Horizonte, MG, Brazil

S. To
Ultra-Precision Machining Centre,
Department of Industrial and
Systems Engineering,
The Hong Kong Polytechnic
University,
Hung Hom, Kowloon, Hong Kong

Frank Wardle
UPM Ltd,
Mill Lane,
Stanton,
Fitzwarren,
Swindon, SN6 7SA, UK

Jiwang Yan
Department of Nanomechanics,
Tohoku University,
Aoba 6-6-01,
Aramaki,
Aoba-ku,
Sendai 980-8579, Japan

1

Introduction

Kai Cheng

Advanced Manufacturing and Enterprise Engineering (AMEE) Department
School of Engineering and Design
Brunel University
Uxbridge, Middlesex UB8 3PH, UK

1.1 Scope of the Subject

Machining processes are industrial processes in which typically metal parts are shaped by removal of unwanted materials. They are still the fundamental manufacturing techniques and it is expected to remain so for the next few decades. According to the International Institution of Production Research (CIRP), machining accounts for approximately half of all manufacturing techniques, which is a reflection of the achieved accuracy, productivity, reliability and energy consumption of this technique.

Future machine tools have to be highly dynamic systems to sustain the required productivity, accuracy and reliability. Both the machine tool system (Machine/Tool-holder/Tool/Workpiece/Fixture) and machining processes are necessary to be optimized for their usability, cutting performance or the process capability to meet the productivity, precision and availability requirements of the end user. Furthermore, the machine dynamics and machining process dynamics are two indispensably integrated parts which should be taken into account simultaneously in optimizing the machine system, as illustrated in Figure 1.1.

The machining and machine dynamics within the machine system should be well understood, optimized and controlled, because they have the following direct effects:

- They may degrade machining accuracy and the machined surface texture and integrity.
- They may lead to chatter and unstable cutting conditions.
- They may cause accelerated tool wear and breakage.
- They may result in accelerated machine tool wear and damage to the machine and part.
- They may create unpleasant noises and sounds on the shopfloor because of the chatter and vibrations.

A number of analytical and experimental methods have been developed to study the dynamics of the machining system, with the two basic objectives [1, 2]:

(1) to identify rules and guidelines to design stable and robust machine tools, and

(2) to develop rules, models and algorithms for undertaking dynamically stable machining processes in an optimal and adaptive manner.

Machining dynamics are a major factor affecting many production operations, especially high speed machining. Taking account of machining dynamics is particularly important in fine finishing operations, such as grinding, diamond turning, and increasingly, nano/micro machining. As a subject, it is multidisciplinary covering cutting mechanics, tribology, sensor and instrumentation, machine design, tooling, process optimization and control, and manufacturing metrology. The subject combines anayltical and experimental work seamlessly together.

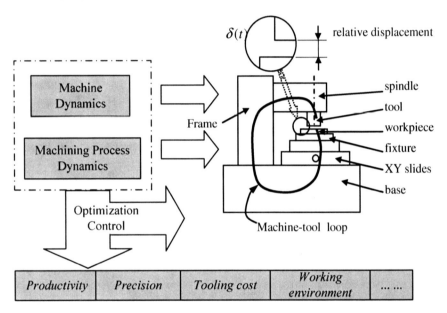

Figure 1.1. The effects of machine and machining dynamics on a machining system

1.2 Scientific and Technological Challenges and Needs

The achievable quality of the precision machined surfaces is affected by four main issues as shown in Figure 1.2. They are the machining process, machine tool performance, workpiece material property and tooling geometry. A scientific approach is needed for building up a theoretical basis to bridge the gap between the surface machined and the determining factors from these four main issues, and to further explore that basis with respect to the desired surface integrity and intended functional performance through machining. It would therefore be of great significance to investigate the fundamentals of high precision surface generation from the manufacturing science viewpoint, which is essential for achieving high precision manufacturing with repeatability, predictability, producibility and productivity. The ultimate goals of manufacturing science and technology are to achieve modelling, simulation, optimization and control of precision machined surfaces including their surface texture, topography, integrity and functionality generation and formation in production processes.

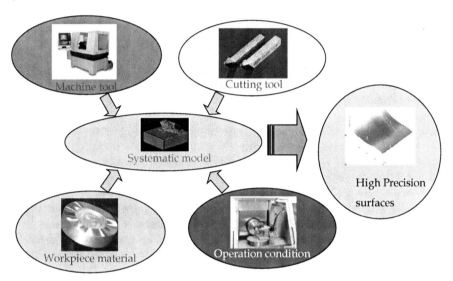

Figure 1.2. Four main issues affecting the precision surface generation

Machining dynamics is the essential and fundamental part for developing the manufacturing science base. They are increasingly important for engaging high speed machining, ultraprecision machining, and nano and micro manufacturing [3, 4, 5].

The advances in computational modelling, sensors, diagnostic equipment and analysis tools, surface metrology, and manufacturing science particularly during the past decade have enabled academia and engineers to research machining dynamics from a new dimension and therefore to have the potential for great industrial benefit, for instance, including:

- Analysis of the material removal dynamics, particularly the effects of cutting speeds and tooling geometry on the stress and temperature conditions at the tool-workpiece interface and thus the surface integrity and functionality.
- Multi-body dynamic analysis of the machine tool structure including the dynamic properties of interfaces between components such as spindles, slideways and drive systems, etc.

- Design of machine tool structures for dynamic repeatability, which is important in predictive control of the machine dynamic performance.
- Dynamic modelling of the machine systems (machine and machining processes) and on line/real time identification of the system modal parameters.
- Development of analytical solutions for the stability of complex contours machining and nonlinear models of interrupted machining.
- Development of novel algorithms (integrated with existing CAD/CAM/CAE tools) for compensation control of machining errors at real time.
- Machining dynamics and micro chatter in ultraprecision machining, and nano and micro cutting.

1.3 Emerging Trends

Increasing demands on manufacturing precision products require the development of precision machines for engaging high value manufacturing. A trend in developing precision machines is that machine tool developers are expected to not only concentrate on the optimization of the machine tool itself in terms of maximum speeds and acceleration of machine axes, but to also take full account of machining dynamics in processes. Therefore, when designing precision machines, it is essential to consider the mechanical structures, control system dynamics, and machining process dynamics simultaneously [6, 7]. An integrated dynamics-driven approach is highly needed for designing precision machines as illustrated in Figure 1.3.

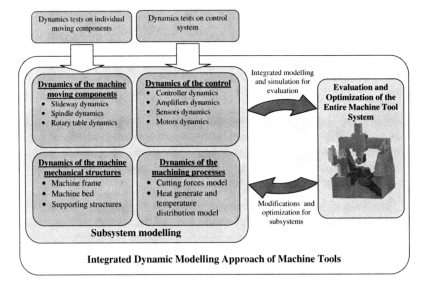

Figure 1.3. Schematic of the integrated dynamics-driven machine design approach

High-accuracy mechanical miniaturized components with dimensions ranging from a few hundred microns to a few millimetres or features ranging from a few to a few hundreds of microns are increasingly in demand for various industries, such as aerospace, biotechnology, electronics, communications, optics, etc. [8]. Advanced high precision machines have the unique advantage of manufacturing high-end miniaturized components in terms of the accuracy, surface finish and geometrical complexity in a wide range of engineering materials. Nevertheless, the micro and functional features on the machined surfaces are becoming dominant particularly for the miniature and micro components and products. Therefore, the detailed and in-depth understanding of the intricate relationships among machines, processes, tooling and materials are increasingly demanded and indispensable for implementing high precision and nano/micro manufacturing. As illustrated in Figure 1.4, machining dynamics driven modelling and simulation can be utilized as the commencing point to comprehensively investigate the complex relationships and phenomenon including:

- Prediction of the surface texture, integrity and functionality generation in machining processes.
- Optimization and control of machining processes against the functionality and performance requirements of the components and products.
- Implementation of the industrial-feasible control algorithms for engaging intelligent, adaptive and high throughput manufacturing.

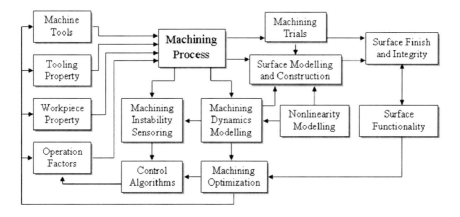

Figure 1.4. Modelling, simulation, optimization and control of the machining process based on machining dynamics

Finite Element Analysis (FEA) is the most practically useful approach for analyzing machining systems because it can be used not only for dynamics analysis, but also for static and thermal analysis. In more recent practice, automeshers using either tetragonal or cubic elements have been increasingly applied because the machining process and associated machining system are the truly dynamically changing process and system and the meshers should thus adaptively change accordingly. Furthermore, multiscale modelling based on combining FEA, micro-mechanics or molecular dynamics (MD) is being used for modelling the formation of surface integrity such as surface roughness, residual stress, micro hardness, microstructure change and fatigue. Throughout the past decade, there have been tremendous research and development for achieving the ultimate goals as illustrated in Figure 1.4 [9, 10, 11, 12], although it would be a continuous long-lasting process.

References

[1] Stephenson, DA and Agapiou, JS. Metal Cutting Theory and Practice. 2nd Edition, 2006, New York: Taylor & Francis.
[2] Altintas, A. Manufacturing Automation: Metal Cutting Mechanics, Machine Tool Vibrations and CNC Design. 2000, Cambridge: Cambridge University Press.
[3] Erdel, BP. High Speed Machining. 2003, Dearborn, Michigan: Society of Manufacturing Engineers.
[4] Ehmann, KF, Bourell, D, Culpepper, ML, Hodgson, TJ, Kurfess, TR, Madou, M, Rajurkar, K and De Vor, RE. International assessment of research and development in micromanufacturing. 2005. World Technology Evaluation Center, Baltimore, Maryland.
[5] Liu, X, DeVor, RE, Kapoor, SG, and Ehmann, KF. The mechanics of machining at the microscale: Assessment of the current state of the science. Journal of Manufacturing Science and Engineering, Transactions of the ASME, 2004. 126 (4): p. 666–678.
[6] Zaeh, M. and Siedl, D. A new method for simulation of machining performance by integrating finite element and multi-body simulation for machine tools. Annals of the CIRP, 2007. 56(1): p. 383–386.
[7] Maj, R, Modica, F and Bianchi, G. Machine tools mechatronic analysis. Proceedings of the Institution of Mechanical Engineers, Part B: Journal of Engineering Manufacture, 2006. 220(3): p. 345-353.
[8] Luo, X, Cheng, K, Webb, D and Wardle, F. Design of ultraprecision machine tools with applications to manufacture of miniature and micro components. Journal of Materials Processing Technology, 2005. 167(2-3): p. 515–528.
[9] Cheung, CF and Lee, WB. Modelling and simulation of surface topography in ultraprecision diamond turning. Proceedings of the Institution of Mechanical Engineers, Part B: Journal of Engineering Manufacture, 2000. 214(4): p. 463–480.
[10] Salisbury, EJ, Domala, KV, Moon, KS, Miller, MH and Sutherland, JW. A three-dimensional model for the surface texture in surface grinding, part 1: surface generation model. Transactions of the ASME: Journal of Manufacturing Science and Engineering, 2001. 123: p. 576–581.
[11] Luo, XC. High Precision Surfaces Generation: Modelling, Simulation and Machining Verification. PhD Thesis, 2004. Leeds: Leeds Metropolitan University.
[12] Huo, D. and Cheng, K. A dynamics-driven approach to the design of precision machine tools for micro-manufacturing and its implementation perspectives. Proceedings of the Institution of Mechanical Engineers, Part B: Journal of Engineering Manufacture, 2008. 222(1): p. 1–13.

2

Basic Concepts and Theory

Dehong Huo and Kai Cheng

Advanced Manufacturing and Enterprise Engineering (AMEE)
School of Engineering and Design
Brunel University
Uxbridge, Middlesex UB8 3PH, UK

2.1 Introduction

This chapter starts with an introduction of the machine-tool-workpiece loop stiffness and deformation, and then fundamentals of vibrations and followed by the definition and categories of machining chatter. It is not the purpose of this chapter to present the general theory of vibration and chatter in depth as there are a number of excellent books and papers available on these subjects. It is intended from the machining system's viewpoint to provide the basic concept and formulations and the necessary theory background for the following up chapters. Furthermore, the generic concept and classification of machining instability are proposed based on the analysis of various machining instable behaviors and their features.

2.2 Loop Stiffness within the Machine-tool-workpiece System

2.2.1 Machine-tool-workpiece Loop Concept

From the machining point of view, the main function of a machine tool is to accurately and repeatedly control the contact point between the cutting tool and the uncut material - the 'machining interface'. Figure 2.1 shows a typical machine-tool-workpiece loop. The machine-tool-workpiece loop is a sophisticated system which includes the cutting tool, the tool holder, the slideways and stages used to move the tool and/or the workpiece, the spindle holding the workpiece or the tool, the chuck/collet, and fixtures, etc. If the machine tool is being taken as a dynamic loop, the internal and external vibrations, and machining processes should be also integrated into this loop as shown in Figure 2.2.

Stiffness can normally be defined as the capability of the structure to resist deformation or hold position under the applied loads. Whilst the stiffness of individual components such as spindle and slideway is important, it is the loop stiffness in the machine-tool system that determines machining performance and dimensional and forming accuracy of the surface being machined, *i.e.*, the relative position between the workpiece and the cutting tool directly contributes to the precision of a machine tool and correspondingly leads to the machining errors.

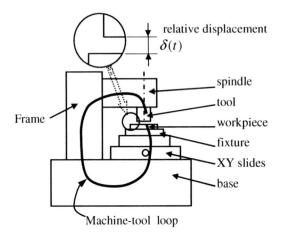

Figure 2.1. A typical machine-tool loop

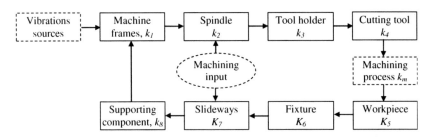

Figure 2.2. The machine-tool-workpiece loop taking account of machining processes and dynamic effects

2.2.2 Static Loop Stiffness

Static loop stiffness in machine tools refers to the performance of the whole machine-tool loop under the static or quasi-static loads which normally come from gravity and cutting forces in machine tools.

A simplified analogous approach to obtaining the static loop stiffness is to regard the machine tool individual elements as a number of springs connected to each other in series or in parallel, so that the static loop stiffness can be derived based on the stiffness of each individual element [1]:

$$x_{static\_loop} = \frac{F}{k_{static\_loop}} = \underbrace{\frac{1}{k_{s1}} + \frac{1}{k_{s2}} + \ldots + \frac{1}{k_{sn}}}_{\text{connected in series}} + \underbrace{\frac{1}{k_{p1} + k_{p2} + \ldots + k_{pn}}}_{\text{connected in parallel}} \quad (2.1)$$

Typically, a well designed machine-tool-workpiece system may have a static loop stiffness of around 50N/μm; a figure of 500 N/μm is well desired for heavy cutting machine tools in particular. While a loop stiffness of about 10N/μm seems not rigid enough, it is quite common in precision machines. Static loop stiffness can be predicted at the early design stage by analytical or numerical methods and thus design optimization and improvement are essential; also, a continuous process because of the increasing demands from the various applications.

2.2.3 Dynamic Loop Stiffness and Deformation

Apart from the static loads, machine tools are subjected to constantly changing dynamic forces and the machine tool structure will deform according to the amplitude and frequency of the dynamic excitation loads, which is termed dynamic stiffness. Dynamic stiffness of the system can be measured using an excitation load with a frequency equal to the damped natural frequency of the structure.

Equations 2.2-2.5 provide a rough approximation of dynamic stiffness k_{dyn} and deformation x_{dyn}:

$$x_{dyn} = \frac{\tilde{F}}{k_{dyn}} \quad (2.2)$$

$$k_{dyn} = \frac{k_{static}}{Q} \quad (2.3)$$

where $\tilde{F}$ is the dynamic load applied to the machine tool, k_{static} is the static stiffness of the machine tool, and Q is the amplification factor which can be calculated from:

$$Q = \frac{1}{2\zeta} = \frac{1}{2\frac{c}{2M\omega_0}} = \frac{M\omega_0}{c} \quad (2.4)$$

where M and c is the mass and damping:

$\omega_0 = \sqrt{\frac{k_{static}}{M}}$ is the natural frequency

$\zeta = \frac{c}{2M\omega_0}$ is the damping ratio

Therefore,

$$x_{dyn} = \frac{\tilde{F}}{k_{dyn}} = \tilde{F}\frac{1}{c\omega_0} = \tilde{F}\frac{1}{c}\sqrt{\frac{M}{k_{static}}} \quad (2.5)$$

In order to accurately predict and calculate dynamic loop stiffness or the behaviour of a whole machine-tool system, a dynamic model including all elements in the machine-tool loop needs to be developed. The finite element method has been widely used to establish the machine tool dynamics model and provide the solution with reasonable accuracy, but it would take more computational time because of the complexity of the machine tool system. On the other hand, some alternative analysis techniques to predict dynamics of machines have been proposed. For example, Zhang *et al.* proposed a receptance synthesis method-based approach to predict the dynamic behaviours of the whole machine-tool system [2], although the approach has the limitation of modelling accuracy.

2.3 Vibrations in the Machine-tool System

Vibrations in the machine-tool system are a well-known fact in causing a number of machining problems, including tool wear, tool breakage, machine spindle bearings wear and failure, poor surface finish, inferior product quality and higher energy consumption.

Vibrations can be classified in a number of ways according to a number of possible factors. For instance, vibrations can be classified as free vibrations, forced vibrations and self-excited vibrations based on external energy sources. It is useful to identify vibrations types in machine tools. The basic principles of the three vibrations above can be found in most textbooks in the subject area [3-4], but the contents discussed below are a formulation in the context of machine tools and provide fundamental concepts for the following up chapters.

2.3.1 Free Vibrations in the Machine-tool System

If an external energy source is applied to initiate vibrations and then removed, the resulting vibrations are free vibrations. In the absence of non-conservative forces, free vibrations sustain themselves and are periodic.

The vibrations of machine tools under pulsating excitations can be regarded as free vibrations. The origins of pulsating excitations in machine tools include:

- Cutter-contact forces when milling or flying cutting
- Inertia forces of reciprocating motion parts
- Vibrations transmitting from foundations
- Imperfects of materials

For instance, taking a single-point diamond turning a part as an example, the part has some material defects such as cavities, as shown in Figure 2.3a. If the cutting tool is taken as the object to be investigated, it can be simplified as a single DOF mass-spring free vibration system as shown in Figure 2.3b, although this is an idealized model and the real system is far more complicated.

Firstly, consider the case of an undamped free vibration system. The general form of the differential equation for undamped free vibrations is:

$$M\ddot{x} + Kx = 0 \tag{2.6}$$

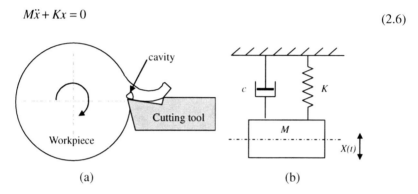

Figure 2.3. a Turning process with material defects b Single DOF free vibration system

Where M and K are the mass and stiffness which are determined during the derivation of the differential equation. Equation 2.6 is subject to the following initial conditions of the form:
$x(0) = x_0$
$\dot{x}(0) = \dot{x}_0$
The solution of Equation 2.6 is:

$$x(t) = x_0 \cos\omega_n t + \frac{\dot{x}_0}{\omega_n} \sin\omega_n t \tag{2.7}$$

where x is displacement at time t:
  x_0 is the initial displacement of the mass
$\omega_n = \sqrt{\dfrac{K}{M}}$ is the undamped natural frequency

There is a slight increase in system complexity while a damping element is introduced to the spring-mass system. Here only viscous damping is taken into account. The general form of the differential equation for the displacement of damped free vibrations becomes:

$$M\ddot{x} + c\dot{x} + Kx = 0 \tag{2.8}$$

where c is the damping of the system. Dividing Equation 2.8 by M gives:

$$\ddot{x} + \frac{c}{M}\dot{x} + \frac{K}{M}x = 0 \tag{2.9}$$

The general solution of Equation 2.9 is obtained by assuming:

$$x(t) = Be^{\alpha t} \tag{2.10}$$

The substitution of Equation 2.10 into Equation 2.9 gives the following quadratic equation for α:

$$\alpha^2 + \frac{c}{M}\alpha + \frac{K}{M} = 0 \tag{2.11}$$

The quadratic formula is used to obtain the roots of Equation 2.11:

$$\alpha_{1,2} = -\frac{c}{2M} \pm \sqrt{\left(\frac{c}{2M}\right)^2 - \frac{K}{M}} \tag{2.12}$$

The mathematical form of the solution of Equation 2.9 and the physical behaviour of the system depend on the sign of the discriminant of Equation 2.12. The case when the discriminant is zero is a special case and occurs only for a certain combination of parameters. When this occurs the system is to be critically damped. For fixed values of K and M, the value of c which causes critical damping is called the critical damping coefficient, c_c:

$$c_c = 2\sqrt{KM} \tag{2.13}$$

The non-dimensional damping ratio, ζ, is defined as the ratio of the actual value of c, to the critical damping coefficient:

$$\zeta = \frac{c}{c_c} = \frac{c}{2\sqrt{KM}} \tag{2.14}$$

The damping ratio is an inherent property of the system parameters. Using Equations 2.13 and 2.14, Equation 2.12 is rewritten in terms of ζ and ω_n as:

$$\alpha_{1,2} = -\zeta\omega_n \pm \omega_n\sqrt{\zeta^2 - 1} \tag{2.15}$$

Therefore, the general solution of Equation 2.9 is:

$$x(t) = e^{-\zeta\omega_n t}(C_1 e^{\omega_n\sqrt{\zeta^2-1}t} + C_2 e^{-\omega_n\sqrt{\zeta^2-1}t})\tag{2.16}$$

where C_1 and C_2 are the arbitary constants of integration. From Equation 2.16, it is evident that the nature of the motion depends on the value of ζ; Equation 2.9 then becomes:

$$\ddot{x} + 2\zeta\omega_n\dot{x} + \omega_n^2 x = 0\tag{2.17}$$

This is the standard form of the differential equation governing the free vibrations with damping.

There are different conditions of damping: critical, overdamping, and underdamping. Detailed discussions of these three cases can be found in most of the subject textbooks [3, 4].

2.3.2 Forced Vibrations

If vibrations occur during the presence of an external energy source, the vibrations are called forced vibrations. The behaviour of a system undergoing forced vibrations is dependent on the type of external excitation. There are a few types of external forces including harmonic, periodic but not harmonic, step, impulse and arbitrary force, etc. If the excitation is periodic, the forced vibrations of a linear system are also periodic.

Considering the internal grinding process as shown in Figure 2.4a in which the spindle is out of balance, the resulted unbalance force is assumed in a harmonic form, $F\sin(\omega t+\varphi)$. This force will vibrate the grinder relative to the workpiece and result in forced vibrations.

Again, an undamped mass-spring system under harmonic forces is considered as shown in Figure 2.4b. The differential equation for undamped forced vibrations subjected to an excitation of harmonic force is:

$$\ddot{x} + \omega_n^2 x = \frac{F}{M}\sin(\omega t + \varphi)\tag{2.18}$$

If excitation frequency ω is not equal to ω_n the following equation is used to obtain the particular solution of Equation 2.18:

$$x_p(t) = \frac{F}{M(\omega_n^2 - \omega^2)}\sin(\omega t + \varphi)\tag{2.19}$$

The homogeneous solution is added to the particular solution with the initial conditions applied, yielding:

$$x(t) = \left[x_0 - \frac{F\sin\varphi}{M(\omega_n^2 - \omega^2)}\right]\cos(\omega_n t)$$
$$+ \frac{1}{\omega_n}\left[\dot{x}_0 - \frac{F\omega\cos\varphi}{M(\omega_n^2 - \omega^2)}\right]\sin(\omega_n t) + \frac{F}{M(\omega_n^2 - \omega^2)}\sin(\omega t + \varphi)$$
(2.20)

In a damped forced vibration system with harmonic excitation the standard form of the differential equation is:

$$\ddot{x} + 2\zeta\omega_n \dot{x} + \omega_n^2 x = \frac{F}{M}\sin(\omega t + \varphi)$$
(2.21)

The particular solution of Equation 2.21 is:

$$x_p(t) = \frac{F}{M[(\omega_n^2 - \omega^2) + (2\zeta\omega\omega_n)^2]}[-2\zeta\omega\omega_n \cos(\omega t + \varphi)$$
$$+ (\omega_n^2 - \omega^2)\sin(\omega t + \varphi)]$$
(2.22)

Equation 2.22 can be rewritten in the following alternative form:

$$x_p(t) = A\sin(\omega t + \varphi - \phi)$$
(2.23)

where
$$A = \frac{F}{M\sqrt{(\omega_n^2 - \omega^2) + (2\zeta\omega\omega_n)^2}}$$

$$\phi = \tan^{-1}\left(\frac{2\zeta\omega\omega_n}{\omega_n^2 - \omega^2}\right)$$

A is the amplitude of the forced response and ϕ is the phase angle between the response and the excitation.

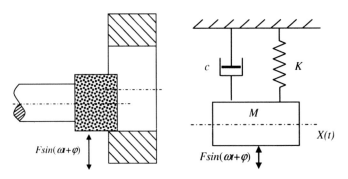

Figure 2.4. a Internal grinding process b Single DOF forced vibration system

Forced vibrations in machine tools can be generated from two kinds of energy sources, which are internal and external vibration sources. External vibration sources, such as seismic waves, usually transfer vibrations to the machine tool structure via the machine base. The design and use of effective vibration isolators will be able to eliminate or minimize forced vibrations caused by external vibration sources. There are many internal vibration sources which cause forced vibrations. For instance, an unbalanced high speed spindle, an impact force in machining processes, and inertia force caused by a reciprocal motion component such as slideways, etc.

2.4 Chatter Occurring in the Machine Tool System

2.4.1 Definition

Apart from free and forced vibrations, self-excited vibrations exist commonly in machine-tool system. A self-excited vibration is a kind of vibration in which the vibration resource lies inside the system. In machining self-excited vibrations usually result in machine tool chatter vibration. It should be noted that chatter vibration can also be caused by the forced vibration, but it is usually not a major problem in machining because the external force or the dynamic compliance of the machine structure can be reduced to reasonable levels when the external force causing the chatter is identified [5].

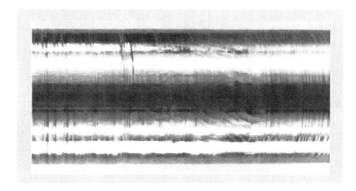

Figure 2.5. Poorly machined surface resulted from chatter (Courtesy: GE Company)

Chatter occurs mainly because one of the structural modes of the machine tool-workpiece system is initially excited by cutting forces. Chatter is a problem of instability in the machining process, characterized by unwanted excessive vibration between the tool and the workpiece, loud noise, and consequently a poor quality of surface finish. It also has a deteriorating effect on the machine and tool life, and the reliability and safety of machining operation [6]. The problem has affected the manufacturing community for quite some time and it is a popular topic for academic and industrial research. Therefore, it is very important to identify and to get a better understanding of the machine structural dynamic performance at both the machine design and production stage. Figure 2.5 shows a poorly machined surface resulting from chatters, and more information about chatters is available in Chapters 3 and 4 of this book.

2.4.2 Types of Chatters

There are mainly three forms of self-excited chatters. The first one is the velocity dependent chatter or Arnold-type chatter, named after the man who discovered it, which is due to a dependence on the variation of force with the cutting speed. The second form is known as the regenerative chatter, which occurs when the unevenness of the surface being cut is due to consequent variations in the cutting force when on the previous occasion the tool passed over that location, causing detrimental degeneration of the cutting force. Depending on the phase shift between the two successive wave surfaces, the maximum chip thickness may exponentially grow while oscillating at a chatter frequency that is close to but not equal to the dominant structural mode in the system. The growing vibrations increase the cutting forces and produce a poor and wavy surface finish [7]. The third form of chatter is due to mode coupling when forces acting in one direction on a machine-tool structure cause movements in another direction and vice versa. This results in simultaneous vibrations in two coupling directions. Physically it is caused by a number of sources, such as friction on the rake and clearance surfaces [8] and mathematically described by Wiercigroch [6].

Most of the chatters occurring in practical machining operations are regenerative chatter [9], although other chatters are also common in some cases. These forms of chatters are interdependent and can generate different types of chatter simultaneously. However, there is not a unified model capable of explaining all chatter phenomena observed in machining practice [10].

2.4.3 The Suppression of Chatters

After identifying chatters occurring in the machine-tool system, a number of approaches for reducing chatters have been proposed. Classical approaches usually use the stability diagrams to avoid the occurrence of chatters [9, 11-12]. The following approach formulates some general methods for the reduction of chatters both on the design and the production stage:

- Selecting the optimal cutting parameters
- Selecting the optimal tooling geometry
- Increasing the stiffness and damping of the machine tool system
- Using the vibration isolator as necessary
- Altering the cutting speed during the machining process
- Using a different coolant

More recently, modern control and on-line chatter detection techniques were applied to suppress chatters [13, 14, 15, 16]. Furthermore, a change of tool geometry is also an industrial feasible approach to chatter control [17], for instance, through the application of cutting tools with irregular spacing or variable pitch cutters [18].

2.5 Machining Instability and Control

2.5.1 The Conception of Machining Instability

In the previous sections, many aspects of self-excited machine tool vibrations or chatters have been briefly discussed. In practice, however, many problems of poor work surface finish are due to forced vibrations and the methods of reducing forced vibrations should thus well be understood. Forced vibrations are usually caused by an out-of-balance force associated with a component integrated with, or external to, the machine tool, whereas a self-excited vibration is spontaneous and increases rapidly from a low vibratory amplitude to a large one; the forced vibration results in an oscillation of constant amplitude. An exploration into chatter vibrations enables a better understanding of machining instability in practice.

From the machining point of view, with the designed machining conditions, a desired surface finish will be produced under a stable machining process. But as a complicated dynamic system, various mechanisms inherent in the machining process may lead the innately stable machining system to work at a dynamically unstable status which invariably results in unsatisfactory workpiece surface quality [19]. The machining instability coined here is a new generalized concept, which includes all phenomena making the machining process departure from what it should be. For instance, a variety of disturbances affect the machining system such as self-excited vibration [20], thermomechanical oscillations in material flow [21], and feed drive hysteresis [10], but the most important is self-excited vibrations resulting from the dynamic instability of the overall machine-tool/machining-process system [22-23]. However, sometimes the machining process is carried out with a relative vibration between the workpiece and the cutting tool, especially in heavy cutting and rough machining, in order to obtain high material removal rates. The relative vibration is not necessarily a sign of the machining instability for the designed machining conditions and prescribed surface finish. In another extreme case, such as in ultra-precision machining or micro/nano machining, the relative vibration between the workpiece and the cutting tool is too small to be measured, but the machining is sensitive to environmental disturbances. The surface generated may be unsatisfactory because of the disturbance, even though the machining system itself operates in the stable state. Therefore, the machining instability is related to the level of the surface quality required and the designed machining conditions.

Table 2.1 The classification of machining instability [25]

		Machining Instability						
		Chatter vibrations			Random and free vibrations			Forced Vibration
		Regenerative (Dominate)	Frictional	Mode coupling	Tool dependent	Workpiece dependent	Environment dependent	
Location		Between cutting edge and workpiece	Tool flank-workpiece; chip-tool rake face	In cutting and thrust force directions	Tool flank-workpiece; chip-rake face	Cutting zone	Whole cutting process	Whole cutting process
Causes		Overlapping cut	Rubbing on the flank face and the rake face	Friction on the rake and clearance faces; chip thickness variation, shear angle oscillation.	Tool wear and breakage; BUE, etc.	Material softening and hardening; hard grain and other kinds of flaws	Environmental disturbances	Off-balance of moving components, such as the spindle
Features		Self-excited vibration; left a wavy surface on workpiece	Self-excited vibration; amplitude depends on the system damping	Mode coupling vibration; Simultaneous vibration in two directions	Random and chaotic; depends on cutting conditions	Random and chaotic; depends on material property and its heat treatment	Radom and chaotic; depends on work environment	Forced vibration
Suppression method		Select proper depth of cut and spindle speed according to regenerative stability chart	Select proper clearance and rake angles	Change the tool path; Select proper cutting variables	Select high quality tool materials and proper cutting parameters	Select proper cutting tool and cutting parameters	If needed, isolate the machine tool	Well balance moving component in machine tools

2.5.2 The Classification of Machining Instability

Based on the conception above, Cheng *et al.* summarize all kinds of machining instability and their features as listed in Table 2.1 [24-25]. The instability is classified as the chatter vibration, the random or free vibration and forced vibration. The random or free vibration usually includes any shock or impulsive loading on the machine tool. A typical random vibration is the tool vibration, for instance, when the tool strikes at a hard spot during the cutting process. The tool will bounce or vibrate relative to the workpiece, which is the beginning of the phenomenon of a self-excited vibration. The initial vibration instigated by the hard spot is heavily influenced by the dynamic characteristics of the machine tool structure which must be included in any rational chatter analysis.

Acknowledgements

The authors are grateful for the support of the EU 6th Framework NMP Program under the contract number NMP2-CT-2-4-500095. Thanks are due to all partners at the MASMICRO project consortium, and the RTD 5 subgroup in particular.

References

[1] Weck, M. Handbook of Machine Tools, Volume 2: Construction and Mathematical Analysis, Wiley, London, 1980
[2] Zhang, G. P., Huang, Y. M., Shi, W. H. and Fu, W. P. Predicting dynamic behaviours of a whole machine tool structure based on computer-aided engineering. International Journal of Machine Tools and Manufacture, 2003, 43: 699–706
[3] Benaroya, H. Mechanical Vibration – Analysis, Uncertainties, and Control. Marcel Dekker, New York, 2004
[4] Rao, S. S. Mechanical Vibrations, Prentice Hall, New Jersey, USA, 2003
[5] Merrit, H. E. Theory of self-excited machine-tool chatter-contribution to machine tool chatter research. Transactions of the ASME: Journal of Engineering for Industry, 1965, 87(4): 447–454
[6] Wiercigroch, M. Chaotic vibrations of a simple model of the machine tool-cutting process system. Transactions of the ASME: Journal of Vibration Acoustics, 1997, 119: 468–475
[7] Altintas Y. Manufacturing Automation: Metal Cutting Mechanics, Machine Tool Vibrations and CNC Design. Cambridge University Press, Cambridge, UK, 2000
[8] Cook, N. H. Self-excited vibration in metal cutting. Transactions of the ASME: Journal of Engineering for Industry, 1959, 81: 183–186
[9] Tobias, S. A. Machine Tool Vibration, Blackie and Son, London, 1965
[10] Wiercigroch, M. and Budak, E. Sources of nonlinearities, chatter generation and suppression in metal cutting, Philosophical Transactions: Mathematical, Physical and Engineering Sciences, 2001, 359(A): 663–693
[11] Sweeney, G. Vibration of machine tools, The Machinery Publishing Co. Ltd, UK, 1971
[12] Tobias, S. A. and Fishwick, W. The chatter of lathe tools under orthogonal cutting conditions, Transactions of the ASME: B, 1958, 80: 1079–1088

[13] Altintas, Y. and Chan, P. K. In-process detection and suppression of chatter in milling. International Journal of Machine Tools and Manufacture, 1992, 32(3): 329–347
[14] Tewani, S. G., Rouch, K. E. and Walcott, B. L. A study of cutting process stability of a boring bar with active dynamic absorber, International Journal of Machine Tools and Manufacture, 1995, 35: 91–108
[15] Li, X. Q., Wong, Y. S. and Nee, A. Y. C. Tool wear and chatter detection using the coherence function of two crossed accelerations. International Journal of Machine Tools and Manufacture, 1997, 37(4): 425–435
[16] Bayly, P. V., Metzler, S. A., Schaut, A. J. and Young, K. A. Theory of torsional chatter in twist drills: model, stability analysis and composition to test. Transactions of the ASME: Journal of Manufacturing Science and Engineering, 2001, 123: 552–561
[17] Liu, C. R. and Liu, T. M. Automated chatter suppression by tool geometry control, Transactions of the ASME: Journal of Engineering for Industry, 1985, 107: 95–98
[18] Budak, E. Improving productivity and part quality in milling of titanium based impellers by chatter suppression and force control, the Annals of CIRP, 2000, 49(1): 31–36
[19] Shaw, M. C. Metal Cutting Principles, Oxford University Press, Oxford, 1984
[20] Stepan, G. Modelling nonlinear regenerative effects in metal cutting, Philosophical Transaction: Mathematical, Physical and Engineering Sciences, 2001, 359(A): 739–757
[21] Davies, M. A. and Burns, T. J. Thermomechanical oscillations in material flow during high-speed machining, Philosophical Transactions: Mathematical, Physical and Engineering Sciences, 2001, 359(A): 821–846
[22] Budak, E. and Altintas, Y. Analytical prediction of chatter stability in milling, Part I: General formulation, Transactions of the ASME: Journal of Dynamic Systems, Measurement and Control, 1998, 120(1): 22–30
[23] Budak, E. and Altintas, Y. Analytical prediction of chatter stability in milling, Part II Application of the general formulation to common milling systems, Transactions of the ASME: Journal of Dynamic Systems, Measurement and Control, 1998, 120(1): 31–36
[24] Luo, X. K., Cheng, K. and Luo, X. C. A simulated investigation on machining instability and non-linear aspects in CNC turning processes, Proceedings of the 18th NCMR Conference, Leeds, UK, 10-12 September 2002: 405–410
[25] Luo, X. K., Cheng, K., Luo, X. C. and Liu, X. W. A simulated investigation on machining instability and dynamic surface generation, International Journal of Advanced Manufacture Technology, 2005, 26(7-8): 718–725

3

Dynamic Analysis and Control

Erhan Budak

Faculty of Engineering and Natural Sciences
Sabanci University
Tuzla, Istanbul 34956
Turkey

All machining processes are subject to dynamic effects due to transient or forced vibrations, and dynamic mechanisms inherent to the process such as regeneration. If not controlled, they may result in high amplitude oscillations, instability and poor quality. Dynamic rigidity of the structures involved in the machining is very important in determining the dynamic behaviour of the process. Structural rigidity is also critical for deformations, and the dimensional quality of machined parts. In this chapter, important aspects of the machining process dynamics are discussed, and the methods that can be used for the analysis and modelling of the machine tool structural components and the processes are presented. Chatter stability and suppression methods will also be explained with applications.

3.1 Machine Tool Structural Deformations

Machining systems involve a machine tool, a cutting tool and holder, and a workpiece and work holding devices as structural elements. Depending on their relative rigidity, one or more components may dominate the total deformation at the tool-workpiece contact point contributing to the form errors and the dynamics of the process which may yield instability. For example, for machining centers which are composed of a bed, linear and rotary axes, a column, a spindle etc., the spindle-holder-tool assembly is usually the most flexible part of the whole system due to the slender geometries of these components and multiple interfaces between them. In some applications, on the other hand, the workpiece flexibility can outweigh the flexibility of the machine tool and the tooling such as in the case of the milling of compressor blades. In either case, the cutting process forces are the main cause of the structural deformations. In this section, analytical methods which can be used for the modelling of cutting processes and structural deformations are presented.

3.1.1 Machining Process Forces

Cutting forces are the main cause of the deformations of machine tool structures and workpieces resulting in form errors and tolerance violations. Although they may affect the structural components of a machine tool distributed in a large space, cutting forces are generated in a very small area at the work-tool interface. Practical cutting operations such as turning and milling have complex geometries involving 3 dimensional cutting actions due to tool angles and cutting edges. However, the basic mechanics of the process can be understood by using an orthogonal cutting model as shown in Figure 3.1. In this simplest machining process model, the cutting edge is perpendicular to the relative cutting velocity between the tool and the workpiece. The tool with rake angle α moves along the work material removing an uncut chip thickness of h. The work material, i.e., the uncut chip, goes through a plastic deformation with a very high strain and strain rate resulting in high temperatures, and meets the rake face where it flows up under plastic and elastic contact conditions. The complicated mechanics of the process is simplified by assuming 2 deformation zones: the primary deformation or the shear zone, and the secondary deformation or the rake contact. In shear zone models, which are more realistic for high cutting speeds, the material is assumed to shear along the *AB* plane as shown in Figure 3.1, and form a chip. The various forces encountered in this process are shown in Figure 3.2, where R is the resultant force acting on the tool. Note that 3 sets of forces acting on different planes shown are all equivalent to the resultant force R. F_t and F_r are the tangential and feed forces in the cutting velocity and feed directions, whereas N and F are the normal and frictional forces on the rake face. F_s and F_n are the shear and normal forces acting on the shear plane. The shear angle ϕ_s is the most fundamental parameter in a cutting process, as it is needed to perform the force analysis. This is why it has been one of the focal points of machining research for more than half a century. In his pioneering work, Merchant [1] used a minimum energy condition to determine the value of the shear angle in an orthogonal cutting process. He assumed a perfectly sharp tool (no rubbing or ploughing), a two dimensional deformation (no side spread) and a uniform stress distribution in the shear plane which is equal to the yield shear strength of the work material to arrive at his famous shear angle relationship:

$$\phi_s = \frac{\pi}{4} - \frac{1}{2}(\beta - \alpha) \tag{3.1}$$

where β is the friction angle, i.e., $\beta = \tan^{-1}(\mu)$, μ being the dry friction coefficient on the rake face. This model is an approximation as the contact mechanics on the rake face is much more complicated. There are two zones of contact between the chip and the rake face: the sticking zone (plastic flow) and the sliding zone (elastic contact). In the sticking zone, the chip is in the plastic state, the top layer being bounded to the rake face and the rest flowing over it. As a result, the friction in the sticking zone is lower than the dry friction coefficient. This can also be explained by analyzing the stresses in those regions. The tangential force, i.e., the frictional force, in the sticking zone is to the shear yield stress of the work material whereas it is lower than this in the sliding zone as the material is in the elastic state. On the other hand, the normal stress is very high in the plastic zone and reduces to zero at the end of the total contact in an exponential manner. As a result, the ratio of the tangential force to the normal force, *i.e.*, the friction coefficient, in the sticking zone is much lower than the one in the sliding zone. Therefore, the overall or the average friction coefficient on the rake face depends on the friction coefficients and the lengths of both zones. The length proportions of both zones depend on the material characteristics as well as the cutting conditions. Thus, different friction coefficients can be obtained for the same work-tool material pair under different cutting conditions. For example, as the pressure on the rake face is increased due to the lower shear angle, which may be caused by decreasing the rake angle, the plastic zone becomes longer resulting in a lower overall frictional coefficient.

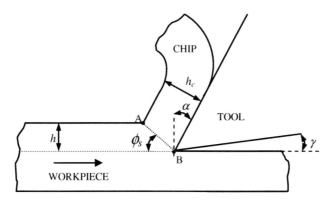

Figure 3.1. Orthogonal cutting geometry

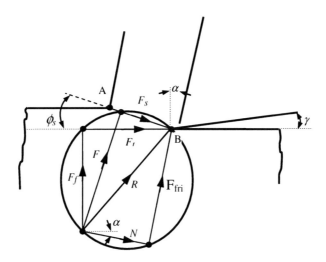

Figure 3.2. Forces in the orthogonal cutting process

The forces in the tangential, F_t, and the feed directions, F_f, can be expressed as:

$$F_t = hwK_t \quad ; \quad F_f = hwK_f \tag{3.2}$$

where h is the uncut chip thickness, w is the width of cut, and K_t and K_f are the cutting force coefficients. The cutting force coefficients can be obtained from the force analysis of the orthogonal cutting as follows:

$$K_t = \tau \frac{\cos(\beta - \alpha)}{\sin \phi_s \cos(\phi_s + \beta - \alpha)} \quad ; \quad K_f = \tau \frac{\sin(\beta - \alpha)}{\sin \phi_s \cos(\phi_s + \beta - \alpha)} \tag{3.3}$$

where t is the shear stress of the work material in the shear plane.

The accuracy of the orthogonal force model is limited due to the assumptions described above. The predictions can be improved by considering a more realistic deformation zone and material models such as a finite shear zone thickness, non-uniform rake face contact stress distribution and material flow characteristics. They can also be improved by using experimental approaches. The cutting parameters can be identified through orthogonal cutting tests performed on a lathe where a tube can be used as the work piece in order to eliminate the rubbing forces. The shear angle, the shear stress and the friction coefficient can be obtained from orthogonal cutting tests as follows:

$$\tan \phi_s = \frac{r \cos \alpha}{1 - r \sin \alpha} \quad , \quad \tau = \frac{\left(F_t \cos \phi_s - F_f \sin \phi_s\right) \sin \phi_s}{wh} \quad , \quad \tan \beta = \frac{F_f + F_t \tan \alpha}{F_t - F_f \tan \alpha} \tag{3.4}$$

where r is the cutting ratio or the ratio of the uncut chip thickness to the chip thickness.

In Equation 3.2, the edge forces are also included in the cutting force coefficient which is usually referred to as the exponential force model. They are separated from the cutting force coefficients in the edge force or the linear-edge force model as follows:

$$F_t = w(K_{tc}h + K_{te}) \quad ; \quad F_f = w(K_{fe}h + K_{fe}) \tag{3.5}$$

where subscripts (e) and (c) represent edge force and cutting force coefficients, respectively. If the linear-edge force model is to be used then the edge cutting forces must be subtracted from the cutting forces measured in each direction using linear regression. Then, the edge force coefficients are identified from the edge cutting forces. After the orthogonal cutting tests are repeated for a range of cutting speed, rake angle and uncut chip thickness, an orthogonal cutting database is generated for a certain tool and work material pair [2-4].

Most of the real machining processes are oblique as the cutting edge is not always perpendicular to the cutting speed direction due to the inclination angle i. In oblique cutting models, there are several important planes which are used to measure tool angles and perform analysis [2]. The normal plane, which is perpendicular to the cutting edge, is commonly used in the analysis. After several assumptions, the following expressions are obtained for the cutting force coefficients in an oblique cutting process [2]:

$$K_{tc} = \frac{\tau}{\sin\phi_n} \frac{\cos(\beta_n - \alpha_n) + \tan\eta_c \sin\beta_n \tan i}{c} \;;\; K_{rc} = \frac{\tau}{\sin\phi_n \cos i} \frac{\sin(\beta_n - \alpha_n)}{c}$$

(3.6)

$$K_{ac} = \frac{\tau}{\sin\phi_n} \frac{\cos(\beta_n - \alpha_n)\tan i - \tan\eta_c \sin\beta_n}{c}$$

where $c = \sqrt{\cos^2(\phi_n + \beta_n - \alpha_n) + \tan^2\eta_c \sin^2\beta_n}$. K_{tc}, K_{rc} and K_{ac} are the cutting force coefficients in the cutting speed, feed and axial direction.

In Equation 3.6, (τ) is the shear stress in the shear plane, ϕ_n is the shear angle in the normal plane, i is the angle of obliquity and η_c is the chip flow angle measured on the rake face. The chip flow angle can be solved iteratively based on the equations obtained from force and velocity relations [3-4]. However, for simplicity, Stable's rule [5] may also be used which states that $\eta_c \approx \beta$. β_n and α_n are the friction and the rake angle in the normal plane, respectively.

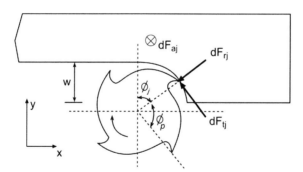

Figure 3.3. Cross-sectional view of an end mill showing differential forces

Cutting process models are general and can be used to predict forces in a variety of machining processes. As an example, the milling force modelling will be presented here. Milling is one of the most commonly used machining process, and has a relatively more complex geometry due to its rotating tool, multiple cutting edges and intermittent cutting action, as shown in Figure 3.3. Two different methods will be presented for the force analysis: the mechanistic and mechanics of cutting models which differ in the way the cutting force coefficients are determined.

In the mechanistic force model, the cutting force coefficients are calibrated for certain cutting conditions using the experimental data. Consider the cross-sectional view of a milling process shown in Figure 3.3. For a point on the (j^{th}) cutting tooth, the differential milling forces corresponding to an infinitesimal element thickness (dz) in the tangential, dF_t, radial, dF_r, and axial, dFa, directions can be given as:

$$dF_{t_j}(\phi, z) = K_t h_j(\phi, z) dz \; ; \; dF_{r_j}(\phi, z) = K_r dF_{t_j}(\phi, z) \; ; \; dF_{a_j}(\phi, z) = K_a dF_{t_j}(\phi, z) \quad (3.7)$$

where ϕ is the immersion angle measured from the positive y-axis as shown in Figure 3.3. The axial force component, F_a, is in the axial direction of the cutting tool, which is perpendicular to the cross-section shown in Figure 3.3. For the edge force or linear-edge force model, the differential forces can be expressed similar to Equation 3.5. The radial (w) and axial depth of cut (a), the number of teeth (N), the cutter radius (R) and the helix angle (i) determine what portion of a tooth is in contact with the work piece for a given angular orientation of the cutter, $\phi = \Omega t$, where t is the time, Ω is the angular speed in (rad/sec) or $\Omega = 2\pi n/60$, n being the (rpm) of the spindle. The chip thickness at a certain location on the cutting edge can be approximated as follows:

$$h_j(\phi, z) = f_t \sin \phi_j(z) \quad (3.8)$$

where f_t is the feed per tooth and $\phi_j(z)$ is the immersion angle for the flute (j) at the axial position z. Due to the helical flute, the immersion angle changes along the axial direction as follows:

$$\phi_j(z) = \phi + (j-1)\phi_p - \frac{\tan i}{R} z \quad (3.9)$$

where the pitch angle is defined as $\phi_p = 2\pi/N$. The tangential, radial and axial forces given by Equation 3.7 can be resolved in the feed, x, normal, y, and the axial direction, z, and can be integrated within the immersed part of the tool to obtain the total milling forces applied on each tooth. For the exponential force model, the following is obtained after the integration:

$$F_{x_j}(\phi) = \frac{K_t f_t R}{4 \tan \beta} \left[-\cos 2\phi_j + K_r \left(2\phi_j(z) - \sin 2\phi_j(z) \right) \right]_{z_{jl}(\phi)}^{z_{ju}(\phi)}$$

$$F_{y_j}(\phi) = -\frac{K_t f_t R}{4 \tan \beta} \left[\left(2\phi_j(z) - \sin 2\phi_j(z) \right) + K_r \cos 2\phi_j(z) \right]_{z_{jl}(\phi)}^{z_{ju}(\phi)} \quad (3.10)$$

$$F_{zj}(\phi) = -\frac{K_a K_t f_t R}{\tan \beta} \left[\cos \phi_j(z) \right]_{z_{jl}(\phi)}^{z_{ju}(\phi)}$$

where $z_{jl}(\phi)$ and $z_{ju}(\phi)$ are the lower and upper axial engagement limits of the in cut portion of the flute j. The total milling forces can then be determined as:

$$F_x(\phi) = \sum_{j=1}^{N} F_{x_j}(\phi) \quad ; \quad F_y(\phi) = \sum_{j=1}^{N} F_{y_j}(\phi) \quad ; \quad F_z(\phi) = \sum_{j=1}^{N} F_{z_j}(\phi) \quad (3.11)$$

For the linear-edge force model, the forces are obtained similarly by using Equation 3.5, and integrating within the engagement limits as follows [6]:

$$F_{xj}(\phi) = \frac{R}{\tan \beta} \{ K_{te} \sin \phi_j(z) - K_{re} \cos \phi_j(z) +$$

$$\frac{f_t}{4}[K_{rc}(2\phi_j(z) - \sin 2\phi_j(z)) - K_{tc} \cos 2\phi_j(z)]\}_{z_{jl}}^{z_{ju}}$$

$$F_{xj}(\phi) = \frac{R}{\tan \beta} \{ -K_{re} \sin \phi_j(z) - K_{te} \cos \phi_j(z) + \quad (3.12)$$

$$\frac{f_t}{4}[K_{tc}(2\phi_j(z) - \sin 2\phi_j(z)) - K_{rc} \cos 2\phi_j(z)]\}_{z_{jl}}^{z_{ju}}$$

$$F_{xj}(\phi) = \frac{R}{\tan \beta}[K_{ae}\phi_j(z) - f_t K_{ac} \cos \phi_j(z)]_{z_{jl}}^{z_{ju}}$$

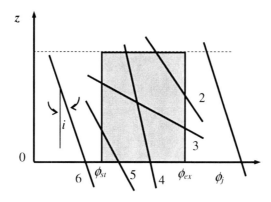

Figure 3.4. Helical end mill and cutting zone intersection cases

The engagement limits depend on the cutting and the tool geometries. An unwrapped end mill surface shown in Figure 3.4 can be used to demonstrate how the limits can be determined. The dark area in the figure represents the cutting zone between ϕ_{st} and ϕ_{ex} in the angular direction, and between 0 and a in the axial direction. ϕ_{st} and ϕ_{ex} are the start and exit immersion angle which can be expressed as:

$$\phi_{st}(z) = \pi - \cos^{-1}\left(1 - \frac{w}{R}\right) \quad \text{(down milling)}$$

$$\phi_{ex}(z) = \cos^{-1}\left(1 - \frac{w}{R}\right) \quad \text{(up milling)} \quad (3.13)$$

Note that ϕ_{ex} is always π in down milling and ϕ_{st} is always 0 in up milling, according to the convention used in Figure 3.3. The helical cutting edges of the tool can intersect this area in 6 different ways based on the immersion angle of each flute at $z=0$, ϕ_j. The limits corresponding to each case are given in Table 3.1 and can be used in Equation 3.12 in order to determine the milling forces per tooth. The forces given by Equations 3.10 and 3.12 can be used to predict the cutting forces for a given milling process if the milling force coefficients are known. In the mechanistic force model, milling force coefficients K_t, K_r and K_a can be determined from the average force expressions [6] as follows:

$$K_r = \frac{P\overline{F_y} - Q\overline{F_x}}{P\overline{F_x} + Q\overline{F_y}} \ ; \ K_t = \frac{\overline{F_x}}{f_t(P - QK_r)} \ ; \ K_a = \frac{\overline{F_z}}{f_t K_t T} \qquad (3.14)$$

where:

$$P = \frac{aN}{2\pi}[\cos 2\phi]_{\phi_{st}}^{\phi_{ex}} \ ; \ Q = \frac{aN}{2\pi}[2\phi - \sin 2\phi]_{\phi_{st}}^{\phi_{ex}} \ ; \ T = \frac{aN}{2\pi}[\cos\phi]_{\phi_{st}}^{\phi_{ex}} \qquad (3.15)$$

Table 3.1. Engagement limits of helical flutes with the cutting zone

	Condition	In/out	z_{jl}	z_{ju}
1	$\phi_j > \phi_{ex}$ and $(\phi_j - a\frac{\tan\beta}{R}) > \phi_{ex}$	Out	NA	NA
2	$\phi_j > \phi_{ex}$ and $\phi_{st} < (\phi_j - a\frac{\tan\beta}{R}) < \phi_{ex}$	In cut	$\frac{R}{\tan\beta}(\phi_j - \phi_{ex})$	a
3	$\phi_j > \phi_{ex}$ and $(\phi_j - a\frac{\tan\beta}{R}) < \phi_{st}$	In cut	$\frac{R}{\tan\beta}(\phi_j - \phi_{ex})$	$\frac{R}{\tan\beta}(\phi_j - \phi_{st})$
4	$\phi_{st} < \phi_j < \phi_{ex}$ and $\phi_{st} < (\phi_j - a\frac{\tan\beta}{R}) < \phi_{ex}$	In cut	0	a
5	$\phi_{st} < \phi_j < \phi_{ex}$ and $(\phi_j - a\frac{\tan\beta}{R}) < \phi_{st}$	In cut	0	$\frac{R}{\tan\beta}(\phi_j - \phi_{st})$
6	$\phi_j < \phi_{st}$ and $(\phi_j - a\frac{\tan\beta}{R}) < \phi_{ex}$	Out	NA	NA

The average forces, $\overline{F_x}, \overline{F_y}$ and $\overline{F_z}$ can be obtained experimentally from the milling tests. In the exponential force model, the chip thickness affects the force coefficients. Since the chip thickness varies continuously in milling, the average chip thickness, h_a, is used:

$$h_a = f_t \frac{\cos\phi_{st} - \cos\phi_{st}}{\phi_{ex} - \phi_{st}} \tag{3.16}$$

In calibration tests, the usual practice is to conduct the experiments at different radial depths and feed rates in order to cover a wide range of h_a for a certain tool-material pair. The force coefficients can then be expressed as the following exponential functions:

$$K_t = K_T h_a^{-p} \; ; \; K_r = K_R h_a^{-q} \; ; \; K_a = K_A h_a^{-s} \tag{3.17}$$

where K_T, K_R, K_A, p, q and s are determined from the linear regressions performed on the logarithmic variations of K_t, K_r, K_a with h_a.

In the linear-edge force model the total cutting forces are separated into two parts. The edge force represents the parasitic part of the forces which are not due to cutting, and thus do not depend on the uncut chip thickness, whereas the cutting forces do. Then, the average forces can be described similarly as follows:

$$\overline{F_q} = \overline{F_{qe}} + f_t \overline{F_{qc}} \quad (q = x, y, z) \tag{3.18}$$

where the edge and cutting components of the average forces ($\overline{F_{qe}}, \overline{F_{qc}}$) are determined using the linear regression of the average measured milling forces. The milling force coefficients for the linear-edge force model can be obtained from the average forces similar to the exponential force model as follows:

$$K_{tc} = 4\frac{\overline{F_{xc}}P + \overline{F_{yc}}Q}{P^2 + Q^2} \quad K_{rc} = \frac{K_{tc}P - 4\overline{F_{xc}}}{Q} \quad K_{ac} = \frac{\overline{F_{zc}}}{T}$$

$$K_{te} = -\frac{\overline{F_{xe}}S + \overline{F_{ye}}T}{S^2 + T^2} \quad K_{re} = \frac{K_{te}S + \overline{F_{xe}}}{T} \quad K_{ae} = -\frac{2\pi}{aN}\frac{\overline{F_{ze}}}{\phi_{ex} - \phi_{st}} \tag{3.19}$$

where P, Q, and T are given by Equation 3.15, and:

$$S = \frac{aN}{2\pi}[\sin\phi]_{\phi_{st}}^{\phi_{ex}} \tag{3.20}$$

In the mechanistic approach, the cutting force coefficients must be calibrated for each tool-material pair covering the conditions that are of interest. The oblique cutting model can be used to predict these coefficients reducing the number of tests significantly. In the mechanics of the milling approach, proposed by Armarego and Whitfield [3], and later by Budak et al. [6-7], the required data are obtained from the orthogonal cutting tests in order to reduce the number of variables and the number of tests, and also to generate a more general database which can be used for other processes as well. These data can then be used to determine the cutting force coefficients using the oblique model given by Equation 3.6.

As a demonstration of the force models presented here, a titanium (Ti6Al4V) milling example is considered. First of all, an orthogonal cutting database is generated using carbide tools with different rake angles, at different speeds and feed rates [6, 8]:

$\tau = 613 \text{ MPa}$, $\beta = 19.1 + 0.29 \, \alpha$

$r = r_0 h^a$, $r_0 = 1.755 - 0.028\alpha$, $a = 0.331 - 0.0082\alpha$

$K_{te} = 24 \text{ N/mm}$, $K_{re} = 43 \text{ N/mm}$

A milling example is considered here where a half-immersion up milling test is performed using a 30^0 helix, a 19.05 mm diameter and a 4-fluted end mill with 12^0 rake angle. The axial depth of cut is 5 mm, and 0.05 mm/tooth feed was used at 30 m/min cutting speed. The measured and the predicted cutting forces using the force coefficients identified from the milling tests and calculated using the oblique model in all 3 directions are shown in Figure 3.5. As it can be seen from this figure, the predictions are very close to the measured forces.

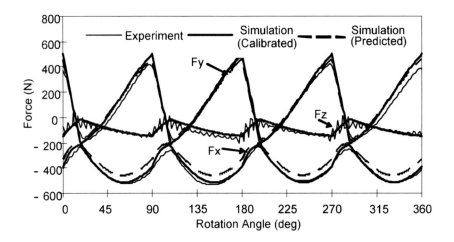

Figure 3.5. An example of the predicted and the measured milling forces

3.1.2 The Deformations of Machine Tool Structures and Workpieces

Machine tool structural components and workpieces deform under thermal, inertial and cutting loads. These deformations result in variations of the intended tool and workpiece positions, and the generated surfaces, causing dimensional errors and tolerance violations. Cutting forces are applied on the workpiece and the cutting tool, and transmitted to the rest of the machine tool. Depending on the machining system, one or more components may contribute to the resultant deflection under the cutting forces. In the case of very flexible parts such as in the case of the milling of thin walled components or the turning of slender shafts, the workpiece deflections may be the major source of dimensional errors, whereas in end milling or boring applications with long tools, the tool deflections may be the main contributors. In order to eliminate or reduce these errors several approaches can be taken. An obvious one is to reduce the cutting forces, and thus the deflections, by decreasing the feed rate or the depth of cut. Another idea is to employ additional semi-finishing or finishing passes on the surface to remove or reduce the deflections and the form errors. Both methods would result in longer machining times, and thus reduced productivity. Another usual practice in industry is to use "sizing cuts" where the deviations are measured on a test part, and then compensated in the CNC codes. Although this is a practical solution in mass production, in low volume or one-part production such as the case in the die and mould industry, this is not a viable solution. On the other hand, the prediction of the errors through structural and cutting force models can be a very effective method of dealing with form errors. In addition, these models can be used to determine the optimal cutting conditions where the dimensional errors are minimized without losing productivity. In this section, form error prediction will be demonstrated for two cases: turning and milling. However, similar methodologies can be used for other machining processes as well.

Consider a turning application where the outside diameter of a slender part is machined as shown in Figure 3.6. These parts are usually supported at the tail stock as well for increased rigidity, but even then the deflections can be substantial under the radial force. The part can be modelled as a beam based on its original diameter as usually very small stock is removed in finish cuts. Then, the following beam equation can be used for the prediction of the form errors:

$$y(x) = F_r x^2 \frac{3L\left[(L-x)^2 - L^2\right] + x\left[3L^2 - (L-x)^2\right]}{12EIL^3} \qquad (3.21)$$

where x is the position of the tool along the part measured from the fixed end, and E and L are the modulus of elasticity and the length of the part, respectively. I is the area moment of the part based on its original diameter; however, the average of the original and the final diameters can also be used. In cases where the difference between the original and the final diameters is significant, a segmented beam model can also be used to increase the accuracy of the predictions as it will be shown for the deflection analysis of slender end mills later in this section. Note that the deflection given by Equation 3.21 varies as the tool travels along the part resulting in different errors. The maximum error occurs at $x/L=0.6$. The deflections of the tool, tail stock and the spindle under the same radial cutting force can be added to the part deflection if they are significant.

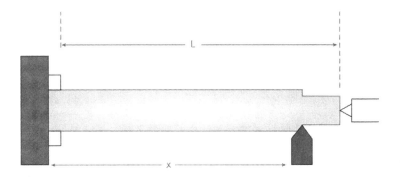

Figure 3.6. Turning of a slender workpiece resulting in form errors

Form error modelling in peripheral milling is presented next. In peripheral milling the work piece surface is generated as the cutting teeth intersect the finish surface. These points are called the *surface generation points* as shown in Figure 3.7. As the cutter rotates, these points move along the axial direction due to the helical flutes, completing the surface profile at a certain feed position along the *x*-axis. The surface generation points z_{cj} corresponding to a certain angular orientation of the cutter, ϕ, can be determined from the following relation:

$$\phi_j(z_{cj}) = \phi + j\phi_p - \frac{\tan\beta}{R} = \begin{cases} 0 & \text{for up milling} \\ \pi & \text{for down milling} \end{cases} \quad (3.22)$$

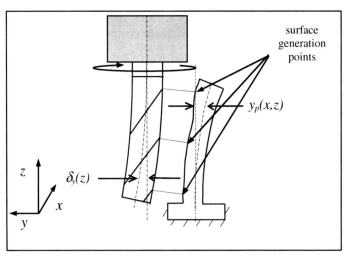

Figure 3.7. Surface generation in peripheral milling

The surface generation points can then be resolved from the above equation as follows:

$$z_{cj}(\phi) = \frac{R(\phi + j\phi_p)}{\tan \beta} \quad \text{for up milling}$$

$$z_{cj}(\phi) = \frac{R(\phi + j\phi_p - \pi)}{\tan \beta} \quad \text{for down milling}$$

(3.23)

As the surface is generated point by point by different teeth resulting in *helix marks* on the surface, in helical end milling the surface finish is not as good as the finish that would be obtained by a zero-helix tool. In case of non helical end milling, the whole surface profile at a certain feed location is generated by a single tooth as the immersion angle does not vary along the axial direction. Thus, the helix marks do not exits with zero-helix end mills, and a better surface finish is obtained. However, helical flutes result in much smaller force fluctuations, lower peak forces, and thus a smoother cutting action with reduced impacts. In addition, helical flutes improve chip evacuation.

The deflections of the tool and the workpiece in the normal direction to the finish surface are imprinted on the surface resulting in form errors. The form error can be defined as the deviation of a surface from its intended, or nominal, position. In the case of peripheral milling, the deflections of the tool and the part in the direction normal to the finished surface cause the form errors as shown in Figure 3.7. Then, the total form error at a certain position on the surface, $e(x,z)$, can be written as follows:

$$e(x,z) = \delta_y(z) - y_p(x,z) \tag{3.24}$$

where $\delta_y(z)$ is the tool deflection at an axial position z, and $y_p(x,z)$ is the work deflection at the position (x,z).

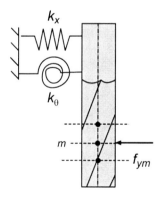

Figure 3.8. View of an end mill showing differential forces

The end mill deflections can be predicted using a cantilever beam model with clamping stiffness as shown in Figure 3.8. k_x and k_θ represent the linear and torsional clamping stiffness at the holder-tool interface. They can be identified experimentally for a certain tool-holder pair [8, 9].

The cutting tool is divided into n elements along the axial direction. The normal force in the mth element, f_{ym}, can be written as:

$$f_{ym}(\phi) = -\frac{K_t f_t R}{4\tan\beta} \sum_{j=1}^{N} \left[\left(2\phi_j(z) - \sin 2\phi_j(z)\right) + K_r \cos 2\phi_j(z) \right]_{z_{m-1}}^{z_m} \quad (3.25)$$

where z_m represents the axis boundary of the cutter in element m shown in Figure 3.8. The elemental cutting forces are equally split by the nodes m and $(m-1)$ bounding the tool element $(m-1)$. The deflection at a node k caused by the force applied at the node m is given by the cantilever beam formulation as [8, 9]:

$$\delta_y(k,m) = \frac{f_{ym} z_m^2}{6EI}(3v_m - v_k) + \frac{f_{ym}}{k_x} + \frac{f_{ym} v_m v_k}{k_\theta} \quad \text{for } 0 < v_k < v_m$$

$$\delta_y(k,m) = \frac{f_{ym} v_m^2}{6EI}(3v_k - v_m) + \frac{f_{ym}}{k_x} + \frac{f_{ym} v_m v_k}{k_\theta} \quad \text{for } v_m < v_k \quad (3.26)$$

where E is the Young's Modulus, I is the area moment of inertia of the tool, $v_k = L - z_k$, L being the gauge length of the cutter. The total static defection at the nodal station k can be calculated by the superposition of the deflections produced by all $(n+1)$ nodal forces:

$$\delta_y(k) = \sum_{m=1}^{n+1} \delta_y(k,m) \quad (3.27)$$

The tool deflections at the surface generation points can be determined from Equation 3.27 and substituted into Equation 3.24 to determine the form errors.

The area moment of inertia must take the affect of the flutes into account for improved predictions. The use of an equivalent tool radius, $R_e = sR$, where $s=0.8$ for common end mill geometries was demonstrated to yield reasonably accurate predictions by Kops and Vo [10]. An improved method of tool compliance modelling is given by Kivanc and Budak [11] where the end mill deflections were approximated using a segmented beam model. For such a case if a load is applied at the tip of the tool the maximum deflection is given by [11]:

$$y_{max} = \frac{FL1^3}{3EI1} + \frac{1}{6}\frac{FL1(L2-L1)(L2+2L1)}{EI2} + \frac{1}{6}\frac{FL2(L2-L1)(2L2+L1)}{EI2} \quad (3.28)$$

where, $L1$ is the flute length, $L2$ is the overall length, F is the point load, $I1$ is the moment of inertia of the part with flute and $I2$ is the moment of inertia of the part without flute. In the case of distributed forces and the existence of clamping stiffness, a formulation similar to Equation 3.28 can be derived. Due to the complexity of the cutter cross-section along its axis, the inertia calculation is the most difficult aspect of the static analysis. The cross sections of some end mills are shown in Figure 3.9.

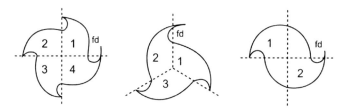

Figure 3.9. Cross sections of 4-flute, 3-flute and 2-flute end mills

In order to determine the inertia of the whole cross section, the inertia of region 1 is first derived, and the inertia of other regions are obtained by transformation [12]. The total inertia of the cross section is then obtained by summing the inertia of all regions. The inertia of region 1 is derived by computing the equivalent radius R_{eq} in terms of the radius r of the arc and position of the centre of the arc (a) [12]:

$$R_{eq4-flute}(\theta) = a.\sin(\theta) + \sqrt{(r^2 - a^2) + a^2.\sin^2(\theta)} \qquad 0 < \theta \leq \pi/2$$

$$R_{eq3-flute}(\theta) = a.\cos(\theta + \frac{\pi}{3}) + \sqrt{(r^2 - a^2) + a^2.\cos^2(\theta + \frac{\pi}{3})} \qquad 0 < \theta \leq 2\pi/3 \quad (3.29)$$

$$R_{eq2-flute}(\theta) = -a.\cos(\theta) + \sqrt{(r^2 - a^2) + a^2.\cos^2(\theta)} \qquad 0 < \theta \leq \pi$$

The moment of inertia of region 1 of a 4-Flute end mill about x- and y- axes can be written as:

$$I_{xx4-flute} = \left[\int_0^{\frac{\pi}{2}} \int_0^{R_{eq4-flute}(\theta)} \rho^3.\sin^2(\theta).d\rho.d\theta\right] - \left[\frac{1}{8}\pi(\frac{fd}{2})^4 + \frac{\pi(\frac{fd}{2})^2}{2}.(r + a - \frac{fd}{2})^2\right]$$

$$I_{yy4-flute} = \left[\int_0^{\frac{\pi}{2}} \int_0^{R_{eq4-flute}(\theta)} \rho^3.\cos^2(\theta).d\rho.d\theta\right] - \left[\frac{1}{8}\pi(\frac{fd}{2})^4\right]$$

(3.30)

where $0 < \rho \leq R_{eq}(\theta)$. The same formulation can be written for region 1 of the 3-flute and 2-flute tool. After transforming the inertia of region 1, the total inertias are found as follows:

$$I_{xx4-flute,TOT} = I_{yy4-flute,TOT} = 2(I_{xx4-flute} + I_{yy4-flute})$$
$$I_{xx3-flute,TOT} = I_{yy3-flute,TOT} = 1.5(I_{xx4-flute} + I_{yy4-flute}) \quad (3.31)$$
$$I_{xx2-flute,TOT} = 2(I_{xx2-flute}), I_{yy2-flute,TOT} = 2(I_{yy2-flute})$$

Finally, using finite element analysis simplified equations were also derived to predict deflections of tools for given geometric parameters and density:

$$deflection_{max} = C \cdot \frac{F}{E} \left[\frac{L1^3}{D1^4} + \frac{(L2^3 - L1^3)}{D2^4} \right]^N \quad (3.32)$$

where F is the applied force E is the modulus of elasticity (MPa) of the tool material, $D1$ and $D2$ are the mill and the shank diameters respectively. The geometric properties of the end mill are in mm. The constant C is 9.05, 8.30 and 7.93 and N is 0.950, 0.965 and 0.974 for 4-flute, 3-flute and 2-flute cutters, respectively.

After the tool is modelled accurately, the clamping stiffness must also be known for the total tool deflection. Depending on the tool and clamping conditions, the contribution of the clamping flexibility to the total deflection of the tool can be significant. Since the contact area is a function of tool diameter and contact length, the magnitudes of the interference displacements can be related to them. The elastic displacement in the connection can be determined as [13]:

$$\delta = \frac{2cq}{\pi d} \quad (3.33)$$

where c is the contact compliance coefficient, $q=F/L$ is the force per contact length and d is the tool diameter. Therefore, the clamping stiffness can be expressed as:

$$k = \frac{F^m / n}{\delta} = \frac{\pi L d}{2c} \quad (3.34)$$

where n is a constant used to compensate the effect of the small lengths of contact. Experiments showed that it is between 3 to 2, being close to 3 for smaller contact lengths (L<30 mm). The effect of tool materials is represented by m. For carbide it is close to 1 and for HSS tools it is 0.9. Coefficient c should be experimentally determined for a connection. After many static deflection tests with different clamping conditions, c was determined to be approximately 0.07 for holders without collets such as power chucks and shrink fit holders whereas for collet type holders c changes substantially (0.05-0.15) depending on the type of the collet.

Workpieces also deflect under cutting forces contributing to form errors. In general, the Finite Element Method (FEM) can be used to determine the structural deformations of the workpiece. The elemental cutting forces in the normal y-direction given by Equation 3.25 are to be used as the force vector. For the cases where the part is very thin such as a turbine blade or a thin plate, the change in the structural properties of the workpiece due to removed material can be very important for accurate prediction of the deflections [8, 14]. In addition, the tool-work contact, and thus the force application points change as the tool moves along the feed direction. Therefore, the form error prediction due to workpiece deflections in milling requires that the FEM solutions be repeated many times in order to consider these special effects, *i.e.*, varying part thickness and force location.

Figure 3.10 shows the work piece model which is used by Budak [8, 14] for deflection calculations. The part thickness is reduced from t_u to t_c at the cutting zone where the cutter enters the part at point *B* and exits at point *A* in the down milling mode. The nodal forces on the tool are applied in the opposite direction on the corresponding nodes in the cutting zone. For a down milling case, the cutting teeth on an end mill with a positive helix angle enters the cut at the bottom of the part where it is the most rigid. However, the tool deflections at these points are high. As the tool rotates, the contact points move along the axial direction where tool deflections are much smaller. For a plate-like structure, however, these are the most flexible sections of the part resulting in high workpiece deflections. Therefore, depending on the application, both part and tool deflections can be very significant and must be included in the calculations. The form error calculations at a certain location of the tool result in the surface profile at that position. After repeating this at many locations along the feed direction, the complete form error map of the surface is obtained.

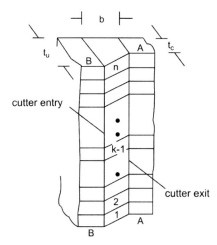

Figure 3.10. Work piece model for used in deflection and form error analysis

The deflections will change the radial depth of cut, and thus the start and exit immersions angles, ϕ_{st} and ϕ_{ex}, as follows:

$$\phi_{st}(z) = \pi - \cos^{-1}\left(1 - \frac{w_f(z)}{R}\right) \quad \text{down milling} \tag{3.35}$$

$$\phi_{ex}(z) = \cos^{-1}\left(1 - \frac{w_f(z)}{R}\right) \quad \text{up milling}$$

The effective start and exit angles vary along the axial, z-direction as the actual value of the desired width of cut, w, changes due to deflections:

$$w_f(z) = w + \delta_y(z) - y_p(x,z) \tag{3.36}$$

The effective start and exit angles given by Equation 3.35 must be used in the force and form error predictions for accurate results when the deflections are comparable to the radial depth of cut. The force model which includes these effects was named as the *flexible force model* by Budak [8, 14]. The solution is an iterative one as the deflections and forces depend on each other.

Peripheral milling of a cantilever plate made out of titanium (Ti6Al4) is considered as an example application. The plate is very flexible with dimensions of 64x34x2.45 mm. Its flexibility is further increased by reducing the thickness by 0.65 mm down to 1.8 mm in a single milling pass, *i.e.*, the axial depth of cut is 34 mm. A 19 mm carbide end mill with a 30° helix and single flute was used in order to eliminate run out effects. The tool gauge length is 55.6 mm and the linear clamping stiffness was measured to be 19.8 kN/mm where the torsional clamping stiffness was negligible. In order to constrain deflections and eliminate work-tool separation, a very small feed rate of f_t=0.008 mm/tooth was used. The cutting force coefficients were identified from the milling tests using the exponential force model:

$$K_T = 207\,(\text{MPa}), \quad p = 0.67, \quad K_R = 1.39, \quad q = 0.043 \tag{3.37}$$

The Young's Modulus of the tool and work materials is 620 GPa and 110 GPa, respectively. Experimentally measured and simulated form errors on the plate are shown in Figure 3.11. Only the flexible force model predictions are shown as the deflections are very high compared to the radial depth. The form error at the cantilevered edge of the plate is due to the tool deflection, and it is approximately 30 μm. The maximum form error occurs close to the free end of the plate where it is most flexible. The error increases from the start of the milling at the left side to the end due to the reduced thickness and increased flexibility of the plate. The rigid model, where the deflection-process interaction is neglected, overestimates the form errors significantly, about 150 μm at the maximum error location [8, 14]. The flexile model predictions agree with the measurements as shown in the figure.

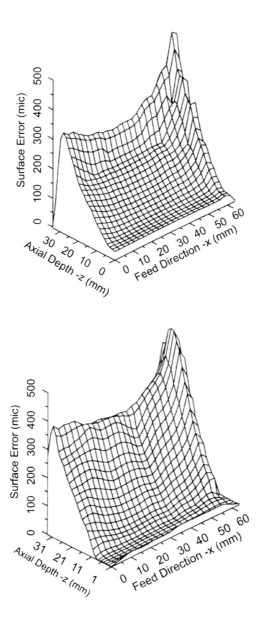

Figure 3.11. Measured and simulated form errors in the peripheral milling of a cantilever plate

3.1.3 The Control and Minimization of Form Errors

In machining operations, the usual practice is to use a constant feed rate for a machining cycle. In cases where there are tolerance violations due to excessive deflections, the feed may have to be reduced in the whole cycle, since even the maximum form error occurs only at a specific location. The *Feed rate scheduling method* proposed by Budak and Altintas [8, 14] can be used to constraint form errors, reducing cycle times. This method is based on an estimation of the maximum allowable feed rate for a specified dimensional tolerance. As in general the form errors vary along the tool path, different feed rates are obtained for different locations. The algorithm is an iterative one since the relation between the deflections and the chip thickness is not linear. For a dimensional tolerance value of (T) and feed location x, the feed rate at an iteration step m can be determined as follows:

$$f_t(x,m) = f_t(x,m-1)\frac{T}{e_{max}(x)} \tag{3.38}$$

where emax(x) is the maximum dimensional error obtained in the iteration step m-1. The starting value of the feed rate can be taken as the feed rate determined in the previous x location. After all feeds are determined along the tool path, the total machining time can be calculated. In a simulation program, different radial depths can be used to identify the milling conditions for the minimum machining time [8].

As a demonstration, the same plate presented in the previous section was machined using the scheduled feed rate for the allowable form error of 250 µm. The scheduled feed rates used in the test and the measured resulting form errors are shown in Figure 3.12. Again, the flexible force model was used in the simulations. The figure also shows that the form errors are kept within the tolerance using the scheduled feed rates.

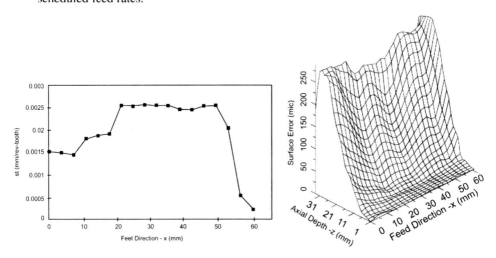

Figure 3.12. Form errors measured on the plate after it is machined with the scheduled feeds as shown

In the precision milling of highly flexible systems deflections can be very high, and in order to maintain tolerance integrity of the part, and slow feed rates, and thus small material removal rates (MRR) may have to be used resulting in low productivity. However, it may be possible to determine certain milling conditions which minimize form errors without sacrificing the productivity significantly [9]. The material removal rate per revolution, MPR, can be used as a measure of the productivity:

$$MPR = awf_t N \tag{3.39}$$

where f_t is the feed rate per tooth, N is the number of teeth on the cutter, and a and w are the axial and radial depth of cuts, respectively. The above equation indicates that the material removal rate is linearly proportional to the feed rate and depth of cuts. The relations between the force and the surface errors, and the cutting conditions, on the other hand, are nonlinear. Therefore, it may be possible to determine w and f_t which produce the required tolerances without sacrificing, perhaps even increasing, the productivity. A suitable index for this purpose is the ratio of the maximum dimensional error, e_{max}, to the MPR which will be called the *specific maximum surface error* (SMSE):

$$SMSE(w, f_t) = \frac{e_{max}(w, f_t)}{MPR(w, f_t)} \tag{3.40}$$

In other words, the SMSE shows the maximum form error generated in order to remove 1 mm³ of material for a certain pair of w and f_t. Therefore, the optimization procedure is simply the identification of w and f_t which minimize SMSE. This method is applicable to both up and down milling in identifying optimal or preferred machining conditions for reduced form errors. However, outstanding results are obtained for up milling due to the increased effect of the radial depth on the SMSE. This will be demonstrated by an example later; however, the mechanism will be explained first. In up milling, the components of the tangential, F_t, and radial, F_r, forces in the normal y-direction are opposite to each other, resulting in cancellations, and thus a reduced total force in that direction, F_y. The same mechanism results in reduced forces in the feed direction, F_x, which do not contribute to the form errors. It is obvious that the radial depth of cut has the highest effect on this mechanism. Thus, if we can determine the radial depth of cut which results in a maximum cancellation of the forces in the y-direction, then we obtain the optimal value of the radial depth. As the force will be close to zero, so will be the deflections, and we can increase the feed rate until the limit of the surface roughness is reached. Since the peak milling force in the y-direction cannot be expressed analytically, the average forces will be used in the analysis. There are two assumptions in this approach. First of all, tool and part deflections, and the form errors, do not depend only on the amplitude of the force, but also its distribution as well. In addition, the peak forces can be quite different than the average forces. However, by keeping all the other parameters constant, *i.e.*, tool geometry and axial depth of cut, one can easily justify that the minimization of the average F_y force by selection of the radial depth of cut would result in the optimal conditions for our purpose. The average force in peripheral milling can be expressed as follows:

$$\overline{F}_y = \frac{Na}{4\pi} K_t f_t (\phi_{ex} - \frac{1}{2}\sin 2\phi_{ex} - K_r \sin^2 \phi_{ex}) \qquad (3.41)$$

The exit angle and the radial depth of cut are related as given in Equation 3.13. The magnitude of the average normal force with respect to the exit angle can be determined by differentiating the square of Equation 3.40 with respect to ϕ_{ex}. It results in an implicit algebraic equation which can be approximately solved by numerical techniques [9], and the following equation is obtained:

$$\phi_{ex}^o = 60.8 K_r - 15.5 K_r^2 \qquad (3.42)$$

where ϕ_{ex}^o is the optimal value of the exit angle which minimizes the average normal force. Note that since K_r in Equation 3.41 depends on the average chip thickness, i.e., the feed rate and the radial depth, and thus the exit angle. Therefore, either an average value of K_r or an iterative solution is to be used. Note that this is only for an analytical demonstration, accurate variations of the form errors by the radial depth of cut can easily be determined using the simulation based on the models presented here.

Up milling experiments were performed on free machining steel by using a 19 mm diameter carbide end mill with 30° helix and 50 mm gauge length to demonstrate the optimal selection of cutting conditions. The work piece was rigid and the form errors were only due to tool deflections. The axial depth of cut was 19 mm in all tests. The linear tool clamping stiffness was measured to be 25 kN/mm. Cutting force coefficients were calibrated as: K_T=1140 (MPa), K_R=0.470, p=0.28, q=0.078. The maximum from error and SMSE were simulated for a range of radial depths and feed rates, and the results indicate that the optimal radial depth is close to 3.3. mm [8, 9]. In order to verify these results, several milling tests were performed with different radial depths and feeds which are shown in Figure 3.13. The figure shows that there is a very good agreement between the experimental and simulation results, and the form error is minimal for radial depth of about 3.35 mm. Compared to 1 mm radial depth, the MRR is more than tripled even though the form errors are almost the same. In addition, at a 3.35 radial depth, the feed rate can be increased without affecting the form errors significantly, which further increases the MRR significantly.

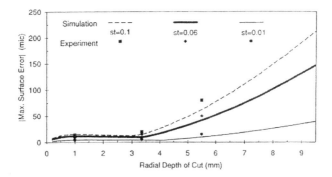

Figure 3.13. Measured and predicted maximum form errors for different radial depths and feed rates

3.2 Machine Tool Dynamics

Dynamic rigidity is one of the most critical characteristics of machine tools especially for high precision and high performance machining applications. It determines the dynamic response of the machine structure to cutting forces and inertial loads during the acceleration and the deceleration of the axes. High amplitude vibrations in response to these loads may result in poor machined part quality and potential damage to the machine. A machine tool's dynamic rigidity depends on many factors such as its configuration, size, construction method, etc. The overall dynamic rigidity in a machining system depends on all of the components involved, *i.e.*, machine tool, tooling, fixtures, workpiece, etc. Therefore, the rigidity of all components in a machining system is critical as the one with the lowest rigidity usually determines the rigidity of the whole system.

Machine tool vibrations exhibit some special characteristics compared to other machinery. The most important type of machine tool vibrations, chatter, is a result of the interaction between the cutting process and the machine structure. The mechanism responsible for this is called the "regeneration of waviness" where the vibration waves imprinted on the cut surface in the previous pass and the present vibrations result in a modulated chip thickness causing periodic cutting forces which further excite the structures. It must be noted that chatter may develop even if there is no external periodic excitation. Thus, chatter is a self-excited vibration type which develops at one of the natural modes of the structures, normally the most flexible one at the point of contact between the cutting tool and the work material. Therefore, the dynamic rigidity, which can be represented by the receptance or frequency response function of the structure at the point of interest, is the fundamental information to be used in the machine tool dynamics and chatter stability analyses. This is different than most other structural analysis applications where the focus is mainly on the modal frequencies in order to predict and avoid the resonance. In the analysis of machine tool dynamics, the amplitude of the dynamic response is the most relevant and important information. In this section, important aspects of experimental methods are reviewed, and a novel analytical method based on structural dynamics methods for machine tool dynamics modelling are presented.

3.2.1 Experimental Methods

The most common method of machine tool dynamic analysis has been through testing, although the application of analytical and numerical methods are also used in modelling dynamic characteristics of machine tools, especially in recent years. The harmonic response of machine tools is of interest almost always as it is the required form for chatter stability analysis. The frequency response function (FRF) or the transfer function (TF) can be generated using several experimental methods. The main idea is to excite the structure at a certain frequency and location, and determine its dynamic response at the same or a different point. When this is repeated for a range of frequencies, the frequency response or the transfer function at the relevant point is obtained. The excitation to the structure can be provided by an electromagnetic or hydraulic shaker where its tip is fastened to the point of interest. Although the response can be measured by many different sensor types such as displacement or velocity transducers, the most common method is to use accelerometers. As the shaker generates a harmonic force at a certain frequency both the force and the response signals are recorded. Then, the frequency response at this particular frequency can be directly determined from the ratio of the response amplitude to the force amplitude. The phase information can also be obtained from the recorded data. Although the data processing part is quite straightforward, this method has several disadvantages. First of all, the shaker can be a costly instrument compared to other alternatives. Also, it needs to be fastened to the structure which requires some sort of permanent modification such as drilling and threading a hole. Furthermore, the test can be quite time consuming as it needs to be repeated for many frequencies in the range of interest.

A more practical approach is to use the Fourier analysis where an impulsive force instead of a harmonic one can be used to excite a wide range of frequencies in one test. This can be done since theoretically an ideal impulse function with infinitesimal impulse period, *i.e.*, 0^+, contains an infinite number of frequencies which can be shown by Fourier analysis. Thus, the idea is to use the Fourier transform of input and output signals to determine the entire frequency response function in one test. This approach has been commonly used in impact tests where a point on the structure is excited using an instrumented hammer and the response at the same or another point is measured by a sensor, usually an accelerometer. A schematic view of a typical impact test setup is shown in Figure 3.14. The impact hammer is instrumented with a force sensor to capture the applied impact force. Piezoelectric crystals are commonly used sensor types for the force and acceleration measurements, and they may need charge amplification before the data can be acquired by a computer through analogue to digital conversion.

There are several important practical considerations which may affect the measurement accuracy and quality significantly. First of all, in order to increase the measurement quality and increase signal-to-noise ratio (SNR) low noise cables with the minimum possible length should be used. Sensor selection is another critical step in dynamic measurements. The accelerometer size affects its sensitivity and dynamic bandwidth as well as its weight, and thus has to be selected properly. An accelerometer with a large mass will affect the dynamics of the structure during testing due to the added mass yielding erroneous measurements, whereas a very small accelerometer may not have the required sensitivity especially in the low frequency range. The hammer size and its tip geometry are other important decisions in the dynamic tests. The hammer size and the tip must be selected properly to provide sufficient excitation to the structure for the required frequency range. A small hammer may not generate enough excitation to generate a response with good SNR whereas heavy hammers or soft tips increase the contact time at the impact test point reducing the frequency bandwidth of the force, and thus the frequency range of the transfer function. Although there are general guidelines such as the ones summarized here, a certain experience is required in order to obtain good dynamic measurements on machine tools.

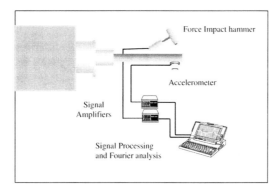

Figure 3.14. Impact test supup used for frequency response measurements

In order to obtain the TF, the measured input and the output are transferred to the frequency domain using the Fourier analysis. As a result, the TF between the response measurement point (i) and the force application point (j) can be obtained as follows:

$$G_{ij}(\omega) = \frac{X_i(\omega)}{F_j(\omega)} \tag{3.43}$$

When the response and the excitation at the same point (i) are considered, then the direct TF, $G_{ii}(\omega)$, at this point is obtained. Note that since usually the transfer function in terms of displacement, X, is needed in the dynamic analysis, the accelerometer signal needs to be integrated. In addition, several signal processing techniques such as windowing and filtering are used to improve the measurement quality [7].

Structural dynamic parameters, *i.e.*, mass, stiffness and damping, at the point of interest can be identified from the measured TF using modal analysis methods [16]. In general, in the TF given by Equation 3.42 there can be contributions of many structural modes of the measured structure. Therefore, for a general case, multi mode identification methods using curve fitting and error minimization techniques must be used [16]. However, for most structures these modes are well separated, and only a couple of modes (sometimes even just one) dominate the dynamic response of the structure at the point of interest. This is why single mode or single-degree-of-freedom (SDOF) system methods are used in many cases. This is even more so for machine tools as the amplitude of the response is the most important information for cutting stability, and usually one mode due to a certain component such as a long slender tool dominate the TF. In the following section the single mode identification methods are briefly reviewed.

The real part of the frequency response of a SDOF system is shown in Figure 3.15. In frequency response measurements, the real, imaginary, or the total response can be used for the identification of structural parameters. As it can be seen from the figure, the low frequency response, *i.e.*, for frequencies much lower than the natural frequency, is controlled by the stiffness similar to static deflection. Thus, in general, the low frequency response can be used to identify the stiffness of the system. However, in most TF measurements the low frequency signal is not very reliable if an accelerometer is used for the response measurement. In those cases, identification of the stiffness and damping from the resonance peak is much more reliable. The damping ratio ζ can be obtained from the half power points as shown in the figure where $\Delta\Omega = 2\zeta$, and the stiffness from the peak amplitude. It must be mentioned that the stiffness and damping at a particular point can be very much affected by the interfaces on a machine tool between the tooling and the machine, and among the machine tool components. The interface damping can significantly increase the overall damping of a machine tool whereas the reverse is true for stiffness. Once the TF at a point is obtained, the dynamics rigidity of the machine tool and its stability against chatter vibrations can be evaluated.

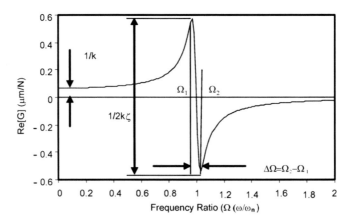

Figure 3.15. Real part of the frequency response of a single mode system

Although for chatter stability analysis only the transfer function at the tool tip is required, the measurements at different points, and between different points, should be performed to determine the full dynamic behaviour of the machine tool. By performing the impact test at different points either by moving the response or the impact points the TF matrix, [G], can be obtained where:

$$\{X\} = [G]\{F\} \qquad (3.44)$$

The TF matrix is also useful to determine the mode shapes. In many cases it is necessary to determine the component which is responsible for vibrations. This cannot be determined through a single TF measurement whereas it can easily be identified from mode shapes. This can be done using a modal analysis software or for most practical purposes by just using G_{ij}. If for example, the excitation is given at point 1, the amplitudes of G_{11}, G_{12}, G_{13} etc. at the first resonance peak can be used to determine the first mode shape, and in the second peak for the second mode shape etc.

Figure 3.16 shows the mode shapes of the tool assembly on a horizontal machining center. The assembly includes a long tool adapter which is necessary for accessibility to deep pockets on the part but makes the system very flexible. As it can be seen from the figure, the first mode (260 Hz) is the spindle mode as the displacement in the rest of the assembly is relatively small. The second and the third modes belong to the holder and the adapter, respectively. The last mode shown at 5740 Hz is the tool mode where the elastic displacement of the tool can be clearly seen.

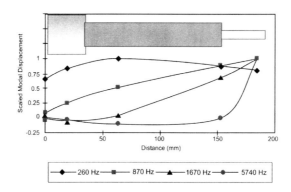

Figure 3.16. Mode shapes of a tool assembly on a horizontal machining center

3.2.2 The Analytical Modelling of Machine Tool Dynamics

Machine tool dynamics can be measured using the appropriate devices and performing proper analysis. However, for a different combination of the system components, *e.g.*, when the overhang length of the tool is changed, a new test will be required since the system dynamics will change. Therefore, the use of experimental modal analysis may not be always very practical, especially for production applications where many combinations are possible. In order to reduce experimentation, the receptance coupling theory of structural dynamics can be used. This method is based on the coupling of component dynamics analytically, and can be used effectively for modelling the dynamics of complex structures provided that the component dynamics are known. It can also be applied as a semi-analytical method where some of the component dynamics are measured such as in [17, 18] where the dynamics of the spindle-holder subassembly is obtained experimentally. Then, it is coupled with the dynamics of the tool, which is obtained analytically by considering the tool as a beam. In a recent study, Ertürk et al. [19-21] presented an analytical model to predict the tool point FRF by modelling all components of the spindle-holder-tool assembly analytically. They used the Timoshenko beam theory, receptance coupling and structural modification techniques. Due to the high length to diameter ratios of many spindle and tool sections, the Euler-Bernoulli beam model may result in considerable errors in prediction of modal frequencies which is significantly improved using the Timoshenko beam formulation. Their model is briefly given in this section.

Spindle, holder and tool are modelled as multi-segment beams by using the Timoshenko beam theory. The individual multi-segment components are formed by coupling the end point receptances of uniform beams rigidly. The determination of the end point receptances of a uniform Timoshenko beam with free end conditions is given in [19] in detail. The end point receptance matrix of a beam A can be represented as:

$$[A] = \begin{bmatrix} [A_{11}] & [A_{12}] \\ [A_{21}] & [A_{22}] \end{bmatrix} \tag{3.45}$$

where submatrices of the above matrix include the point and transfer receptance functions of the segment end points (1) and (2). For example, the point receptance matrix of node $A1$ in beam A is given as:

$$[A_{11}] = \begin{bmatrix} H_{A1A1} & L_{A1A1} \\ N_{A1A1} & P_{A1A1} \end{bmatrix} \tag{3.46}$$

The receptance functions, which are denoted by letters H, N, L and P, are defined as follows:

$$y_j = H_{jk} \cdot f_k \quad \theta_j = N_{jk} \cdot f_k$$
$$y_j = L_{jk} \cdot m_k \quad \theta_j = P_{jk} \cdot m_k$$

where y and θ represent the linear and angular displacements, respectively, and f and m are the forces and the moments, respectively, at the points i and j. Two beams, A and B, can be coupled dynamically using rigid receptance coupling and the receptance matrix of the resulting two-segment beam C can be obtained as follows:

$$[C] = \begin{bmatrix} [C_{11}] & [C_{12}] \\ [C_{21}] & [C_{22}] \end{bmatrix} \quad (3.47)$$

where:

$$[C_{11}] = [A_{11}] - [A_{12}] \cdot [[A_{22}] + [B_{11}]]^{-1} \cdot [A_{21}]$$
$$[C_{12}] = [A_{12}] \cdot [[A_{22}] + [B_{11}]]^{-1} \cdot [B_{12}]$$
$$[C_{21}] = [B_{21}] \cdot [[A_{22}] + [B_{11}]]^{-1} \cdot [A_{21}]$$
$$[C_{22}] = [B_{22}] - [B_{21}] \cdot [[A_{22}] + [B_{11}]]^{-1} \cdot [B_{12}]$$

(3.48)

By following the same formulation, one might continue coupling more segments like a chain to form an n-segment beam. In order to include the dynamics of bearings, the structural modification technique presented by Özgüven [21] can be used as shown in detail in [20]. In this case, the dynamic structural modification matrix represents the translational and rotational, stiffness and damping information of the bearings. The final step is to couple the main system components to obtain the tool point FRF. However, these components should be coupled elastically due to the flexibility and damping at the contacts. When the end point receptances of the spindle on bearings (S) are coupled with those of the holder (H), the end point receptance matrices of the spindle-holder assembly (SH) can be obtained from:

$$[SH_{11}] = [H_{11}] - [H_{12}] \Big[[H_{22}] + [K_{sh}]^{-1} + [S_{11}] \Big]^{-1} [H_{21}] \quad (3.49)$$

$[K_{sh}]$ is the complex stiffness matrix representing the spindle-holder interface dynamics. Note that the receptance matrix $[SH_{11}]$ is very similar to $[C_{11}]$ given in Equation 3.48 with the addition of $[Ksh]^{-1}$ only. Finally, the tool (T) can be added to the spindle-holder (SH) system to obtain the end point FRFs of spindle-holder-tool (SHT) assembly. The FRF required for the stability lobe diagram of a given spindle-holder-tool assembly is the one that gives the relation between the transverse displacement and force at the tool tip, which is the first element of the following:

$$[SHT_{11}] = [T_{11}] - [T_{12}] \Big[[T_{22}] + [K_{ht}]^{-1} + [SH_{11}] \Big]^{-1} [T_{21}] \quad (3.50)$$

As an example application, consider the geometry of the spindle-holder-tool combination shown in Figure 3.17. Each of the system components, *i.e.*, spindle, holder and tool are composed of several sections with different diameters and lengths which are modelled as multi-segment beams. The dimensions of the components, bearing and interface dynamical properties are given [19]. In order to verify the results of the model, the vibration modes of this assembly were calculated using the finite element method using ANSYS® 9.0. The beam element, which is based on the Timoshenko beam theory, is used by restricting the degrees of freedom other than motion in one transverse direction and flexural rotation so that the finite element model is consistent with the model. In order to represent the dynamics of bearings and spindle-holder and holder-tool interfaces, combination element COMBIN14 (Spring-Damper) of ANSYS® 9.0 is used. The natural frequencies obtained by the analytical model and the finite element solution are tabulated in Table 3.2. As can be seen from the table, the natural frequencies of the assembly obtained by the model presented in this paper and those obtained by using the finite element software are in good agreement and the maximum difference observed for the first seven modes is about 5%.

It is well known that rotary inertia and especially shear deformation are very important for non-slender components and at high frequencies. Thus, the Euler-Bernoulli beam model may result in modelling errors and Timoshenko model must be used for accurate results. The deficiency of the Euler-Bernoulli model in such a case is illustrated in Figure 3.18.

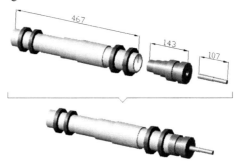

Figure 3.17. Components of the assembly used in the example

Table 3.2. Natural frequencies of the assembly used in the case study

Mode	Model [Hz]	FEA [Hz]	Diff. [%]
1	71.7	71.6	0.14
2	195	193.8	0.62
3	877.8	867.5	1.19
4	1438.3	1424.3	0.98
5	1819.5	1752.6	3.82
6	3639.3	3442.5	5.72
7	3812.5	3634.8	4.89

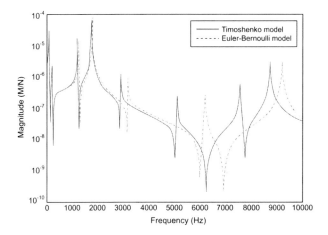

Figure 3.18. The tool point FRF comparing Timeshenko and Euler-Bernouilli beam model predictions

3.2.2.1 Connection Parameters
One important element of machine tool dynamics modelling, or the receptance coupling method in general, is the connection parameters between the components. They may affect the overall response; however there is no accurate modelling methods to predict them. Even experimental identification of these parameters can be quite challenging due to the existence of many unknown parameters in an assembly. However, this task can be simplified by analyzing the effects of parameters in each individual connection on the total response. This provides a deeper insight into the affects of connection parameters which can also be used in their faster and easier identification. For example, Figure 3.19 shows that the bearing stiffness values have a considerable effect on the first two modes of the system shown in Figure 3.17, which are the rigid body modes, whereas they have almost no effect on the remaining (elastic) modes. It was observed [20] that the dynamics of the front bearings primarily control the first rigid body mode whereas the rear bearings mainly affect the second rigid body mode. Therefore, for the system used, the spindle geometry and bearing properties have the most important effect on the first two modes. This also implies that if chatter develops in one of the first two modes, changing the holder or the tool may not help. In a very similar way, the sensitivity of tool point FRF to the spindle-holder interface dynamics is studied. It is observed that the translational stiffness at the spindle-holder interface dominantly affects the first elastic mode of the FRF. It is also observed that rotational stiffness at the same interface has almost a negligible effect on the FRF [20]. A similar analysis is performed in order to study the sensitivity of FRF to the holder-tool interface dynamics. It is observed that the translational stiffness strongly controls the second elastic mode. Similar to the spindle-holder interface, the rotational stiffness at this connection has a negligible effect on FRF. Therefore, the observations made so far indicate that, for the first elastic mode, the spindle-

holder interface is the most important link in the chain, whereas the same is true for the holder-tool interface for the second elastic mode in this case study. The connection damping values have similar effects, but on the FRFs peak amplitudes instead of the frequencies. For example, it is observed that the front bearing damping affects the FRF values at the first rigid body mode, the translational contact damping at the spindle-holder interface mainly alters the peak value of the first elastic mode, etc. [20]. The above conclusions can be used in the parametric identification of the connection dynamics of a given spindle-holder-tool assembly from experimental measurement of tool point receptance much more easily and accurately compared to previous approaches used. With the information of which connection parameters affect which mode, identification should be performed by extracting the parameters of interest from their relevant modes.

3.2.2.2 An Example Application

The measured tip point FRF of the suspended spindle shown in Figure 3.20 and the model predictions of the same FRF using both beam theories are given in Figure 3.21. Note that the inaccuracy associated with using the Euler-Bernoulli theory increases at higher frequencies. A BT40 type holder in which a carbide tool of 12.7 mm diameter and 175 mm length is inserted with an overhang length of 74 mm is assembled to the free spindle shown in Figure 3.20. Then, the tool point FRF of the free assembly is measured by an impact test. The measured FRF and the model simulation of the tool point FRF are given in Figure 3.22. In performing the effect analysis, it is observed that the spindle-holder interface controls the first mode and the holder-tool interface controls the second mode. The interface parameters identified this way are given in Table 3.3. Note that, in the parametric identification process, mainly the translational parameters are tuned and average values are used for the rotational interface dynamics.

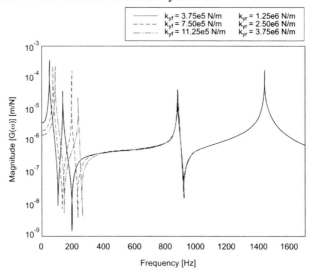

Figure 3.19. The combined effect of bearing stiffness values on the tool point FRF

Dynamic Analysis and Control 53

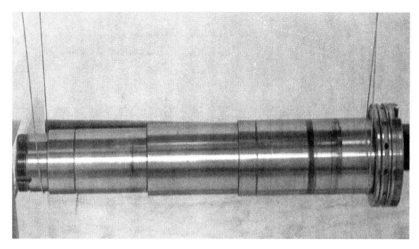

Figure 3.20. Spindle suspended for free-free measurement

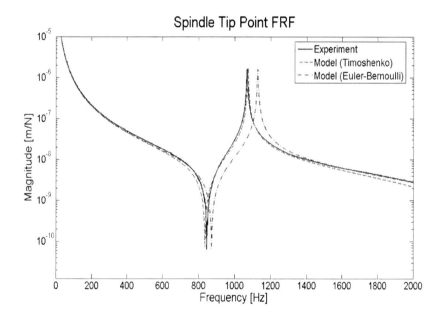

Figure 3.21. Measured and predicted FRF for the spindle

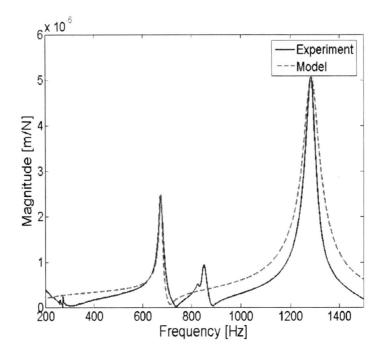

Figure 3.22. Measured and predicted tool point FRF of the assembly

Table 3.3. Connection parameters identified from the measurements

	Spindle-holder interface	Holder-tool interface
Translational stiffness	$2.65 \times 10^7 \, N/m$	$8.75 \times 10^6 \, N/m$
Translational damping	$140 \, Ns/m$	$97 \, Ns/m$
Rotational stiffness	$1.5 \times 10^6 \, Nm/rad$	$1.5 \times 10^6 \, Nm/rad$
Rotational damping	$50 \, Nms/rad$	$50 \, Nms/rad$

3.3 The Dynamic Cutting Process

Vibration in metal cutting is a significant problem as it causes damage to the finished surface, shortens tool life, degrades machine tool components and produces noise contamination. In addition, the mechanics of the cutting process is also affected by the vibrations changing the contact conditions between the tool and the material. The problem of dynamic cutting has been studied by many different groups since 60s, and it has proved to be one of the most complex problems in the area of metal cutting [22, 23]. As a result, the extent of analytical methods that can be used for the analysis and modelling of the dynamic cutting process mechanics is very limited. Therefore, in this section an overview of the dynamic effects on the cutting process will be reviewed with some general observations derived from these studies.

3.3.1 Mechanic of Dynamic Cutting

The process of dynamic cutting is shown in Figure 3.23. The tool is removing a chip from an undulated surface which was generated during the previous pass when the tool vibrated with an amplitude y0 (outer modulation or wave removing). Simultaneously, the tool is vibrating with an amplitude y (inner modulation or wave generation). The process can be visualized as a superposition of these two distinct mechanisms, *i.e.*, wave removing and wave generation. The physics of dynamic cutting can be explained in terms of shear, rake and clearance oscillations, and variation of friction forces and flank contact which is briefly outlined in the following section.

3.3.1.1 Shear Angle Oscillations
As it was demonstrated in the first section of this chapter, the shear angle is perhaps the most important dependent parameter in metal cutting. It has been observed through high speed photography that the shear angle oscillates during dynamic cutting. Although it can be considered as a result of the dynamic cutting process, the oscillatory shear angle will also result in oscillatory cutting forces. Several reasons for the oscillation of shear angle have been proposed - some of which are listed below. Depending on many factors one or more of them may cause the oscillations.

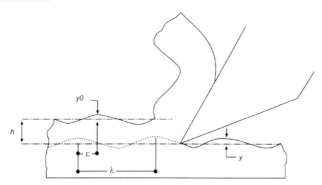

Figure 3.23. Dynamic cutting process

Consider the case of a wave removing shown in Figure 3.24 where the surface if the chip is inclined by angle δ with respect to the direction of cutting speed. The cutting forces for increasing (positive δ) and decreasing (negative δ) chip thickness are different compared to static cutting with constant chip thickness. It was also observed that a similar situation is valid for wave generation as well. This is attributed to the effect of δ on the shear angle ϕ, and expressed as follows:

$$\phi = \phi_m + C\delta \qquad (3.51)$$

where ϕ_m is the value of the shear angle during static cutting and the value of C depends on the ratio of mean chip thickness to the wavelength. This effect could generate negative damping in the process as a decreasing chip thickness leads to a larger force due to a smaller shear angle (Equation 3.51), and vice versa.

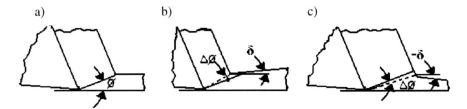

Figure 3.24. Effect of the surface slope on the shear angle

Figure 3.25 shows the wave generation process where it can easily be seen that the angle between the rake face and the normal to the cutting speed direction continuously varies. As a result, it can be concluded that the shear angle also oscillates in respond to changes in the rake angle. This can be considered another source of negative damping as an increasing rake angle due to larger amplitude vibrations leads to a decreasing cutting force.

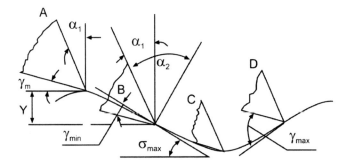

Figure 3.25. Wave generation process

Another reason for the shear angle oscillation can be attributed to the variation of the friction conditions on the rake face due to vibrations. The friction coefficient on the rake face depends on the cutting speed and the rake angle both of which continuously change during dynamic cutting. Most shear angle models such as the Merchant's equation given in Equation 3.1 indicate that the shear angle depends on the friction coefficient, and variation of the friction on the rake due to variations in the relative speed between the tool and the chip, and in the rake angle would result in variations in the shear angle. Since there is an inverse relationship between the velocity and the friction, the rubbing action may result in a negative damping in the process. Additional experimental observations [24] showed that the vibration frequency and cutting speed are two important factors affecting the shear angle oscillations.

Therefore, almost all of the parameters in cutting are interrelated which changes the cutting mechanics completely during dynamic cutting due to a strong interaction between these parameters. The foregoing discussion on the dynamic effects on the shear angle indicates that the complex mechanisms of dynamic cutting may generate sources of negative process damping which may result in process instability. Although this may be the source of chatter in some special cases, as it will be demonstrated in the next sections of this chapter, the regeneration mechanism is a much stronger source of instability and is responsible for the majority of the chatter cases.

3.3.1.2 Process Damping
The forces acting on the clearance surface do not take part in the chip formation process, and therefore they are parasitic loads on the cutting tool. Experimental and theoretical investigations showed that wear land geometry, clearance angle, tool nose radius, material properties and the coefficient of friction affect the flank forces. As some or all of these conditions vary during dynamic cutting, so do the flank forces. This mechanism has been used as an explanation for the low speed stability observed in cutting, and is referred as process damping in the cutting literature. The mechanism of low speed stability or cutting was first explained by Tlusty [22] by analyzing the variation of the clearance between the flank face of the tool and the cut surface shown in Figure 3.25. At the top (A) and bottom (B) points of the wave the clearance angle is the nominal, γ_m, being equal to the standard clearance angle of the tool. In the middle of the downward slope the clearance angle is minimum, γ_{min}, and in the middle of the upward slope it is maximum, γ_{max}. It has been shown that the decrease of the clearance angle leads to an increase of the thrust component of the cutting force. Therefore, during the half cycle from (A) to (C) the normal force is greater than that during the upward motion from (C) to (D). This variation of the thrust force is 90° out of phase with the displacements and this is why it represents damping. This damping is larger for short waves because they have steeper slopes resulting in more interfacing between the tool and the surface on the flank. The wave length of the undulations produced on the cut surface is as follows:

$$\omega = \frac{v}{f} \qquad (3.52)$$

where v is the cutting speed and f is the frequency of vibration. Therefore, a small wavelength can also be present at high speeds as well if the vibration frequency is large enough.

The flank forces can be formulated using the ploughing theory where the flank force is assumed to be proportional to the volume of the material displaced by the tool motion [25]:

$$f_y = \frac{Kw}{v_o} \frac{d}{dt}(y) \tag{3.53}$$

where f_x is the flank force normal to the workpiece surface, K is the damping coefficient, w is the width of cut, v_o is the cutting speed and y is the displacement normal to the workpiece surface. The ploughing force in the other direction, i.e., in the cutting speed direction, can be determined using the friction coefficient in the flank contact zone:

$$f_x = \mu f_y \tag{3.54}$$

3.3.1.3 Dynamic Cutting Force Coefficients (DCFC)
Dynamic cutting forces can be analyzed and modelled using the DCFC which was a very popular approach 1960s and 1970s. The main idea is that the dynamic cutting forces cannot be represented using only the real cutting force coefficient, i.e., forces in phase with the displacements, and complex terms are also added to the formulation. The outer and inner modulations can be superposed and the corresponding transfer function of the dynamic cutting process can be written as follows:

$$\begin{aligned} F_N &= w(K_{di}Y + K_{do}Y_o) \\ F_T &= w(K_{ci} + K_{co}Y_o) \end{aligned} \tag{3.55}$$

where Y is the amplitude of the vibration of the tool normal to the cut surface, F_N is the component of the cutting force in the normal direction and F_T is the tangential component of the force. Subscripts d and c are the direct and cross effects of vibrations in the Y direction on the forces in normal and tangential directions, respectively. Subscripts i and o are inner and outer modulations, respectively. There are 8 DCFC as each of K_{mn} in Equation 3.55 is considered complex:

$$K_{mn} = \mathrm{Re}(K_{mn}) + j\,\mathrm{Im}(K_{mn}) \tag{3.56}$$

The real components represent the components of the forces in phase with displacements, and can be regarded as the stiffness in the process. The imaginary components, on the other hand, are out of phase with the displacements representing the damping in the process. The determination of the DCFC is not an easy task, and the resulting data is quite scattered. There are two main experimental methods: static and dynamic. The static methods are based on the measurement of static force components whereas in the dynamic methods the cutting tool is excited by an impact or a sinusoidal signal. The main characteristics of the measured DCFC data can be summarized as follows [26]:

- The effects of cutting speed and feed on the normal force (direct coefficients) are much stronger than on the tangential force (cross coefficients). This is especially true for the imaginary coefficient $Im(K_{di})$ which is much larger than $Im(K_{ci})$.
- In general, the real parts of inner and outer modulation coefficients are equal.
- The real parts of DCFC are very little affected by the wearland whereas the imaginary components are almost in direct proportion with the flank wear.
- $Im(K_{di})$ the component is almost directly proportional to the frequency of oscillations.
- The cutting speed has a significant effect on all 8 coefficients for all the materials tested.
- The mechanical and physical properties of the workpiece material have a strong influence on the coefficients.

As a result, $Im(K_{di})$ has a special importance on the dynamic cutting process as it represents most of the process damping.

3.3.2 The Dynamic Chip Thickness and Cutting Forces

Regeneration of chip thickness due to successive cuts on a surface with a vibratory cutting tool is the most important mechanism responsible for instability in machining [27, 28]. As a result of the inner and outer modulations, and the phase between them a modulated chip thickness is obtained. The periodically varying chip thickness results in dynamic cutting forces which further excite the structures and may cause instability. Therefore, dynamic chip thickness is the fundamental source of dynamic cutting forces, and instability. In this section dynamic chip thickness and cutting force formulations for turning and milling processes will be presented. Turning represents a simpler case of dynamic cutting due to a stationary tool whereas the dynamics of milling is more complicated due to a rotating tool and an interrupted cutting action. The dynamic chip thickness and cutting force formulations will be used as the starting points in the stability analysis in the next section.

3.3.2.1 Dynamics of the Turning Process
In order to formulate the relationship between dynamic turning forces and the dynamic chip thickness, all components of the dynamic problem are transformed into the machine axes as shown in Figure 3.26b. Observing Figure 3.26a and 3.26b one can deduce that the dynamic displacements at cutting (z) direction do not affect the dynamic chip thickness. By this observation, the dynamic problem is reduced to a 2D model.

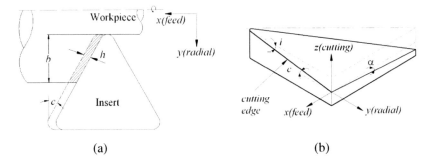

Figure 3.26. (a) Chip thickness in turning (b) 3D view of the three cutting angles on the insert

The modulated chip thickness resulting from vibrations of the tool and workpiece can be written as follows [29]:

$$h(t) = f\cos c + \left(x_c^1 + x_w^1 - x_c^0 - x_w^0\right)\cos c + \left(-y_c^1 - y_w^1 + y_c^0 + y_w^0\right)\sin c \quad (3.57)$$

where, f represents the feed per revolution, c is the side cutting edge angle, (x_c^1, x_w^1) and (y_c^1, y_w^1) are the cutter and workpiece dynamic displacements for the current pass, respectively, and (x_c^0, x_w^0) and (y_c^0, y_w^0) are the cutter and workpiece dynamic displacements for the previous pass in x and y directions, respectively. The feed term represents the static part of the chip thickness, and can be neglected since the static chip thickness does not contribute to the regeneration mechanism. Therefore, the dynamic chip thickness in turning can be defined as follows:

$$h(t) = \Delta x \cos c - \Delta y \sin c \quad (3.58)$$

where

$$\Delta x = x_c^1 + x_w^1 - x_c^0 - x_w^0$$
$$\Delta y = y_c^1 + y_w^1 - y_c^0 - y_w^0 \quad (3.59)$$

Although the dynamic problem can be considered as 2D, the cutting process is 3D due to the existence of the inclination angle. Then, the forces at the cutting edge need to be modelled by an oblique cutting model as explained in Section 3.1.1. The total force acting on the cutting edge is divided into three components: one parallel to the cutting velocity direction, F_t, one perpendicular to the plane formed by the cutting velocity, or F_t and the cutting edge, F_f, and the last one perpendicular to the other two, F_r, (Figure 3.27). Then, the dynamic cutting forces on the tool can be expressed as follows:

$$\begin{Bmatrix} F_f \\ F_r \end{Bmatrix} = -\frac{b}{\cos c} \begin{bmatrix} K_f \\ K_r \end{bmatrix} [\cos c \quad -\sin c] \begin{Bmatrix} \Delta x \\ \Delta y \end{Bmatrix} \qquad (3.60)$$

where K_f, and K_r are the corresponding cutting force coefficients and b is the depth of cut. Note that F_t is not included in the formulation as it does not take part in the regeneration mechanism. However, it is affected by the regeneration, and if needed it can be determined using the value of the dynamic chip thickness and the force coefficient in the cutting speed direction, K_t. By coordinate transformation the cutting forces can be written in the lathe coordinates as follows:

$$\begin{Bmatrix} F_x \\ F_y \end{Bmatrix} = \begin{bmatrix} \cos c & \sin c \\ -\sin c & \cos c \end{bmatrix} \begin{Bmatrix} -F_f \\ F_r \end{Bmatrix} \qquad (3.61)$$

where F_x and F_y are the cutting force components in x and y directions, respectively. Substituting Equation 3.61 into Equation 3.60 the following relationship is obtained:

$$\{F\} = b[A]\{\Delta d\} \qquad (3.62)$$

Where $\{F\}$ is the force vector and $\{\Delta d\}$ is the dynamic displacement vector both defined in the machine coordinates. The directional coefficients matrix can be expressed as:

$$[A] = \begin{bmatrix} -\cos c & \sin c \\ \sin c & \cos c \end{bmatrix} \begin{bmatrix} K_f \\ K_r \end{bmatrix} [\cos c \quad -\sin c] \qquad (3.63)$$

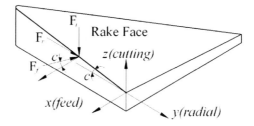

Figure 3.27. Three components of the total cutting force acting on the insert

3.3.2.2 The Dynamics of the Milling Process
In this analysis, both the milling cutter and the workpiece are considered to have two orthogonal modal directions as shown in Figure 3.28. Milling forces excite both the cutter and the workpiece causing vibrations which are imprinted on the

cutting surface. Each vibrating cutting tooth removes the wavy surface left from the previous tooth resulting in a modulated chip thickness which can be expressed as follows:

$$h_j(\phi) = s_t \sin\phi_j + (v_{j_c}^o - v_{j_w}^o) - (v_{j_c} - v_{j_w}) \tag{3.64}$$

where the feed per tooth f_t represents the static part of the chip thickness, and $\phi_j = (j-1)\phi_p + \phi$ is the angular immersion of tooth (j) for a cutter with constant pitch angle $\phi_p = 2\pi/N$ and N teeth as shown in Figure 3.28. $\phi = \Omega t$ is the angular position of the cutter measured with respect to the first tooth and corresponding to the rotational speed Ω (rad/sec). v_j and v_j^o are the dynamic displacements in the chip thickness direction due to tool and work piece vibrations for the current and previous tooth passes, for the angular position ϕ_j, and can be expressed in terms of the fixed coordinate system as follows:

$$v_{j_p} = -x_p \sin\phi_j - y_p \cos\phi_j \qquad (p=c,w) \tag{3.65}$$

where p and c indicate part and cutter, respectively. Similar to the turning, the static part in Equation 3.64, $(s_t \sin\phi_j)$, is neglected in the stability analysis, as it does not contribute to the regeneration mechanism.

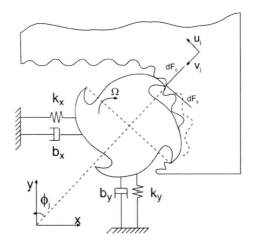

Figure 3.28. Cross-sectional view of an end mill showing differential forces

If Equation 3.65 is substituted in Equation 3.64, the following expression is obtained for the dynamic chip thickness in milling:

$$h_j(\phi) = [\Delta x \sin\phi_j + \Delta y \cos\phi_j] \tag{3.66}$$

where:

$$\Delta x = (x_c - x_c^o) - (x_w - x_w^o) \quad ; \quad \Delta y = (y_c - y_c^o) - (y_w - y_w^o) \quad (3.67)$$

where (x_c, y_c) and (x_w, y_w) are the dynamic displacements of the cutter and the work piece in the x and y directions, respectively. The dynamic cutting forces on tooth (j) in the tangential and the radial directions can be expressed as follows:

$$F_{t_j}(\phi) = K_t a h_j(\phi) \quad ; \quad F_{r_j}(\phi) = K_r F_{t_j}(\phi) \quad (3.68)$$

where a is the axial depth of cut, and K_t and K_r are the cutting force coefficients. After substituting h_j from the equation, the dynamic milling forces can be resolved in x and y directions as follows:

$$\begin{Bmatrix} F_x \\ F_y \end{Bmatrix} = \frac{1}{2} a K_t \begin{bmatrix} a_{xx} & a_{xy} \\ a_{yx} & a_{yy} \end{bmatrix} \begin{Bmatrix} \Delta x \\ \Delta y \end{Bmatrix} \quad (3.69)$$

where a_{xy} are the directional coefficients. The directional coefficients depend on the angular position of the cutter which makes Equation 3.69 time-varying :

$$\{F(t)\} = \frac{1}{2} a K_t [A(t)] \{\Delta(t)\} \quad (3.70)$$

The dynamic milling forces given by Equation 3.70 contain many frequencies. First of all, the coefficient matrix [A] has time varying directional coefficients which are periodic with the frequency being equal to the tooth passing frequency. In addition, the dynamic displacements contain the vibration or chatter frequency. As a result of the interactions between these periodic terms with different frequencies dynamic milling forces include many other frequencies in addition to the chatter frequency which is the only response frequency in turning.

3.4 The Stability of the Cutting Process

Chatter vibrations develop due to dynamic interactions between the cutting tool and the workpiece, and result in a poor surface finish and a reduced tool life. Tlusty et al. [27] and Tobias [28] identified the most powerful source of self-excitation which is associated with the structural dynamics of the machine tool and the feedback between the subsequent cuts on the same cutting surface resulting in regeneration of waviness on the cutting surfaces, and thus modulation in the chip thickness. Under certain conditions the amplitude of vibrations grows and the cutting system becomes unstable. Although chatter is always associated with vibrations, in fact it is fundamentally due to instability in the cutting system. In a particular cutting process, for a certain cutting speed and width of cut, there is a limiting depth of cut above which the system becomes unstable, and chatter develops. Chip thickness, or feedrate, has very little, if any, affect on the stability limit. Additional operations, mostly manual, are required to clean the chatter marks left on the surface. Thus, chatter vibrations result in reduced productivity, increased cost and inconsistent product quality. In this section, the dynamic forces formulated in the previous section will be used to develop the stability conditions for the turning and milling systems.

3.4.1 The Stability of Turning

The response of the cutter and the workpiece at the chatter frequency, ω_c, can be expressed as follows:

$$\{d_j(i\omega_c)\} = [G_j(i\omega_c)]\{F\}e^{i\omega_c t} \quad j=c,w \tag{3.71}$$

where $\{F\}$ represents the dynamic milling force and the transfer function matrix is given as:

$$[G_j] = \begin{bmatrix} G_{j_{xx}} & G_{j_{xy}} \\ G_{j_{yx}} & G_{j_{yy}} \end{bmatrix} \quad j=c,w \tag{3.72}$$

where G_{jxx} and G_{jyy} are the transfer functions in x and y directions, respectively, and G_{jxy} and G_{jyx} are the cross transfer functions. The vibrations in the previous pass, at time $(t-\tau)$, where τ is the delay term which is equal to one spindle revolution period, can be defined as follows:

$$\{d_j^0\} = e^{-i\omega_c \tau}\{d(i\omega_c)\} \quad j=c,w \tag{3.73}$$

By substituting Equation 3.71 into Equation 3.62:

$$\{F\}e^{i\omega_c t} = b\left(1-e^{-i\omega_c \tau}\right)[A][G(i\omega_c)]\{F\}e^{i\omega_c t} \tag{3.74}$$

The geometry of the tool and the workpiece in most of the turning operations are symmetrical and beam like structures; thus for many cases, the cross transfer functions are negligible. Then, the transfer matrix can be further simplified as follows:

$$G(i\omega_c) = G_c(i\omega_c) + G_w(i\omega_c) = \begin{bmatrix} G_{xx} & 0 \\ 0 & G_{yy} \end{bmatrix} \tag{3.75}$$

Dynamic Analysis and Control   65

where, G_{xx} and G_{yy} are the system's total transfer functions in x and y directions. Equation 3.74 has a non-trivial solution if and only if its determinant is zero:

$$\det\left[[I]+\Lambda\left[G_0(i\omega_c)\right]\right]=0 \qquad (3.76)$$

where $\left[G_0(i\omega_c)\right]=[A]\left[G(i\omega_c)\right]$, and the eigenvalue Λ is defined as follows:

$$\Lambda = b\left(e^{-i\omega_c\tau}-1\right) \qquad (3.77)$$

The solution of Equation 3.76 results in the following:

$$\Lambda = 1/\left[G_{yy}\left(K_f\sin^2 c\cos c + K_r\sin c\cos^2 c\right)+G_{xx}\left(K_f\cos^3 c - K_r\cos^2 c\right)\right] (3.78)$$

From Equation 3.77, on the other hand, the stability limit, b_{lim}, at a certain chatter frequency can be obtained as follows:

$$\Lambda_R + i\Lambda_I = b_{lim}\left(\cos\omega_c\tau - i\sin\omega_c\tau - 1\right)$$
$$b_{lim} = \frac{\Lambda_R + i\Lambda_I}{\cos\omega_c\tau - i\sin\omega_c\tau - 1} \qquad (3.79)$$

Since b is a real number, the imaginary part of Equation 3.79 has to vanish, yielding:

$$b_{lim} = -\frac{1}{2}\Lambda_R\left(1+\kappa^2\right) \qquad (3.80)$$

where:

$$\kappa = \frac{\Lambda_I}{\Lambda_R} = \frac{\sin\omega_c\tau}{1-\cos\omega_c\tau} \qquad (3.81)$$

Equation 3.81 can be used to obtain a relation between the chatter frequency and the spindle speed:

$$\varepsilon = \pi - 2\psi, \quad \psi = \tan^{-1}\kappa \qquad (3.82)$$
$$\omega_c\tau = \varepsilon + 2k\pi, \quad n = 60/\tau \qquad (3.83)$$

where ε is the phase difference between the inner and outer modulations, k is an integer corresponding to the number of waves in a period, and n is the spindle speed in rpm. The stable depth of cut of the system can be obtained from Equation 3.82 for different chatter frequencies. These frequencies can be searched around the natural frequency of the most flexible structural mode of the system. Then, the corresponding spindle speeds can be determined from Equation 3.83 for different lobes, *i.e.*, for $k=1,2,3, \ldots$, etc. Thus, the stability lobe diagram of the dynamic system can be obtained by plotting the stable depth of cut vs. the corresponding spindle speeds for different lobes.

An Example Application

The turning stability model is demonstrated by an example application. In order to obtain the Frequency Response Functions (FRF) of the cutter and workpiece, a modal test setup, and to measure the chatter frequency a microphone data acquisition setup is used (Figure 3.29).

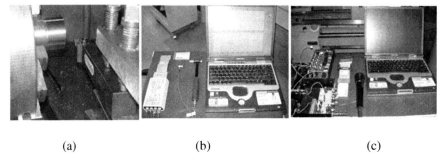

(a) (b) (c)

Figure 3.29. a Turning setup **b** Modal test setup with a CutPro **c** Frequency measurement setup

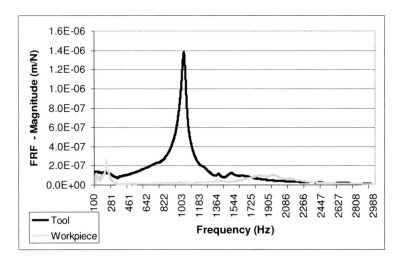

Figure 3.30. Measured transfer functions of workpiece and tool

Dynamic Analysis and Control 67

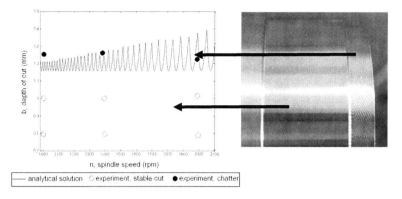

Figure 3.31. Chatter test results and the cut surface of a stable vs. an unstable cut

The workpiece material used during the tests was a medium carbon steel (AISI 1040), and zero rake carbide inserts were used as a cutting tool. All tests were conducted using a feed rate of 0.8 mm/rev. An existing orthogonal cutting database was used for the prediction of cutting coefficients. The measured tool and workpiece transfer functions prior to the tests can be seen in Figure 3.30, where the workpiece is rigidly clamped and the tool is much more flexible. Figure 3.31 shows the comparison of the analytical solution and the experimental results. The depth of cuts tested during chatter experiments are selected in order to verify the stable and unstable cut zones, and the absolute stability limit. Stable cut zones are tested using a depth of cuts fairly under the predicted absolute stability limit. Similarly, unstable cut zones are tested using a depth of cuts above the absolute stability limit. In order to confirm the absolute stability limit prediction, a close enough value is selected. In the cutting tests, one direct way of understanding whether there is chatter or not is to observe the finished surface. An example of a finished surface of a stable and an unstable cut for the tests conducted in 2000 rpm can be seen in Figure 3.31.

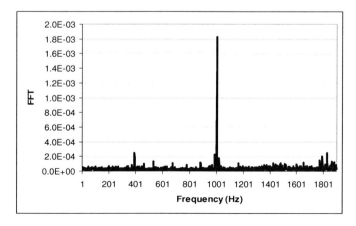

Figure 3.32. Frequency measurement results for the unstable cut

Another method to identify an unstable cut is to determine the frequency of the chatter sound. In Figure 3.32, an example of an FFT measurement can be seen for the same test at 2000 rpm. The chatter frequency can be clearly seen, which was measured to be about 1000 Hz at unstable depths of cuts close to the stability limit, as expected from the transfer function given in Figure 3.30. It can be concluded from these test results that the analytical model can successfully predict the chatter limit.

3.4.2 The Stability of the Milling Process

In Equation 3.70, the directional coefficient matrix $[A(t)]$ is periodic at the tooth passing frequency $\omega = N\Omega$ and can be expanded into its Fourier series:

$$[A(t)] = \sum_{r=-\infty}^{r=\infty} [A_r] e^{ir\omega t} \quad ; \quad [A_r] = \frac{1}{T} \int_0^T [A(t)] e^{-ir\omega t} dt \tag{3.84}$$

Thus, the stability problem has to be solved considering all significant frequencies. Budak and Altintas [8, 30] used two different approaches to the stability solution of milling which will be given separately in the following section.

The Multi Frequency Solution
If the periodic components of $[A(t)]$ are considered in the solution, then the response of the dynamic forces to these variations should also be included:

$$\{F(t)\} = e^{i\omega_c t} \sum_{k=-\infty}^{\infty} \{F_k\} e^{ik\omega t} = \sum_{k=-\infty}^{\infty} \{F_k\} e^{(\omega_c + k\omega)it} \tag{3.85}$$

which is equivalent to Floquet's theorem. The dynamic displacements can also be written as follows using the principle of superposition:

$$\{r_c\} = \sum_{k=-\infty}^{\infty} [G_c(i\omega_c + ik\omega)]\{F_k\} e^{(\omega_c + k\omega)it}$$

$$\{r_w\} = -\sum_{k=-\infty}^{\infty} [G_w(i\omega_c + ik\omega)]\{F_k\} e^{(\omega_c + k\omega)it} \tag{3.86}$$

$[G_c(i\omega)]$ and $[G_w(i\omega)]$ are the structural transfer function matrices of the cutter and the workpiece:

$$[G_p(i\omega)] = \begin{bmatrix} [G_{p_{xx}}(i\omega)] & [G_{p_{xy}}(i\omega)] \\ [G_{p_{yx}}(i\omega)] & [G_{p_{yy}}(i\omega)] \end{bmatrix} \quad p=c,w \tag{3.87}$$

where $[G_{p_{xx}}(i\omega)]$ and $[G_{p_{yy}}(i\omega)]$ are the direct transfer functions in x and y directions, and $[G_{p_{xy}}(i\omega)]$ and $[G_{p_{yx}}(i\omega)]$ are the cross transfer functions. The phase difference between the vibration waves of two successive teeth is $\omega_c T$, where T is the tooth period. The dynamic displacements for previous and current passes can be expressed as:

$$\{r_c^o\} = \{r_c(t-T)\} = e^{-i\omega_c T}\{r_c\}$$
$$\{r_w^o\} = \{r_w(t-T)\} = e^{-i\omega_c T}\{r_w\} \quad (3.88)$$
$$\{\Delta\} = (\{r_c\}-\{r_w\})\left(1-e^{-i\omega_c T}\right)$$

Substituting into Equation 3.70 the following is obtained:

$$\sum_{k=-\infty}^{\infty}\{F_k\}e^{(\omega_c+k\omega)it} = \frac{1}{2}K_t a(1-e^{-i\omega_c t})[A(t)]\sum_{k=-\infty}^{\infty}[G(i\omega_c+ik\omega)]\{F_k\}e^{(\omega_c+k\omega)it}$$

$$\sum_{k=-\infty}^{\infty}\{F_k\}e^{ik\omega t} = \frac{1}{2}K_t a(1-e^{-i\omega_c t})\sum_{r=-\infty}^{\infty}[A_r]e^{ir\omega t}\sum_{k=-\infty}^{\infty}[G(i\omega_c+ik\omega)]\{F_k\}e^{ik\omega t}$$

$$\sum_{k=-\infty}^{\infty}\{F_k\}e^{ik\omega t} = \frac{1}{2}K_t a(1-e^{-i\omega_c t})\sum_{r=-\infty}^{\infty}\sum_{k=-\infty}^{\infty}[A_r][G(i\omega_c+ik\omega)]\{F_k\}e^{i(r+k)\omega t}$$

(3.89)

Where the total transfer function can be determined by summing up the tool and the workpiece transfer functions, i.e., $[G(i\omega)] = [G_c(i\omega)] + [G_w(i\omega)]$. If both sides of Equation 3.88 are multiplied by $1/Te^{-ip\omega t}$, and integrated from 0 to T, using the orthogonality principle, the following is obtained:

$$\{F_p\} = \frac{1}{2}K_t a(1-e^{-i\omega_c t})\sum_{k=-\infty}^{\infty}[A_{p-k}][G(i\omega+ik\omega)]\{F_k\} \quad (p=0,\pm1,\pm2,...) \quad (3.90)$$

The above equation can be written in an infinite matrix form as:

$$\begin{Bmatrix}\{F_o\}\\ \{F_1\}\\ \{F_{-1}\}\\ \dots\end{Bmatrix} = \frac{1}{2}K_t a(1-e^{-i\omega_c t})\begin{bmatrix}[A_o][G(i\omega_c)] & [A_{-1}][G(i\omega_c+i\omega)] & \dots\\ [A_1][G(i\omega_c)] & [A_o][G(i\omega_c+i\omega)] & \dots\\ [A_{-1}][G(i\omega_c)] & [A_{-2}][G(i\omega_c+i\omega)] & \dots\end{bmatrix}\begin{Bmatrix}\{F_o\}\\ \{F_1\}\\ \{F_{-1}\}\\ \dots\end{Bmatrix}$$

(3.91)

Equation 3.91 has nontrivial solutions if the determinant is zero:

$$\det\left[\delta_{pk}[I] - \frac{1}{2}K_t a(1-e^{-i\omega_c T})[A_{p-k}][G(i\omega_c + ik\omega)]\right] = 0 \qquad (3.92)$$

where δ_{pk} is the Kronecker delta. Equation 3.92 defines an infinite determinant which is a characteristic of the periodic systems. A truncated version of this equation must be used to obtain approximate solutions. For the first order approximation, $p, k=0, \pm 1$, the following truncated determinant is obtained:

$$\det\begin{bmatrix} [I]+\Lambda[A_o][G(i\omega_c)] & \Lambda[A_{-1}][G(i\omega_c+i\omega)] & \cdots \\ \Lambda[A_1][G(i\omega_c)] & [I]+\Lambda[A_o][G(i\omega_c+i\omega)] & \cdots \\ \Lambda[A_{-1}][G(i\omega_c)] & \Lambda[A_{-2}][G(i\omega_c+i\omega)] & \cdots \end{bmatrix} \qquad (3.93)$$

where the eigenvalue $\Lambda = -\frac{1}{2}K_t a(1-e^{-i\omega_c T})$. The Fourier coefficients can be determined by integrating Equation 3.84 in the angular domain:

$$[A_q] = \frac{N}{2\pi}\int_0^T [A(\theta)] e^{-iqN\theta} \qquad (3.94)$$

Resulting in the following:

$$\alpha_{xx}^{(q)} = \frac{i}{2}\left[-c_o K_r e^{-iqN\theta} + c_1 e^{-ip_1\theta} - c_2 e^{ip_2\theta}\right]_{\phi_{st}}^{\phi_{ex}}$$

$$\alpha_{xy}^{(q)} = \frac{1}{2}\left[-c_o i K_r e^{-iqN\theta} + c_1 e^{-ip_1\theta} + c_2 e^{ip_2\theta}\right]_{\phi_{st}}^{\phi_{ex}}$$

$$\alpha_{yx}^{(q)} = \frac{1}{2}\left[c_o K_r e^{-iqN\theta} + c_1 e^{-ip_1\theta} + c_2 e^{ip_2\theta}\right]_{\phi_{st}}^{\phi_{ex}} \qquad (3.95)$$

$$\alpha_{yy}^{(q)} = \frac{i}{2}\left[-c_o K_r e^{-iqN\theta} - c_1 e^{-ip_1\theta} + c_2 e^{ip_2\theta}\right]_{\phi_{st}}^{\phi_{ex}}$$

where:

$$p_1 = 2+Nq \quad , \quad p_2 = 2-Nq$$
$$c_o = \frac{2}{Nq} \quad , \quad c_1 = \frac{K_r - i}{p_1} \quad , \quad c_2 = \frac{K_r + i}{p_2} \qquad (3.96)$$

The Single Frequency Solution
In chatter stability analysis the inclusion of the higher harmonics in the solution may not be required for most cases as the response at the chatter limit is usually dominated with a single chatter frequency. Thus, it is sufficient to include only the

average term in the Fourier series expansion of [A(t)] in which case the directional coefficients take the following form [8-10]:

$$\alpha_{xx} = \frac{1}{2}[\cos 2\phi - 2K_r\phi + K_r \sin 2\phi]_{\phi_{st}}^{\phi_{ex}} \quad ; \quad \alpha_{xy} = \frac{1}{2}[-\sin 2\phi - 2\phi + K_r \cos 2\phi]_{\phi_{st}}^{\phi_{ex}}$$
$$\alpha_{yx} = \frac{1}{2}[-\sin 2\phi + 2\phi + K_r \cos 2\phi]_{\phi_{st}}^{\phi_{ex}} \quad ; \quad \alpha_{yy} = \frac{1}{2}[-\cos 2\phi - 2K_r\phi - K_r \sin 2\phi]_{\phi_{st}}^{\phi_{ex}}$$
(3.97)

Then, Equation 3.70 reduces to the following form:

$$\{F(t)\} = \frac{1}{2}aK_t [A_0]\{\Delta(t)\} \tag{3.98}$$

By substituting the response and the delay terms in Equation 3.98, the following expression is obtained:

$$\{F\}e^{i\omega_c t} = \frac{1}{2}aK_t(1-e^{-i\omega_c T})[A_0][G(i\omega_c)]\{F\}e^{i\omega_c t} \tag{3.99}$$

Equation 3.99 has a non-trivial solution only if its determinant is zero:

$$\det\left[[I] + \Lambda[G_0(i\omega_c)]\right] = 0 \tag{3.100}$$

where [I] is the unit matrix, and the oriented transfer function matrix is defined as:

$$[G_0] = [A_0][G] \tag{3.101}$$

and the eigenvalue (Λ) is the same as in the multi frequency solution:

$$\Lambda = -\frac{N}{4\pi}K_t a\left(1 - e^{-i\omega_c T}\right) \tag{3.102}$$

The Evaluation of the Stability Limit
The chatter stability limit can be determined by solving the characteristic equation of the closed loop milling system, *i.e.*, equations 3.92 or 3.100. For the single frequency solution, an analytical determination of the chatter limit is possible whereas for the multi frequency solution an iterative algorithm needs to be used as both the chatter frequency and the spindle speed, which are dependent, appear in the determinant. Therefore, the formulation for the single frequency solution is presented here; however, a similar procedure can be used for a general solution as it was recently shown by Merdol and Altintas [31].

If the cross transfer functions, G_{xy} and G_{yx}, are neglected in Equation 3.100, the eigenvalue can be obtained as follows:

$$\Lambda = -\frac{1}{2a_0}\left(a_1 \pm \sqrt{a_1^2 - 4a_0}\right) \qquad (3.103)$$

where
$$a_0 = G_{xx}(i\omega_c)G_{yy}(i\omega_c)\left(\alpha_{xx}\alpha_{yy} - \alpha_{xy}\alpha_{yx}\right) \qquad (3.104)$$
$$a_1 = \alpha_{xx}G_{xx}(i\omega_c) + \alpha_{yy}G_{yy}(i\omega_c)$$

Since the transfer functions are complex, Λ will have complex and real parts. However, the axial depth of cut (a) is a real number. Therefore, when $\Lambda = \Lambda_R + i\Lambda_I$ and $e^{-i\omega_c T} = \cos\omega_c T - i\sin\omega_c T$ is substituted in Equation 3.102, the complex part of the equation has to vanish yielding:

$$\kappa = \frac{\Lambda_I}{\Lambda_R} = \frac{\sin\omega_c T}{1 - \cos\omega_c T} \qquad (3.105)$$

The above can be solved to obtain a relation between the chatter frequency and the spindle speed [30]:

$$\omega_c T = \varepsilon + 2k\pi \;,\; \varepsilon = \pi - 2\psi \;,\; \psi = \tan^{-1}\kappa \;,\; n = \frac{60}{NT} \qquad (3.106)$$

where ε is the phase difference between the inner and outer modulations, k is an integer corresponding to the number of vibration waves within a tooth period, and n is the spindle speed (rpm). After the imaginary part in Equation 3.103 is vanished, the following is obtained for the stability limit:

$$a_{lim} = -\frac{2\pi\Lambda_R}{NK_t}\left(1 + \kappa^2\right) \qquad (3.107)$$

Therefore, for a given cutting geometry, cutting force coefficients, tool and work piece transfer functions, and chatter frequency ω_c, Λ_I and Λ_R can be determined from Equation 3.103, and can be used in Equations 3.106 and 3.107 to determine the corresponding spindle speed and stability limit. Equation (3.103) provides two Λs for every chatter frequency, however, the one which results in the lowest a_{lim} must be used similar to the other stability problems. When this procedure is repeated for a range of chatter frequencies and the number of vibration waves, k, the stability lobe diagram for a milling system is obtained.

The Analytical Generation of Stability Lobes in Milling
The analytical stability model presented can be used to generate stability lobe diagrams where the variation of the axial stable depth of cut with the spindle speed can be shown. Figure 3.33 shows a sample stability diagram for a milling system analyzed in [32]. As the figure shows, the chatter free material removal rate can be

increased substantially by using the high stability pockets. For example, in the first lobe (close to 12,000 rpm), the stability limit is about 5 times the critical (or minimum) stability limit. Note that the stability pockets become larger at higher speeds which is one of the impacts of high speed machining. The figure also shows that the analytical or single frequency solution, the multi frequency solution and the time domain simulations all converge to the same stability limits for this application.

As a second example, a case considered by Merdol and Altintas [31] is considered, and the stability diagram is given in Figure 3.34. In this case, the stability of a very low radial immersion milling is analyzed. For those cases, the predictions of the single frequency solution may not be as accurate. This is mainly because of the increased Fourier series content in the directional coefficients, and thus in matrix [A], due to a very short contact between the tool and the material, resulting in a very short pulse-width waveform. Thus, higher harmonics in the Fourier series need to be included in Equation 3.90, and the truncated version of Equation 3.91 must include more terms. As a result, many different combinations of the tooth frequency and the chatter frequency affect the system response, *i.e.*, $G(ik\omega \pm \omega c)$, resulting in deviations from the single frequency solution. The stability diagram shown in Figure 3.34 demonstrates one major deviation from the single frequency solution, *i.e.*, the added lobe which occurs only for very small radial immersion cases. However, the multi frequency solution is able to predict this phenomenon precisely.

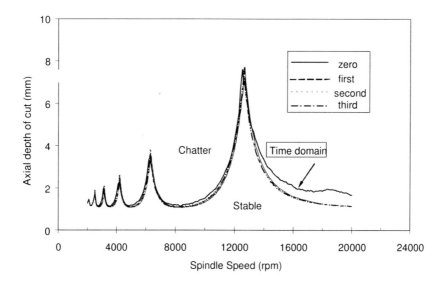

Figure 3.33. A sample stability diagram for a milling application

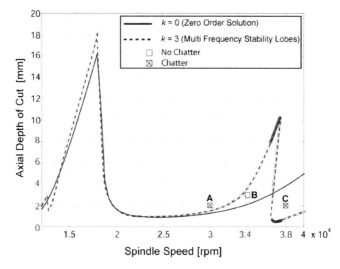

Figure 3.34. Stability diagram for a low radial immersion application

3.4.3. Maximizing the Chatter Free Material Removal Rate in Milling

Stability diagrams in terms of an axial depth of cut for a given radial depth of cut can be generated using a certain spindle speed range. However, for the maximum productivity selection of both axial and radial depths is critical [33]. The radial depth of cut, B, can be defined as follows:

$$B/R = 1 - \cos(\phi_{ex}) \quad \text{up-milling}$$

$$B/R = 1 + \cos(\phi_{ex}) \quad \text{down-milling} \quad (3.108)$$

where ϕ_{st} and ϕ_{ex} are the start and exit angles of the cutting edges to and from the cut, and R is the tool radius. A normalized form of the radial depth of cut, $b=B/2R$ is used in the rest of the formulation for generalization. Thus, b is unitless, and it may only have values in the range of [0, 1]. The stability diagrams can be generated, in terms of either a_{lim} vs. spindle speed (for a fixed b_{lim}) or b_{lim} vs. spindle speed (for a fixed a_{lim}). The common practice is to express stability diagrams in terms of a_{lim} vs. spindle speed. The importance of identifying b_{lim} is two fold. Firstly, in some cases, axial depth is fixed due to the geometry of the part, and thus the maximum stable b must be determined. Secondly, maximum MRR can only be achieved by optimizing both depth of cuts. The procedure to determine b_{lim} starts with selection of an a_{lim} for which the stability diagram, b_{lim} vs. spindle speed, will be generated. Then, by scanning the full range of exit angles (up-milling) or start angles (down-milling), the eigenvalues of the milling system are determined. As the last step, the corresponding value of b_{lim} is determined using Equation 3.108.

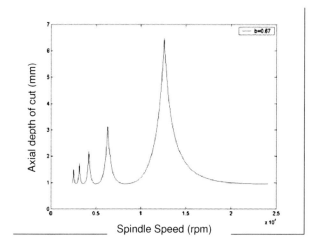

Figure 3.35. Stable axial depth of cut vs. spindle speed for b=0.67

As an example, the system considered by Weck et al. in [32] which involves the milling of an aluminium alloy with a 3-tooth end mill is considered. Figures 3.36 and 3.37 show the stability diagrams generated using the analytical method. One importance of identifying b_{lim} is clearly seen in Figure 3.36. The radial depth of cut limit for the selected axial depth reaches to the maximum value of $b=1$ at some speeds. That is another reason why the maximum material removal rate can only be achieved by optimizing a and b simultaneously.

If we focus on a specific spindle speed where stability limits are the highest, *i.e.*, in the lobes, we see that no decrease in b is necessary for some increase in a; however, after a certain point, a negatively sloped relation exists between the stable limits of axial and radial depths of cut as represented in Figure 3.37.

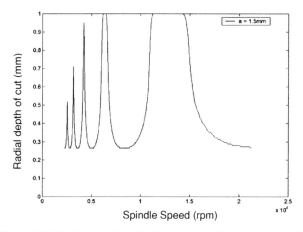

Figure 3.36. Stable radial depth of cut vs. spindle speed for a=1.5 mm

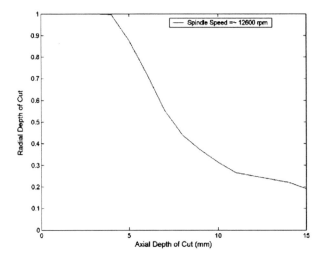

Figure 3.37. Stable radial depth of cut vs. stable axial depth of cut for 12,600 rpm

Since the material removal rate (MRR) is proportional to the multiplication of the axial and radial depths of cut, it is interesting to find out at which combination of axial and radial depths of cut, the maximum value of MRR may be achieved.

$$MRR = a.b.n.N.f_t \qquad (3.109)$$

where a is the axial depth of cut, b is the radial depth of cut, n is the spindle speed, N is the number of cutting teeth, and f_t is the feed per revolution per tooth. In general, the effect of f_t on stability is small and can be neglected. Therefore, a normalized value of MRR is used hereafter:

$$MRR^* = \frac{MRR}{f_f} = a.b.n \qquad (3.110)$$

MRR* for an axial depth of cut is calculated using the b_{lim} corresponding to that a_{lim}. Subsequently the chatter-free MRR* is obtained as shown below:

$$MRR^* = a_{lim}.b_{lim}.n.N \qquad (3.111)$$

The variation of the MRR* with the axial depth of cut at the high stability pocket position of 12,600 rpm is shown in Figure 3.38. The figure indicates that, for this case, the maximum possible MRR* can be obtained for the axial depth of cut of about 5 mm. In some cases the MRR* curve has a peak as in Figure 3.38; in some cases it saturates after a certain depth, depending on the machine tool dynamics. Through simulations [33] it was found that if the natural frequencies of the cutter and workpiece system are different in x and y directions significantly, and the feed is in the direction of lower natural frequency, then it is more likely to see a peak in the MRR* curve. For the example considered here, the modal frequencies of the end mill in the feed and normal directions were 600 Hz and 660 Hz, respectively.

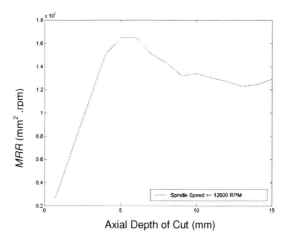

Figure 3.38. Maximum MRR* vs. the stable limit of axial depth of cut

Example Applications

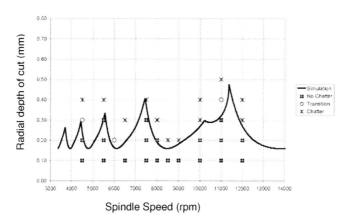

Figure 3.39. Stability diagram for an axial depth of 1.5 mm in the aluminum milling example

An aluminum milling application with 4-teeth milling tool is considered. The cutting coefficients were calibrated as K_t=796 MPa and K_r=0.21 for the feedrate of 0.25 mm/tooth. The modal properties of the tool measured using impact tests are given in [33]. The stability diagram in terms of radial depth for axial depth of cut of 1.5 mm is shown in Figure 3.39. Figure 3.40, on the other hand, shows the variation of the radial depth of cut limit with the axial depth for 11,000 rpm which is useful in identifying the optimal milling conditions for the maximum material removal rate. Results of the chatter tests are also shown in the same figure which are in good agreement with the predictions.

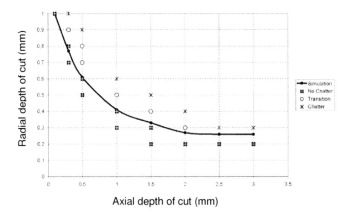

Figure 3.40. Stable radial depth limit vs. the axial depth of cut for 11,000 rpm

Minimal Chatter Free Pocketing Time
Pocketing is a very common operation in milling, *e.g.*, in die, mould, and airframe production, etc. In order to find the minimum pocketing time, one should consider the total number of passes (nop) [29]. The term pass stands for one cutting pass across the pocket length. The total pocketing time (TPT) can be approximately expressed as:

$$TPT = nop \frac{w_p}{f} \qquad (3.112)$$

where w_p and f are the pocket width and the feed rate, respectively. Thus, minimizing the total pocketing time is equivalent to minimizing the number of passes since we keep the feed rate constant throughout the analysis. A proper feedrate value should be used considering the other constraints such as the tool capacity and the surface finish requirement. The number of passes can then be expressed as:

$$nop = ceil\left(\frac{d_p}{a}\right).ceil\left(\frac{l_p}{b}\right) \qquad (3.113)$$

where d_p is the pocket depth, l_p is the pocket length, and ceil is the round up function. The steps of the proposed method are as follows. Once the pocket geometry and the appropriate tool are identified, workpiece and cutter dynamics are to be measured, so that the pairs of stable axial and radial depth of cut limits can be determined using simulations. For the example considered in the previous section, the pairs of stable depths are determined using the method presented here and shown in Table 3.4. The results of the optimization are presented in Table 3.5.

Table 3.4. Pairs of stable limits of the axial and the radial depths of cut

a_{lim} (mm)	b_{lim} (mm)
4.00	1.00
6.00	0.83
8.00	0.65
10.00	0.52
12.00	0.44
14.00	0.38
20.00	0.27

The first two columns of Table 3.5 show the required pocket depth and length. The next two column sets present the resulting number of passes (*nop*) for two different methods. The first method stands for choosing a high radial depth of cut close to full slotting, 0.8 in this case. The second column set, Optimal, presents the results for choosing the optimal pair of depths of cut according to the method presented in this section. The last column shows the percentage improvements in pocketing time attained by the optimal combination.

Table 3.5. Optimal depths of cut for the minimum pocketing time

Pocket		Conventional b = 0.8		Optimal			Improvement
D_p mm	L_p/D	a mm	Nop	a	b	nop	%
4	10	4	13	4	1.0	13	23
6	10	6	13	6	0.83	13	8
8	10	4-4	26	8	0.65	16	38
10	10	5-5	26	10	0.52	20	23
12	10	6-6	26	12	0.44	23	11
14	10	5-5-4	39	14	0.38	27	30
20	10	5-5-5-5	52	20	0.27	38	27

3.4.4 Chatter Suppression-Variable Pitch End Mills

Variable pitch cutters have non uniform pitch spacing between the cutting teeth and may be very effective in suppressing chatter. The non uniform pitch may disturb the regeneration mechanism, and if the angles are selected appropriately the phase between the inner and outer modulation can be minimized. The effectiveness of variable pitch cutters in suppressing chatter vibrations in milling was first demonstrated by Slavicek [34], and later on others [35, 36] proposed different methods to model the stability of these tools [35, 36]. The fundamental difference in the stability analysis of milling cutters with non-constant pitch angle is that the phase delay is different for each tooth:

$$\varepsilon_j = \omega_c T_j \quad (j=1,...,N) \tag{3.114}$$

where T_j is the jth tooth period corresponding to the pitch angle ϕ_{pj}. The dynamic chip thickness and the cutting force relations given for the standard milling cutters apply to the variable pitch cutters as well. The eigenvalue expression will take the following form due to the varying phase:

$$\Lambda = \frac{a}{4\pi} K_t \sum_{j=1}^{N} \left(1 - e^{-i\omega_c T_j}\right) \tag{3.115}$$

The stability limit can be obtained from Equation 3.115 as:

$$a_{lim}^{vp} = -\frac{4\pi}{K_t} \frac{\Lambda}{N-C+iS} \tag{3.116}$$

where:

$$C = \sum_{j=1}^{N} \cos \omega_c T_j \quad ; \quad S = \sum_{j=1}^{N} \sin \omega_c T_j \tag{3.117}$$

As a_{lim} is a real number, the imaginary part of Equation 3.116 must vanish yielding [22]

$$a_{lim}^{vp} = -\frac{4\pi}{K_t} \frac{\Lambda_I}{S} \tag{3.118}$$

The stability limit with variable pitch cutters can be determined using Equation 3.118. However, optimization of the pitch angles for a given milling system has a more practical importance than the stability analysis for an arbitrary variable pitch cutter. Equation 3.118 indicates that in order to maximize the stability limit, $|S|$ has to be minimized. From Equation 3.117, S can be expressed as follows:

$$S = \sin \varepsilon_1 + \sin \varepsilon_2 + \sin \varepsilon_3 + ... \tag{3.119}$$

where $\varepsilon_j = \omega_c T_j$. The phase angle, which is different for every tooth due to the non-constant pitch, can be expressed as follows:

$$\varepsilon_j = \varepsilon_1 + \Delta\varepsilon_j \quad (j = 2,..,N) \tag{3.120}$$

where $\Delta\varepsilon_j$ is the phase difference between tooth j and tooth (1) corresponding to the difference in the pitch angles between these teeth. The pitch angle variation ΔP corresponding to $\Delta\varepsilon$ can be determined as:

$$\Delta P = \frac{\Delta\varepsilon}{2\pi}\theta = \frac{\Omega}{\omega_c}\Delta\varepsilon \tag{3.121}$$

Equation 3.119 can be expanded as follows by using Equation 3.120:

$$S = \sin\varepsilon_1 + \sin\varepsilon_1\cos\Delta\varepsilon_2 + \sin\Delta\varepsilon_2\cos\varepsilon_1 + \sin\varepsilon_1\cos\Delta\varepsilon_3 + \sin\Delta\varepsilon_3\cos\varepsilon_1 + ... \tag{3.122}$$

There are many solutions to the minimization of $|S|$, i.e., $(S=0)$. It can be found out by intuition that $S=0$ for the following conditions [36]:

$$\Delta\varepsilon = k\frac{2\pi}{N} \quad (k = 1,2,...,N-1) \tag{3.123}$$

The corresponding ΔP can be determined using Equation (3.121).

The increase of the stability with variable pitch cutters over the standard end mills can be determined by considering the ratio of stability limits. For simplicity, the absolute or critical stability limit for equal pitch cutters, i.e., the minimum stable depth of cut regardless of the spindle speed, are considered. The absolute stability limit is the minimum stable depth of cut without the effect of lobing which can be expressed as follows:

$$a_{cr} = -\frac{4\pi\Lambda_I}{NK_t} \tag{3.124}$$

Then, the stability gain can be expressed as:

$$r = \frac{a_{\lim}^{vp}}{a_{cr}} = \frac{N}{S} \tag{3.125}$$

where r is plotted as a function of $\Delta\varepsilon$ in Figure 3.41 for a 4-tooth milling cutter with linear pitch variation. The phase ε depends on the chatter frequency, spindle speed and the eigenvalue of the characteristic equation. Therefore, the stability analysis has to be performed for the given conditions. Three different curves corresponding to different ε_1 values are shown in Figure 3.41 to demonstrate the effect of phase variation on r. r is maximized for integer multiples of $2\pi/N$, i.e., for

(1/4, 1/2, 3/4)×2π. $\Delta\varepsilon + k2\pi$ (k=1, 2, 3, ...) are also optimal solutions; however, they result in higher pitch variations which is not desired.

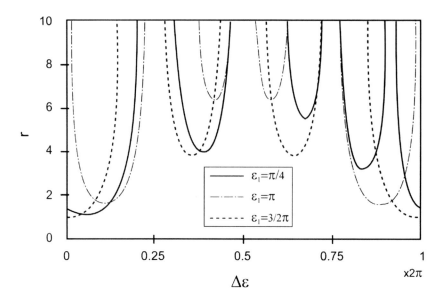

Figure 3.41 Effect of ε_1 on stability gain for a 4-fluted end mill with linear pitch variation

As it can be seen from Figure 3.41 for a 4-tooth end mill with linear pitch variation, a minimum of r=4 gain is obtained for $0.5\pi < \Delta\varepsilon < 1.5\pi$. Thus, the target for $\Delta\varepsilon$ should be π, which is one of the optimal solutions for the cutters with an even number of flutes, but it is also in the middle of the high stability area.

3.5 Conclusions

High productivity and quality in machining strongly depend on the process dynamics and stability. Rigidity of machine tools and the selection of process parameters are two main factors in the dynamic behaviour of cutting operations. In this chapter, methods that can be used for the analysis and modelling of machine tool structures and cutting process dynamics are reviewed. These methods can be used to analyze as well as improve the dynamic behaviour of machining processes. Perhaps the most important application of these methods is the analysis and suppression of chatter which is the most critical vibration type for machining systems. The analytical methods and suppression techniques presented here can be used to improve the stability and productivity of the machining operations. The selected and presented methods have been applied, and are currently being used in industry. The results observed so far indicate that the dynamics in cutting is a complicated problem, but substantial advancements can be obtained using modelling. Therefore, the work in this area is expected to continue in the coming years.

References

[1] Merchant, M. E., Basic mechanics of the metal cutting process, ASME Journal of Applied Mechanics, 66, 1944, 168–175.
[2] Armarego, E.J.A., Brown, R.H., The Machining of Metals, Prentice–Hall, London, 1969.
[3] Armarego, E.J.A., Whitfield, R.C., Computer based modelling of popular machining operations for force and power predictions, Annals of the CIRP 34, 1985, 65–69.
[4] Budak, E. and Altintas, Y., Prediction of milling force coefficients from orthogonal cutting data, ASME 1993 Winter Annual Meeting, New Orleans, USA, Manufacturing Science and Engineering, 1993, 453–459.
[5] Stabler, G.V., Fundamental geometry of cutting tools, Proceedings of the Institution of Mechanical Engineers, 1951, 14–26.
[6] Budak, E., Altintas, Y., Armarego, E.J.A., Prediction of milling force coefficients from orthogonal cutting data, Trans. ASME Journal of Manufacturing Science and Engineering, 118, 1996, 216–224.
[7] Altintas, Y., Manufacturing Automation, Cambridge University Press, 2000.
[8] Budak, E., The mechanics and dynamics of milling thin–walled structures, Ph.D. Dissertation, University of British Columbia, 1994.
[9] Budak, E., Altintas, Y., Peripheral milling conditions for improved dimensional accuracy. International Journal of Machine Tool Design and Research, 34/7, 1994, 907–918.
[10] Kops, L., Vo, D.T., Determination of the equivalent diameter of an end mill based on its compliance, Annals of the CIRP 39, 1990, 93–96.
[11] Kivanc, E., Budak, E., Structural modelling of end mills for form error and stability analysis, International Journal of Machine Tools and Manufacture, 44/11, 2004, 1151–1161.
[12] Nemes, J.A., Asamoah–Attiah, S. and Budak, E., Cutting Load Capacity of End Mills with Complex Geometry, Annals of the CIRP, 50/1, 2001, 65–68.
[13] Rivin, E., Stiffness and Damping in Mechanical Design, Marcel Dekker, NY, 1999.
[14] Budak, E., Altintas, Y., Modelling and avoidance of static deformations in peripheral milling of plates, International Journal of Machine Tool Design and Research 35/3, 1995, 459–476.
[15] Budak, E., Altintas, Y., Flexible milling force model for improved surface error predictions, ASME 1992 European Joint Conference on Engineering Systems Design and Analysis, Istanbul, Turkey, ASME PD–Vol. 47–1, 1992.
[16] Ewins, D.J., Modal Testing: Theory, Practice and Application, Taylor & Francis Group; 2nd edition, London, 2001.
[17] T. Schmitz, R. Donaldson, Predicting high–speed machining dynamics by substructure analysis, Annals of the CIRP 49/1, 2000, 303–308.
[18] T. Schmitz, M. Davies, M. Kennedy, Tool point frequency response prediction for high–speed machining by RCSA, Journal of Manufacturing Science and Engineering 123, 2001, 700–707.
[19] Ertürk, A., Özgüven, H.N. and Budak, E., Analytical modelling of spindle–tool dynamics on machine tools using Timoshenko beam model and receptance coupling for the prediction of tool point FRF, International Journal of Machine Tools and Manufacture, 2008 (in press).

[20] Ertürk, A., Özgüven, H.N. and Budak, E., Effect analysis of bearing and interface dynamics on tool point FRF for chatter stability in machine tools by using a new analytical model for spindle–tool assemblies, International Journal of Machine Tools and Manufacture, 2008 (in press).
[21] Özgüven, H.N., A new method for harmonic response of non–proportionally damped structures using undamped modal data, Journal of Sound and Vibration, 117, 1987, 313–328.
[22] Tlusty, J., Analysis of the state of research in cutting dynamics, Annals of the CIRP 27/2, 1978, 583–589.
[23] Altintas, Y. and Weck, M. Chatter stability of metal cutting and grinding, Annals of the CIRP, 53/2, 2004, 619–642.
[24] Nigm, M.M. and Sadek, M.M., Experimental investigation of the characteristics of dynamic cutting process, Trans. ASME Journal of Engineering for Industry, 1977, 410–418.
[25] Wu, D.W. and Liu, C.R., An analytical model of cutting dynamics, Part I: model building, Trans. ASME Journal of Engineering for Industry, 107, 1985, 107–111.
[26] Tlusty, J. and Rao, S.B., Verification and analysis of some dynamic cutting force coefficients data, Proceedings of the 6th NAMRC, University of Florida, USA, 1978, 420–426.
[27] Tlusty, J., Polacek, M., The stability of machine tools against self excited vibrations in machining, International Research in Production Engineering, ASME, 1963, 465–474.
[28] Tobias, S.A. and Fishwick, W., The chatter of lathe tools under orthogonal cutting conditions, Transactions of ASME, 80, 1958, 1079–1088.
[29] Özlü, E. and Budak, E., Analytical prediction of stability limit in turning operations, Proceedings of the 9th Workshop on the Modelling of Machining Operations, Bled, Slovenia, May 2006.
[30] Budak, E. and Altintas, Y., Analytical prediction of chatter stability in milling–part I: general, Trans. ASME Journal of Dynamic Systems, Measurement and Control, 120, 1998, 22–30.
[31] Merdol, S.D. and Altintas, Y., Multi frequency solution of chatter stability limits for low immersion milling, Trans. ASME Journal of Manufacturing Science and Engineering, 126, 2004, 459–466.
[32] Weck, M., Altintas, Y. and Beer, C., CAD assisted chatter free NC tool path generation in milling, International Journal of Machine Tools and Manufacture, 34, 1994, 879–891.
[33] Tekeli, A. and Budak, E., Maximization of chatter free material removal rate in end milling using analytical methods, Journal of Machining Science and Technology 9, 2005, 147–167.
[34] Slavicek, J. The effect of irregular tooth pitch on stability of milling, Proceedings of the 6th MTDR Conference, Manchester, UK, 1965, 15–22.
[35] Vanherck, P. Increasing milling machine productivity by use of cutters with non–constant cutting edge pitch, 8th MTDR Conference 1966, 947–960.
[36] Budak, E., An analytical design method for milling cutters with non constant pitch to increase stability – Part I: Theory; Part II: Application, Trans. ASME Journal of Manufacturing Science and Engineering, 125, 2003, 29–38.

4

Dynamics Diagnostics: Methods, Equipment and Analysis Tools

Neil D. Sims

Advanced Manufacturing Research Centre with Boeing
Department of Mechanical Engineering
The University of Sheffield
Mappin Street, Sheffield S1 3JD, UK

4.1 Introduction

There are two reasons why it is important to diagnose, identify, or analyse the dynamic behaviour of machine tools. Firstly, the many techniques that can be used to optimise the machining process invariably require some model of the structural dynamics. This is especially true for the problem of avoiding unstable chatter vibrations, and for predicting the surface finish. Knowledge of the structure's natural frequencies, mode shapes, and damping ratios can be used to predict the cutting performance (*i.e.*, chatter stability and geometrical accuracy), and to choose optimal values for the spindle speed and depth of cut. Such techniques could be described as pre-process techniques, because the dynamics testing is carried out before the machining process.

A second reason to analyse dynamic behaviour is that during metal cutting the forces, vibrations, and accompanying sound and acoustic emissions can contain information on the underlying behaviour of the system. In general, this kind of in-process information has been widely used to detect problems such as tool wear and tool breakage. However, in terms of dynamics, the obvious application is once again the issue of regenerative chatter stability.

The aim of this chapter is therefore to review the relevant tools for identifying and diagnosing the dynamic behaviour of machine tool structures and machining processes, with particular emphasis on the problem of regenerative chatter. The other issues that effect productivity (*e.g.*, tool wear and failure, and surface defects), are not specifically considered because they do not arise as a direct consequence of the machine's structural dynamics. Furthermore, there have been a number of publications that review these topics, to which the reader is referred [1–5].

The chapter is organised as follows: first, the relevant modal analysis theory is reviewed. Experimental equipment that is useful for characterising the dynamic behaviour of machine tools is then described. This equipment is classified as either standard equipment (that which could be found in any dynamics laboratory), novel equipment (which is specific to the field of machine-tool testing), or in-process equipment which focuses on equipment for use during metal cutting. The final section of this chapter draws upon the theory and experimental equipment, focussing on techniques for chatter detection during milling and other machining processes.

4.2 Theory

Real machine tool structures are continuous, time-variant, nonlinear elastic structures. Their continuous nature means they have an infinite number of degrees of freedom (*i.e.*, the minimum number of coordinates that can be used to describe their motion). They are nonlinear because of the effect of friction and contact mechanics between various components, *e.g.*, slide-ways and tool holders. Finally, they can be time-variant due to the change in their dynamic behaviour as (for example) the headstock or table move on the slide-way, and as material is removed from the workpiece.

To model such behaviour obviously requires a number of assumptions. The first is to assume time-invariant behaviour – *i.e.*, to neglect any changes of the dynamics in the system during metal cutting. This is normally valid since for the duration of the simulation (*i.e.*, the prediction of behaviour during one part of the cutting process), the change in dynamics will be very small. The second assumption is that the nonlinear behaviour can be linearised, and this can have repercussions when trying to obtain models based upon experimental data. The final assumption is that the response of the machining system can be described with sufficient accuracy by assuming a finite number of degrees of freedom. The choice of the number of degrees of freedom is clearly a trade-off between the model complexity and model accuracy.

With the above assumptions, the structural dynamics can be described by the equation of motion:

$$[M]\{\ddot{x}\}+[C]\{\dot{x}\}+[K]\{x\}=\{f\} \tag{4.1}$$

where [M], [C], and [K] are the NxN mass, damping and stiffness matrices, and N being the number of degrees of freedom. {f} is the vector of time varying externally applied forces (such as the force due to cutting), and {x} is the vector of displacements for the degrees of freedom. Equation 4.1 is sometimes referred to as the spatial model of the structural dynamics, because it is written in terms of spatial, or physical, coordinates.

So how can Equation 4.1 be solved? It transpires [6, 7] that if we begin by neglecting the damping and assuming free vibration (*i.e.*, {f}=0), the behaviour can be described by the modal model:

$$[\omega_r^2] = \begin{bmatrix} \omega_1^2 & 0 & \cdots & 0 \\ 0 & \omega_2^2 & \cdots & 0 \\ \vdots & \vdots & \ddots & \vdots \\ 0 & 0 & \cdots & \omega_N^2 \end{bmatrix}$$

$$[\Psi] = [\{\psi_1\} \ \{\psi_2\} \ \cdots \ \{\psi_N\}] \tag{4.2}$$

Here, ω_r is the natural frequency of mode *r* of the system, ψ_r is its mode shape, and the collected mode shape vectors are referred to as Ψ. The mode shapes are only known within an indeterminate scaling factor [6], and so it is common to normalise them as follows:

$$\{\phi_r\} = \frac{\{\psi_r\}}{\sqrt{m_r}} \tag{4.3}$$

where m_r is the so-called modal mass of mode *r*, and ϕ_r is the mass-normalised mode shape of mode *r*. It can then be shown (*e.g.*, [6, 8, 9]) that the forced response of the system with viscous damping takes the form:

$$\alpha_{jk}(s) = \sum_{r=1}^{N} \frac{\phi_{jr}\phi_{kr}}{s^2 + 2\zeta_r\omega_r s + \omega_r^2} \tag{4.4}$$

which is known as the response model for the structure. The use of modal coordinates has enabled the response to be written as a superposition of the responses due to each individual mode of vibration. The subscript *j* denotes the response location, and the subscript *k* denotes the excitation location. Equation 4.4 has been written in terms of the Laplace operator, *s*, but it could, of course, be presented in the frequency domain. The physical significance of the mass-normalised mode shape becomes clear upon further inspection of Equation 4.4 - the response at a location *j* due to a force at a location *j* is given by:

$$\alpha_{jj}(s) = \sum_{r=1}^{N} \frac{\phi_{jr}\phi_{jr}}{s^2 + 2\zeta_r\omega_r s + \omega_r^2} = \sum_{r=1}^{N} \frac{1}{(m_{jj})_r s^2 + 2\zeta_r\omega_r s + (k_{jj})_r} \tag{4.5}$$

where:

$$(m_{jj})_r = (\phi_{jr})^{-0.5}$$

$$(k_{jj})_r = (m_{jj})_r \omega_r^2$$

which are referred to as the effective mass and effective stiffness, respectively, for mode r at location j. These parameters are unique and physically significant, since they correspond to the mass and stiffness observed in the experimental response [8].

The elegance of Equation 4.4 is that if one can identify the natural frequencies, damping ratios, and mode shapes (at the excitation and measurement locations) for all the modelled modes, then the response of the structure can be readily determined. Furthermore, other forms of the response model, such as state-space models, can be readily determined.

There are a plethora of techniques, tools, and software packages available to assist in the identification of the modal parameters in Equation 4.4. However, perhaps the most easy to visualise approach is the so-called peak-amplitude method [8]. In this case, the analysis begins by assuming that the modes are well-separated in the frequency domain, so that in the region of one particular mode the other modes can be assumed to contribute a constant value to the response. Equation 4.4 can then be re-written as:

$$\alpha_{jk}(s) = \frac{_rA_{jk}}{s^2 + 2\zeta_r\omega_r s + \omega_r^2} + _rD_{jk} \tag{4.6}$$

where $_rA_{jk}$ equals $\phi_{jr}\phi_{kr}$ and is referred to as the modal constant. The constant $_rD_{jk}$ is a complex value associated with mode r, that accounts for the other modes. The peak-amplitude method proceeds as follows:

- Assume that a peak $\left(|\hat{H}_{jk}|\right)$ in the experimentally measured frequency response function $H_{jk}(\omega)$ occurs at the natural frequency ω_r.
- The frequency bandwidth for a response level of $\left(|\hat{H}_{jk}|\right)/\sqrt{2}$ is determined and is referred to as $\Delta\omega$.
- The damping ζ_r can be determined as $\Delta\omega/(2\omega_r)$.
- Finally, the modal constant $_rA_{jk}$ can be determined as $\left(|\hat{H}_{jk}|\right)\omega_r^2 2\zeta_r$

To identify the individual mode shapes, therefore, requires at least one co-located, or driving point, measurement, where $j=k$, so that:

$$_rA_{jj} = \phi_{jr}^2 \tag{4.7}$$

from which the magnitude of the mode shape at location j can be determined.

4.2.1 An Example

To illustrate the application of modal analysis to machining, a simplified example will be investigated. Consider the scenario where a milling tool, with the properties given in Table 4.1, is attached to a spindle assembly that can be assumed rigid.

Table 4.1. Tool properties

Length	200 mm
Diameter	20 mm
Young's modulus	200 GPa
Density	9000 kg/m3

A spatial model (Equation 4.1) with no damping can be developed using finite element (FE) analysis. If one uses three beam elements to represent the beam, and considers motion in only one plane, then six degrees-of-freedom will result (three translational and three rotational). Solving the Eigen value problem leads to the natural frequencies of the model, ω_r, and the mode shapes Ψ. The first three natural frequencies correspond to the bending mode shapes of the beam, and the first two of these mode shapes are plotted in Figure 4.1a. Performing a mass-normalisation of the mode shapes leads to the corresponding unique values of ϕ_r which are plotted in Figure 4.1b. The modal model (ω_r, ϕ_r for each mode) is shown in Table 4.2. The mass-normalised mode shapes can be used to determine the physically meaningful values of effective mass and stiffness at the location of each node in the finite element model, and these are plotted in Figure 4.1c and d, respectively. The mass-normalised mode shapes and natural frequencies can be also used to form a response model (Equation 4.4) which can include viscous (or hysteretic) damping, unlike finite element spatial models. The response model can be presented in state-space form for a time-domain simulation of chatter stability, or alternatively, the real part of the response model can be used directly to predict the analytical chatter stability using Tlusty's method [10].

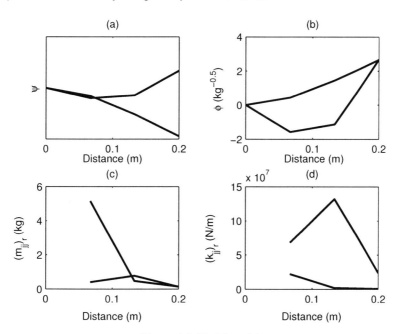

Figure 4.1. Modal model

The main use of modal analysis, however, is to avoid the use of FE formulations and to rely solely on experimental data to develop the response model. This can be advantageous because FE models cannot readily include damping, and often inaccurately predict behaviour due to errors in assumed mass and stiffness properties. Figure 4.2a shows a driving point (or co-located) frequency response function (FRF) which could have been obtained using experimental techniques. Using the peak-picking method, the natural frequencies and their half power points can be identified, and these are illustrated on Figure 4.2a. Identification of the modal constant and damping ratio leads to the values presented in Table 4.3. Because the FRF was obtained using a measurement at the same physical location as the excitation source, the modal constants can be used to determine the values of the corresponding mass-normalised mode shapes. These can then be used to formulate the response model, and also to determine the effective mass and stiffness at the excitation/measurement location. Figure 4.2b shows the two individual components of the response model, along with their sum (the overall response model). It can be seen that the asymptotes of the frequency response for the individual components represent the effective stiffness and mass for the corresponding mode. Meanwhile, the overall response closely matches the original data (Figure 4.2a).

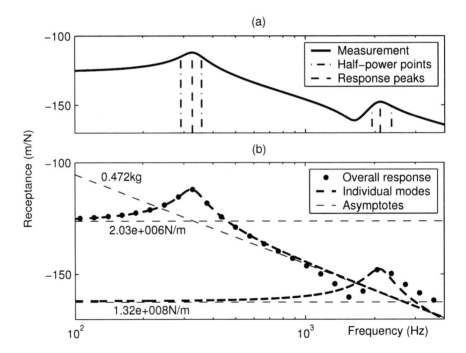

Figure 4.2. Identification of the modal model and response model

The frequency response measurement used in Figure 4.2b was actually derived directly from the finite element model described above, using the node at 0.133 m as the excitation and response location. Consequently the identified modal parameters should match the corresponding values from the original formulation. This agreement can be observed by comparing the highlighted terms in Tables 4.2 and 4.3. There is reasonable agreement between the magnitudes of the terms and more advanced parameter identification techniques can lead to far superior performance [8, 9].

Table 4.2. Model predictions

Natural frequency ω_r (Hz)	Mass-normalised mode shapes ϕ_r (kg)-0.5 at coordinate		
	0.067 m	0.13 m	0.2 m
329.78	0.44035	1.4549	2.6602
2073.2	-1.5785	-1.1332	2.676

Table 4.3. Identified model

Natural frequency (Hz)	Damping ratio	rAjj	Mass-normalised mode shape (kg)-0.5
326	0.102	2.1316	1.46
2108	0.098	1.4161	1.19

4.2.2 The Substructure Analysis

One final concept in modal analysis is worthy of mention at this stage, namely coupled structure analysis. Here, the techniques of modal testing and modelling are applied to separate components of the machine tool structure, from which the overall response of the system is predicted.

There are two possible techniques that can be used to combine, or couple, the individual components together: via the modal models, or via the response models [8]. The modal model approach has the advantage of being a somewhat neater methodology, but it is not well suited to problems that involve non-proportional damping. Furthermore, the modal model should in principle include all of the modes of vibration of the structure, yet for distributed structures (such as machine tools) we know that this is impossible to achieve.

The response model approach does not suffer from such drawbacks, although it can require a larger number of experimental measurements. The method is often referred to as receptance coupling, or impedance coupling, because the receptance/impedance FRFs are directly "coupled" together, using conditions of compatibility and equilibrium.

This approach was used for milling tools by Schmitz [11], with the aim of avoiding the requirement to test tools when mounted on the machine. Using an experimental measurement of the machine tool response at the tool holder, and an FE model of the tool, the overall response was predicted. However, the stiffness and damping behaviour of the tool holder-tool assembly were found to be significant, and Park *et al.* [12] later extended the method to include additional rotational degrees of freedom.

4.3 Experimental Equipment

Having discussed how modal analysis theory can help to develop models of machine-tool structures, it is now necessary to focus on the problem of measuring the experimental data required for the analysis process. At the most basic level, this measurement is used to obtain the driving point FRF at the tip of the cutting tool. This could be used to directly determine the chatter stability [13], or to develop modal/response models suitable for time domain simulation of chatter, surface finish, cutting forces, etc. More complex experiments may be used to determine a number of mode shapes of the machine, tool or workpiece. In any case, the experimental measurement of FRFs is not straightforward, and so the main techniques are summarized below. To begin with, however, the signal processing techniques are briefly described.

4.3.1 The Signal Processing

The aim of experimental FRF measurement is concisely stated with reference to Figure 4.3. Given some excitation $f(t)$ and a measured response $x(t)$, determine the FRF $G(\omega)$, assuming a linear system. The simplest solution (mathematically) is, of course, to use $f=F_0\sin(\omega t)$, measure the response $x(t)=X_0\sin(\omega t+\theta)$ and hence determine $|G(\omega)|=|X_0/F_0|$ and $\angle G(\omega)=\theta$. This is often totally impractical because it requires an actuation system capable of sinusoidal forced vibration, and repetitive testing to sweep through the frequency range, ω, of interest.

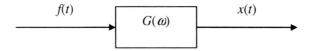

Figure 4.3. Experimental FRF measurement

The alternative approach is to use a non-periodic excitation, f(t), and statistical signal processing techniques. A discussion of these techniques is beyond the scope of the present article - full details can be found in textbooks such as [14], and the application to modal testing is discussed in [6, 8]. The signal processing requires a number of operator decisions: the frequency range, the number of test averages, and choice of windowing procedure. Nowadays, the spectrum analyser that performs the processing is often semi-integrated within a laptop PC.

4.3.2 Excitation Techniques

The most commonly used excitation technique in machining dynamics is the modal hammer. This has an integral force sensor which is used to determine the force exerted on the structure being excited (*e.g.*, the tool). The magnitude of the excitation is governed by the mass of the hammer head and its velocity. The latter is determined by the operator, and since it is not accurately controlled, many modal hammers have the option of using additional masses on the head to modify the impact load.

Meanwhile, the time-domain force pattern, and the corresponding frequency spectrum, are governed by the stiffness of the contacting surfaces and the impactor mass. These stiffnesses and mass introduce a bandwidth on the excitation force spectrum. Consequently, the impactor tip and the head mass must be carefully chosen so that they excite the structure with sufficient energy in the required frequency range. Failure to inject sufficient energy can cause a poor signal-to-noise ratio in the measured vibration signal, giving inaccurate predictions.

The other commonly used excitation technique device is the electromagnetic shaker, but this is less relevant to machining applications and so its discussion is left to other publications [6, 8].

4.3.3 The Measurement Equipment

The measurement of the structure's response can be achieved in a number of ways, but the most common method is the use of piezoelectric accelerometers, shown schematically in Figure 4.4. They operate by using a seismic mass, which during vibration exerts an inertial force on the piezoelectric material. Provided the excitation is sufficiently below the resonant frequency of the device (due to the stiffness of the piezoelectric crystal and the seismic mass), then the seismic mass will move with the same magnitude and phase as the host structure, and the force exerted on the crystal will be proportional to the corresponding acceleration. The piezoelectric crystal produces a charge as a consequence of the applied force. This charge must be conditioned, or amplified, so that it can be properly measured using the signal analyser equipment. The charge amplification is either performed onboard the transducer ("integrated circuit" type devices), or using a separate charge amplifier.

It should be noted that piezoelectric force transducers operate in much the same way, but the load exerted on the crystal matches that transmitted through the device. Such transducers are usually used at the impact tip of modal hammers, described previously.

Piezoelectric accelerometers can be attached to the host structure in a variety of ways. A threaded stud can be cemented onto the host structure, and the accelerometer screwed in place. For magnetisable host structures, magnets can be used to hold the accelerometer in place. Alternatively, a wax can be used to temporarily adhere the transducer to the host structure, and the simplicity of this approach means that it is often used when testing cutting tools. The attachment method can often influence the measurement response, particularly at frequencies above 2-5 kHz. However, typical machine tool vibrations tests involve frequencies below this value.

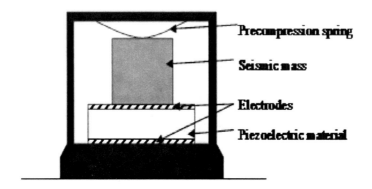

Figure 4.4. Simplified schematic of a piezoelectric accelerometer

The major drawback of using accelerometers is that the mass of the accelerometer can have an effect on the natural frequencies of the structure being tested. This is especially true for very lightweight structures, such as small milling tools. In this case, it can be desirable to use alternative, non-contacting measurement techniques such as laser transducers. The laser Doppler velocometer (LDV) is capable of detecting instantaneous velocity of the surface of an object, by measuring the wavelength of the reflected laser light compared to the original source. The wavelength is changed due to the Doppler effect, as a result of the velocity of the reflecting surface. LDVs can operate to very high frequencies and have a vibration measurement range of between 0.01mm/s and 20m/s, depending upon the model. However, they require a line-of-sight measurement, need to be positioned a certain distance from the target, and can require a smooth, normal surface from which to reflect the laser beam. For machine shop applications, the floor of the building can vibrate which adds to the noise of the velocity reading.

4.3.4 Novel Approaches

Although there is a wide variety of measurement and excitation techniques available for modal analysis of structures, the particular problems associated with machine tool problems has motivated research into specific, novel techniques. Some of the shortfalls of existing equipment are illustrated in Figure 4.5, where a modal hammer and accelerometer are being used on a small milling tool. This measurement will suffer from a number of inaccuracies. Firstly, the mass of the accelerometer will modify the observed natural frequencies of the structure. Secondly, the impact force is not normal to the axis of the tool, and so the impact and measurement coordinates are not aligned. Finally, the flutes on the tool make it extremely difficult to accurately strike the surface, and there is the risk of damaging the cutting edge in the process.

Figure 4.5. Use of a modal hammer on a small tool

In an attempt to overcome these shortfalls, previous work by the author and his colleagues investigated the use of surface-mounted piezoelectric transducers for predicting milling tool stability lobes [15]. With reference to Figure 4.6, the method involved determining the mass-normalised mode shapes at the co-located excitation location ($v_a(t)$ and $v_s(t)$), and (using an LDV or capacitance probe) measuring the mode shape at the tool tip. By calibrating this mode shape with a measurement of the tool's static deflection, the FRF of the tool tip, $x_t(\omega)/f_t(\omega)$ could be predicted. This approach avoided the requirement to excite the structure at the tool tip, which can be particularly problematic for small milling tools.

Another research activity has focussed on an automated approach to determining the FRF of a milling tool, again for the purpose of detecting chatter [16, 17]. Here, a magnetisable dummy tool was mounted in the spindle, and it was shown that by placing a magnet near the tool, a periodic force was imposed as the spindle rotated. The resulting displacement of the tool was measured using a non-contacting optical sensor. The sensor was triggered once per revolution, so that the acquired data indicated the real part of the frequency of periodic excitation. By ramping the spindle speed up, the full FRF (real part) could be evaluated and used to predict stability lobes. One drawback with this technique is that the maximum excitation frequency is limited by the maximum spindle speed of the machine, which may not be sufficient to characterise high frequency modes of vibration. Furthermore, a magnetisable tool is required.

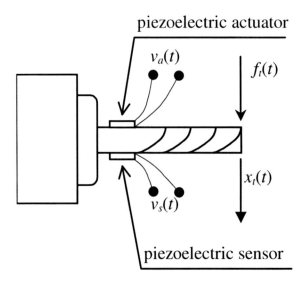

Figure 4.6. Modal testing with piezoelectric transducers

4.3.5 In-process Sensors

The theory and experimental methods described so far have focussed almost entirely on techniques for predicting the dynamic behaviour of a machining process. This may be for the purpose of determining chatter stability (regenerative or otherwise) or to predict the quality of the surface finish and the cutting forces. Another aspect of dynamics diagnostics is that of characterising the behaviour during metal cutting. Undesirable dynamics may be characterised in terms of regenerative chatter stability, mode coupling chatter, primary chatter, or stable behaviour with high levels of forced vibration. Alternatively, measuring the behaviour during machining could serve to calibrate models, or (as a research tool) provide a new understanding of the physics of metal removal. A whole category of machining process monitoring techniques have been developed for other problems, such as identifying tool wear and failure [1-5]. As mentioned in the introduction, these techniques are beyond the scope of the present study. This section will therefore focus on the experimental equipment that is commonly used to diagnose the dynamic behaviour of machining processes. However, some of the devices previously mentioned (*e.g.*, accelerometers and laser Doppler velocometers) are also suitable for in-process measurements.

4.3.6 Dynamometers

Cutting force dynamometers are perhaps the most widely used instrument for characterising the dynamic behaviour of a cutting process. Various configurations are available, such as table mounted devices for milling and drilling, tool holder devices for lathes, or devices mounted on the spindle/spindle bearings [18].

However, one issue with the use of dynamometers is that they have a limited bandwidth, which for example may not be sufficient for measuring the high frequency forces that arise during chatter.

4.3.7 The Current Monitoring

An alternative method to estimate the forces during machining is to process the current signal from the spindle drive. Given the knowledge of the spindle speed, it is possible to determine the resulting torque [18]. In milling/grinding, the diameter of the tool/grinding wheel can then be used to estimate the tangential force on the tool. For turning, the diameter of the workpiece can be used in a similar fashion. However, the effect of the electromagnetic circuit, and the power train, is to filter out the high frequencies of the torque, which limits the bandwidth of the measurement. Despite this problem Soliman and Ismail [19] had some success in identifying the onset of unstable chatter behaviour of milling.

4.2.8 The Audio Measurement

During machining, the chip formation process leads to the release of acoustic energy at high frequencies - typically hundreds of kHz - which can be used to determine events such as excessive tool wear, or tool breakage [4, 20]. The stress waves, or acoustic emission, is clearly at a much higher frequency than the vibrations that arise due to dynamic behaviour such as chatter. These vibrations of the surfaces of the structure lead to sound pressure waves that can be detected by a microphone, or the human ear. Consequently, it is this lower frequency audio signal that is useful for characterising machining dynamics such as chatter. Delio and colleagues [21] performed a detailed review of sensors in machining, and concluded that the microphone is a low cost solution to detecting chatter. The main drawback is that frequencies below 100 Hz cannot be measured easily, and this causes problems if the low-frequency structural modes of the machine are the cause of chatter. Furthermore, the microphone tends to pick up high levels of background noise, particularly in a production environment, due to its omni-directional sensing behaviour.

4.3.9 Capacitance Probes

One research group has developed a capacitance probe system for detecting the vibration of compliant drilling and milling tools during machining [22, 23]. The system used a housing attached to the spindle assembly, so that the probes monitored the vibration near the tip of the tool. The data was used to compare experimental behaviour with predictions from drilling and milling operations. Whilst capacitance probes can provide very high resolution at high frequencies, their performance is particularly sensitive to the earthing of the measurement system. The technique is therefore effective for research purposes, but less relevant for industrial problems.

4.3.10 Telemetry and Slip Rings

A number of experimental devices have employed slip-ring assemblies or telemetry systems in an effort to extract measurement signals from the rotating spindle system. For example, Dohner [24] developed an active vibration control system for milling tools, that required strain gauge sensors on the tool. The strain gauge signals were transmitted to the vibration control system using a telemetry device. A commercially available spindle-mounted dynamometer also uses a telemetry system, whilst Weck [25] developed a slip ring system for monitoring tool vibrations. The main drawback with telemetry and slip ring devices are the additional complexity of the equipment, and the possible noise/interference on the measurement signal.

4.3.11 Fibre-optic Bragg Grating Sensors

A novel alternative to slip rings and telemetry devices has been proposed [26], which uses fibre optic devices to measure the strain on a milling tool. It was proposed that the optical signal could be transmitted from the rotating tool by using fibre-optic rotary joints. Optical signals have the advantage of being immune to contamination by electromagnetic interference, whilst Bragg grating devices can offer a more compact and distributed sensor than the traditional strain gauge.

4.4 Chatter Detection Techniques

In the production environment, chatter can often be detected by an experienced machine tool operator, due to the noise produced during cutting, and the characteristic surface finish. However, a number of factors have motivated research into more advanced chatter detection techniques, such as:

- the need for automation
- the avoidance of human error
- objective, rather than subjective, diagnosis.

Consequently a wide range of approaches have been developed, and Figure 4.7 makes some attempt to classify these various techniques. In this section these methods will be reviewed and some of the more common methods will be illustrated using numerical and experimental examples of milling operations.

The numerical examples use the modelling approach described in reference [27], and assumed a rigid tool with a single workpiece mode of vibration normal to the feed direction. The natural frequency was 922 Hz, the stiffness 1.34 MN/m, and the damping ratio 0.011. The milling tool had 2 vertical flutes, and a 12.7 mm radius. The simulation assumed a 5% radial immersion with a 0.05mm feed per tooth, and the depth of cut and spindle speed were varied depending on the requirements.

The experimental data was collected by Zhang [28], and involved the machining of a flexible workpiece which had a natural frequency at around 420 Hz. A two-flute 30 degree helix 10 mm diameter cutter was used, and the vibration of the workpiece was recorded using a surface mounted piezoelectric strain gauge.

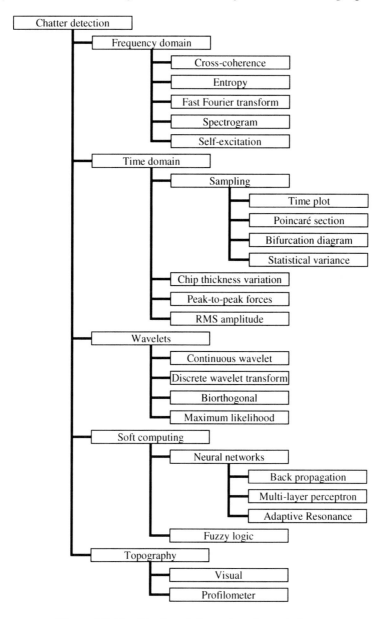

Figure 4.7. Classification of chatter detection techniques

4.4.1 The Topography

Chatter usually leaves an undesirable surface finish which, for demanding applications (such as aerospace components), is unsatisfactory. Consequently, one of the most obvious ways to diagnose or illustrate chatter is through visual inspection of the surface finish. In Figure 4.8 a photograph of the peripheral surface of a component is shown. The cutting sequences near the top of the surface have been unstable, leaving very clear chatter marks on the surface. In contrast, for the stable sequences near the bottom of the workpiece the surface marks can be attributed to feed marks from the tooth engagement.

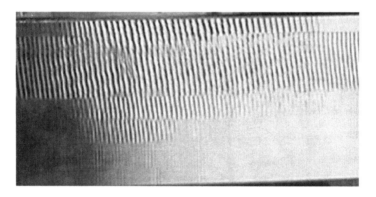

Figure 4.8. Visual identification of chatter on the peripheral-milled surface

Whilst such images help to illustrate the chatter stability to the uniformed reader, it can be more constructive to analyse the surface topography using profileometers or white light interferometers [29]. In this case the surface average roughness (Ra) value can be determined, and given knowledge of the cutting conditions the wavelength of the chatter vibrations can be used to determine the chatter frequency.

At very low depths of cut it can be difficult to photograph or analyse the peripheral surface of the workpiece. Chatter can still be identified by inspecting the face-milled surface (as shown in Figure 4.9), but it is not easy to determine the corresponding frequency of the vibrations.

4.4.2 The Frequency Domain

It was mentioned earlier that machine tool operators can often identify chatter due to its tell-tale audio signature. It follows that a frequency domain analysis of the audio signal should yield the same result. In practice, a variety of measurement signals could be used for the analysis, such as workpiece or tool accelerations/displacements, or force readings from a dynamometer. However, it transpires that the microphone is the most versatile and easy to use approach, because it captures the vibration of more than one component in the machine-tool system [21].

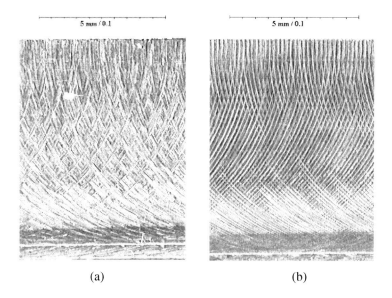

Figure 4.9. Visual identification of chatter on the face-milled surface

Regardless of the measurement used, the underlying theory of frequency domain techniques is that the self-excited chatter vibrations will occur at frequencies incommensurate with the forced vibrations from to the cutting process. For turning, these forced vibrations are simply constant (*i.e.*, zero frequency), whereas in milling the rotation of the tool teeth into the workpiece causes periodic forced vibrations at the tooth passing frequency and its harmonics. In terms of nonlinear dynamics, regenerative chatter in milling is an example of a secondary Hopf, or Neimark, bifurcation [30].

The simplest way to determine the frequency domain information is to take the fast Fourier transform (FFT) of the measured signal, but more advanced statistical signal processing techniques (such as computing the power spectral density with Welch's method [31]) can yield better results. As an example, Figure 4.10 shows the power spectral density of the numerical simulation at 15,000 rpm. The stable cut exhibits vibrations at the force vibration frequency (500 Hz) and its harmonics. In contrast, the unstable cut exhibits vibrations at a frequency incommensurate with the force vibrations. This frequency - around 900 Hz - is close to the natural frequency (922 Hz) of the structure. The combination of the two vibrations (forced and self-excited) leads to peaks in the spectrum, which occur at the frequencies [30]:

$$f = \left| m_f f_f + m_c f_c \right| \tag{4.8}$$

where f_f and f_c are the forced and chatter frequencies, and m_f, m_c are integers. Evaluating Equation 4.8 for a few terms of m_f and m_c leads to the frequencies indicated in Figure 4.11, which match the peaks in the calculated spectrum.

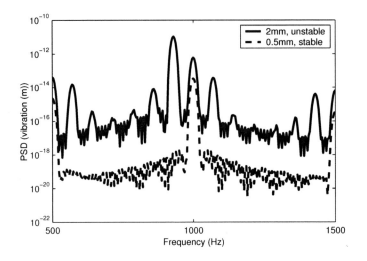

Figure 4.10. Power spectral density of the vibration (Hopf bifurcation)

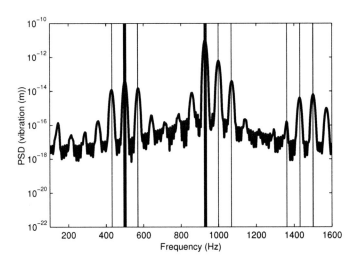

Figure 4.11. Peaks of the forced and self-excited vibrations in Figure 4.10

In very low radial immersion milling, another form of regenerative chatter can occur which is subtly different to the normal case. Whereas classical regenerative chatter results in self-excited vibration at frequencies near a natural frequency of the structure, low radial immersion milling can lead to vibrations at sub-harmonics of the forced vibration, due to the so-called period doubling [32] or flip bifurcation phenomena. The power spectral density plot for this kind of instability is shown in Figure 4.12. It should be noted that similar characteristics can emerge due to the effects of tool runout [32] and so care should be taken when analysing this behaviour.

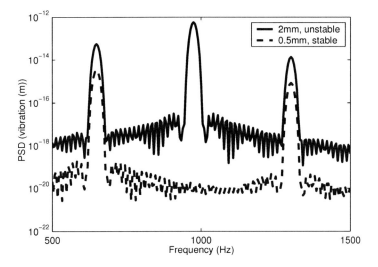

Figure 4.12. Power spectral density of the vibration (flip bifurcation)

The Fourier analysis techniques described above have been widely deployed by both researchers and software producers. For example, the Harmonizer software [33] uses a microphone and laptop-based software to identify chatter frequencies and suggest alternative, more stable, spindle speeds.

It is worth mentioning that these methods are mostly relevant to milling chatter, due to the presence of high frequency force and self-excited vibrations. For turning the forced vibrations tend to be at zero frequency, which can make it easier to identify the onset of the chatter vibration. However, an alternative frequency domain approach has been proposed by Li et al. [34], who used the cross-coherence of two perpendicular acceleration signals, and Kondo [35] who considered the transfer function from the inner to outer chip modulation in turning.

A major drawback with the majority of frequency domain techniques is that they offer no information in the time domain. For example, if chatter occurred irregularly during the measured signal, then the frequency domain would not indicate the region where chatter occurred, and the magnitude of the chatter vibrations would be diluted due to the stable regions of the cut. One solution is to use smaller frames of the measured signal for the Fourier analysis (as used in [33]), but this leads to increased dependency on the operator, and hence more subjectivity. It can also be time consuming, and so an alternative approach is to automate the framing and Fourier analysis by calculating the so-called spectrogram, or short-time Fourier transform, of the signal [14]. An example is shown in Figure 4.13, for the experimental data set. However, the drawback of using a smaller frame for the Fourier analysis is that resolution in the frequency domain is lost. This is further illustrated in Figure 4.13b, where the analysis frame has been reduced to increase the time-domain resolution at the expense of frequency domain resolution. Neither of the graphs in Figure 4.13 has clearly identified chatter vibrations - only the forced vibrations at 420 Hz dominate the response. This time-frequency resolution can only be overcome by wavelet-based techniques, which will be described later.

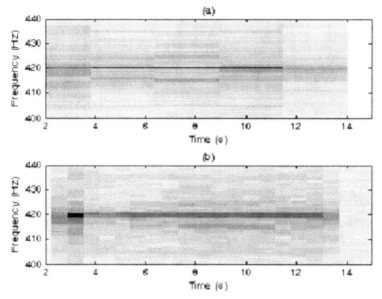

Figure 4.13. Experimental chatter recognition. **a** Spectrogram with FFT length of 215 **b** Spectrogram with FFT length of 212

A final issue with frequency domain approaches is that they tend to need a large number of sample points, and tool revolutions, in order to compute the chatter stability. This is perfectly acceptable for experimental or production scenarios, where 100 or more tool revolutions will easily occur within each CNC program sequence. However, time domain simulations of milling can be computationally intensive, and so in this case it is desirable to develop techniques that require fewer (simulated) tool revolutions. One approach, described in [27], achieved this by determining the growth or decay of the self-excited vibration. It is well known that the viscously-damped free vibration exhibits a logarithmic decay; reference [27] used this concept for the case of chatter vibrations. Figure 4.14 shows an example for the simulating milling scenario for various depths of cut. For stable cutting, the self-excitation decays logarithmically, whereas for unstable cutting the vibration first grows logarithmically and then reaches a steady-state limit cycle due to the nonlinearity of tooth loss-of-contact. The advantage of this approach is that it can be used with relatively few (*e.g.*, 10-20) simulated revolutions, and the chatter threshold (where the vibrations neither decay nor grow) is physically meaningful. This is in contrast to the spectral density approach, where some subjectivity is introduced: the operator must decide what magnitude of self-excited vibration (relative to the forced vibration) is acceptable before it is classified as chatter.

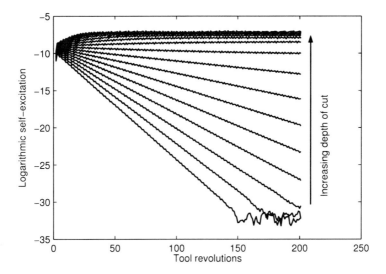

Figure 4.14. Determining the chatter stability by analysing the self-excited vibrations

To conclude the discussion of frequency domain techniques, it is worth mentioning some of the practical and industrial applications of this approach. The MetalMax Harmonizer [33] software is one example, which employs a microphone signal and performs Fourier analysis to suggest alternative spindle speeds. This approach is based upon an iterative method for optimising the spindle speed given knowledge of the chatter and forced vibration frequencies [36]. A similar software system, including tool breakage detection, is described by El Hafsi [37]. Automatic spindle speed regulation has also been proposed [38] and Altintas demonstrated the implementation of the approach as part of a modular intelligent machine [39].

4.4.3 Time Domain

One of the more elegant means to identify unstable chatter in milling has its origins in nonlinear dynamics. Nayfeh and Balanchandran [30] describe how the stability of periodic solutions can be analysed using a Poincaré section, which captures a snapshot of the system's states once every cycle of the periodic solution. For milling, the forced vibrations repeat once every tooth pass, thus defining the periodicity of the solution. Properly speaking, a Poincaré section would sample all the states of the milling process once every tooth pass. However, in practice it has been found [29] that the behaviour can be illustrated with only one or two experimental vibration measurements. These measurements are typically the deflection of the tool or workpiece in mutually perpendicular directions on the plane of the cut. Alternatively, the notion of a delayed coordinate can be used [30, 32]. One of the most straightforward and versatile measurements that can be used is that from a microphone [29].

As an example, Figure 4.15 presents the results from the simulation problem, using the y-direction displacement and velocity for the data samples. It can be seen that the unstable result (Figure 4.15a and b) exhibits a non-constant value of the sampled data, which results in a circular motion on the Poincaré section. In contrast, the unstable cut (Figure 4.15c and d) produces (after an initial transient behaviour) a constant value for the sampled data, which converges to a stable "fixed point" on the Poincaré section.

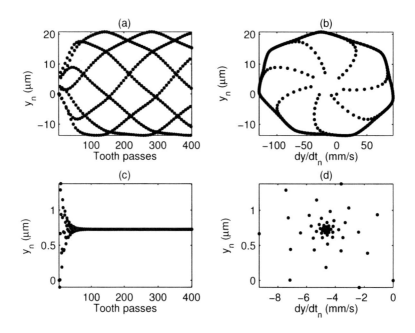

Figure 4.15. Sampled measurements (**a, c**), and Poincaré sections (**b, d**), for a Hopf bifurcation. (**a, b**) 2 mm, unstable; (**c, d**) 0.5 mm, stable.

For the period-doubling bifurcation that can occur during very low radial immersion milling, the Poincaré section and sampled data signal can still be used to detect instability, as shown in Figure 4.16. The term "period doubling" is now justified, as the Poincaré section shows that the solution alternates between two locations.

The sampled data signals and Poincaré sections are an excellent means of illustrating the stable or unstable behaviour, but it can be more useful to display the data against an increasing depth of cut. The resulting bifurcation diagram [30] is so called because the transition from stable cutting to chatter - the bifurcation - can be clearly seen. Figure 4.17 shows the bifurcation diagram for the simulation results at a Hopf bifurcation. The chatter stability margin, or bifurcation point, lies between 1.2 and 1.3 mm depth of cut.

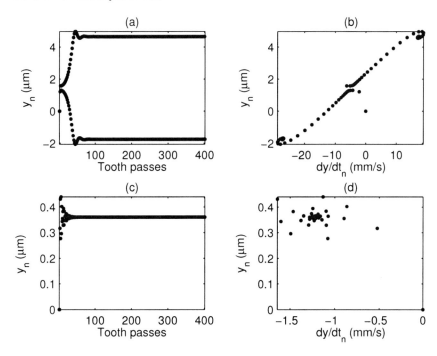

Figure 4.16. Sampled measurements (**a**, **c**), and Poincaré sections (**b**, **d**), for a flip bifurcation. (**a**, **b**) 2 mm, unstable; (**c**, **d**) 0.5 mm, stable.

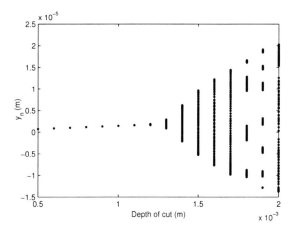

Figure 4.17. Bifurcation diagram for the detection of the (Hopf) chatter stability boundary

It should be noted that the bifurcation diagram must be based upon the quasi-steady data from the sampled data (Figure 4.15), because the transient behaviour at the start of the cut (or simulated cut) would lead to a misleading result. Some researchers have shown hysteresis of the bifurcation point for the case of a nonlinear relationship between the cutting force and the chip thickness [40, 41].

A more convenient way to present the bifurcation data as a function of the depth of cut, is to calculate the variance of the quasi-steady sampled data, for each depth of cut [42]. This approach was used by Schmitz et al. [29], with microphone signals, to identify the onset of chatter. Bayly *et al.* [43] used the variance indicator with a time domain model to determine the stability boundary, but this required a larger number of simulated tool revolutions so that there are sufficient quasi-steady data points for a variance calculation.

The time-domain methods described so far have all been based upon the concept of once-per-revolution, or once-per-tooth, sampling. There are some alternative time-domain approaches which have been proposed, particularly for milling simulations. The peak-to-peak forces method was proposed by Smith and Tlusty [13], who showed that as the limit of stability is reached, the cutting forces tend to drastically increase. They showed that the peak-to-peak forces provided a rapid means of identifying the stability boundary during time-domain computations of milling. More recently, Altintas and co-workers [44] proposed an alternative method that they claimed was more effective for low-radial immersion milling simulations. They recorded the computed chip thickness during the simulation, and compared the value to that for a rigid machine structure. The maximum variation was used, with a suitable threshold value, to indicate the onset of chatter. Whilst this approach has been implemented in commercially available software [45], it is not suitable for practical (*i.e.,* non-simulated) milling since it requires a measurement of the chip thickness.

At this point, it is worth reviewing four of the chatter detection methods that are particularly suited to milling simulations, rather than experiments. These are the variance, peak-to-peak force, chip thickness variation, and self-excitation damping ratio methods. They are presented in Figure 4.18a-d, respectively, for the simulation scenario as plots of the criteria against increasing depth of cut. It can be seen that each method has advantages and disadvantages. The variance method has a value of nearly zero for stable cuts, but this rises sharply after instability. The chip thickness variation has a similar pattern, but with a value of unity for stable cuts. In both cases, the chatter threshold must be chosen to be slightly higher than the stable value (0 or 1) but not so high as to give a stable depth prediction that is too high. For the peak-to-peak forces method, there is no obvious value for the threshold, but the value does start to increase more sharply after the onset of instability. The self-excitation damping ratio has a physically meaningful threshold value (zero), but it requires more signal processing steps than the other methods, and may not be appropriate for variable-pitch tools.

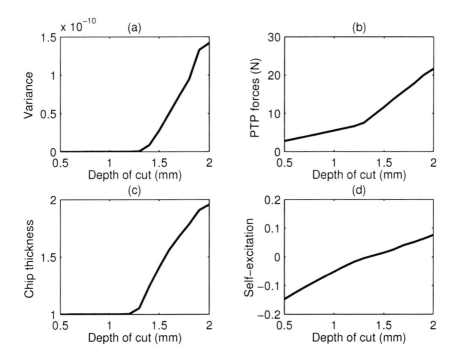

Figure 4.18. Identification of chatter stability (Hopf bifurcation)

4.4.4 Wavelet Transforms

It was mentioned earlier that one of the main drawbacks with the frequency domain chatter detection methods is that they tend to lose resolution in the time domain. This can be particularly troublesome for experimental data, where different regions of a machining sequence can lead to either stable or unstable cutting, perhaps with a change in the mode of vibration that causes chatter, and a corresponding change in the chatter frequency. For this problem, the time domain methods – particularly using sampled data – can still be used to determine the regions of cut that have caused chatter [29], but they offer no information on the frequency of the chatter vibrations.

Wavelet-based techniques can overcome these shortfalls, because wavelet-based signal processing reduces the problem of time versus frequency resolution [46, 47]. Some of the earliest work that described the use of wavelets for chatter problems was that of Khraisheh *et al.* [48]. They investigated the dynamics of primary (*i.e.*, non-regenerative) chatter using the continuous wavelet transform, showing that the effects of friction and built-up-edge formation led to chaotic behaviour. Berger *et al.* [49] used the bi-orthogonal wavelet to investigate chatter in turning, whilst Choi and Shin [50] used the wavelet-based maximum likelihood method. Here, the fractal dimension is used as an indicator of the randomness of the dynamic behaviour, thus indicating the occurrence of chatter. Yoon and Chin [51] used the discrete wavelet transform to investigate the onset of chatter in end-milling. In general, wavelet transforms can offer more insight into the behaviour than time- or frequency-based analyses, but this insight is at the expense of additional computations, and more complex signal processing which may require a judicious choice of processing parameters.

As an example, the experimental data sequence is analysed in Figure 4.19a using the continuous wavelet transform [46]. The result is similar to that for the spectrogram (Figure 4.13), except that there is the possibility of improving the time and frequency (strictly speaking, "scale" rather than "frequency") resolution. Consequently the region where chatter occurred – and the chatter frequency - can now be readily identified. When the cutting first begins (at time 3.8 seconds) there is a sudden increase in the vibration magnitude at both the forced vibration frequency (420 Hz) and a chatter frequency (413 Hz). However, the chatter vibrations seem to stabilise until about 8 seconds, when they reappear at a frequency of 415 Hz, for a period of about 4 seconds.

For comparison purposes, Figure 4.19b plots the approximate once-per-tooth samples for the same data. The sample rate is approximate because a proximity probe was not used to trigger the samples or to determine the exact spindle speed. Consequently the samples appear to drift slightly during the cut. Nevertheless, the region of chatter (between 8 and 12 seconds) can be identified by the higher variance of the data. Compared to the wavelet-based approach, the sampled time method has not required any parameter specification for the signal processing steps (other than the spindle speed). Meanwhile, the continuous wavelet transform requires a number of parameters to be chosen, making it less useful for industrial practitioners.

4.4.5 Soft Computing

One approach to identifying or characterising chatter is through the use of soft computing methods, which aim to mimic the pattern recognition and decision making capabilities of the human mind. This approach is justifiable for chatter problems, because experienced machine operators can often recognise chatter relatively easily, suggesting that "intelligent" computing methods could be used instead.

One of the most popular techniques is that of using artificial neural networks, such as multi-layer perceptions or back propagation neural networks [52]. However, these methods require substantial "training" of the network, and it has been argued that these supervised neural networks are less suitable for chatter detection because the frequency of the chatter vibrations will vary depending on the application [53]. Consequently, unsupervised networks that rely on adaptive resonance theory [53, 54] have also been proposed.

Although neural networks attempt to mimic the decision making capabilities of the mind, they do not harness any of the physical understanding of the problem being considered. Since the dynamics of regenerative chatter are well understood, the case in favour of such a "black box" approach is not so strong. An alternative soft computing approach that can still incorporate physical models is fuzzy logic, which has been implemented on various machining dynamic problems [55-57].

4.4.6 The Information Theory

One final category of chatter detection technique that deserves mention is the use of Information Theory. This has been shown to be well-suited to the detection of grinding chatter [58, 59] using the concepts of entropy, course-grained entropy rate, and course-grained information rate. Other researchers have employed the concept of "mutual information" to determine the onset of chatter in turning [60].

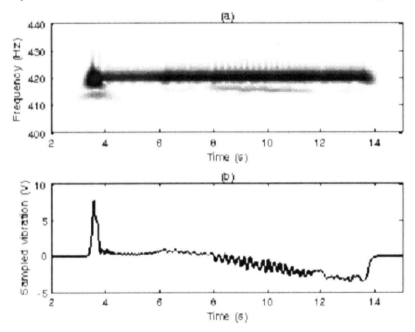

Figure 4.19. Experimental chatter recognition. **a** Continuous wavelet transform **b** Vibration sampled at approximately one sample per tool revolution

4.5 Summary and Conclusions

This chapter has reviewed the various techniques available for diagnosing the dynamic behaviour of machine tools and machining processes. The techniques can be classified as either pre-process techniques, or in-process techniques.

Pre-process techniques aim to develop an understanding of the dynamics of the machine-tool structure (the machine, the tool, and the workpiece) so that along with a model of the process, the dynamic behaviour can be predicted. This enables the optimization of the process, for example by avoiding chatter or improving surface finish. Basic pre-process techniques could involve one simple vibration test, with no subsequent modelling of the dynamics. More advanced approaches involve developing complete modal and response models of the structure, to gain a deeper understanding of its performance. In either case, a range of experimental techniques must be used, along with statistical signal processing, and perhaps modal identification methods.

In-process techniques aim to illustrate the dynamic response of the structure during the machining process. This requires measurements to be taken during the machining process, and although some of the equipment used for pre-process measurements is still relevant (*e.g.*, accelerometers) the experimental approach is generally quite different. A wide range of specialist signal-processing procedures are available to elucidate the dynamic behaviour, in particular the chatter stability, based upon the experimental data. Some of these methods provide an almost instantaneous analysis of the behaviour, and as such are suitable for use in intelligent or adaptive machining applications. Other methods require information that is not readily available from experimental machining processes, and are therefore only useful for characterising the dynamics of simulated machining processes (*i.e.*, during modelling).

The wide range of techniques available for characterising the dynamics of the structure, either before or during the machining process, indicates that no one technique has gained widespread acceptance, either in research or in industrial practice. This is not surprising, since it is clear that each method suffers from numerous advantages and disadvantages. The final decision on what methods are appropriate for a particular application rests with the user.

Acknowledgements

The author is grateful for the support of the EPSRC under grant references GR/S49841/01 and GR/S49858/01, and the Advanced Manufacturing Research Centre with the Boeing Company, at the University of Sheffield. Thanks are due to Mr Yuanming Zhang who collected the experimental data used in this study.

References

[1] Shiraishi, M. Scope in in–process measurement, monitoring and control techniques in machining processes – part 1: In–process techniques for tools. Precision Engineering, 1989. 10(4): p. 179–189.

[2] Shiraishi, M. Scope in in–process measurement, monitoring and control techniques in machining processes – part 2: In–process techniques for workpieces. Precision Engineering, 1989. 11(1): p. 27–37.
[3] Shiraishi, M. Scope in in–process measurement, monitoring and control techniques in machining processes – part 3: In–process techniques for cutting processes and machine tools. Precision Engineering, 1989. 11(1): p. 39–47.
[4] Axinte, DA, Gindy, N, Fox, K, and Unanue, I. Process monitoring to assist the workpiece surface quality in machining. International Journal of Machine Tools and Manufacture. 2004. 44(10): p. 1091–1108.
[5] Cho, D–W, Lee, SJ, and Chu, CN. The state of machining process monitoring research in Korea. International Journal of Machine Tools and Manufacture, 1999. 39(11): p. 1697–1715.
[6] Maia, NMM and Silva, JMM, Theoretical and Experimental Modal Analysis. 1997, London: Research Studies Press Ltd.
[7] Altintas, Y, Manufacturing Automation: Metal Cutting Mechanics, Machine Tool Vibrations, and CNC Design. 2000, Cambridge: Cambridge University Press.
[8] Ewins, DJ, Modal Testing: Theory, Practice, and Application. 2000, London: Research Studies Press.
[9] Balmès, E. Structural Dynamics Toolbox. SDTOOLS, Paris, 2002.
[10] Tlusty, J, Manufacturing Processes and Equipment. 1999, New Jersey: Prentice Hall.
[11] Schmitz, TL. Predicting high–speed machining dynamics by substructure analysis. CIRP Annals – Manufacturing Technology, 2000. 49(1): p. 303–308.
[12] Park, SS, Altintas, Y, and Movahhedy, M. Receptance coupling for end mills. International Journal of Machine Tools and Manufacture, 2003. 43(9): p. 889–896.
[13] Smith, S and Tlusty, J. Update on High–Speed Milling Dynamics. Journal of Engineering for Industry, 1990. 112: p. 142–149.
[14] Oppenheim, AV and Schafer, RW, Discrete–Time Signal Processing. 1989, Prentice–Hall: Englewood Cliffs, NJ. p. 713–718.
[15] Sims, ND, Bayly, PV, and Young, KA. Piezoelectric sensors and actuators for milling tool stability lobes. Journal of Sound and Vibration, 2005. 281(3–5): p. 743–762.
[16] Snyder, JP, Davies, M, Pratt, J, and Smith, S. A New Stable Speed Test Apparatus for Milling. Transactions of NAMRI/SME, 2001. 24: p. 153–157.
[17] Schmitz, TL, Davies, MA, Medicus, K, and Snyder, J. Improving high–speed machining material removal rates by rapid dynamic analysis. CIRP Annals – Manufacturing Technology, 2001. 50(1): p. 263–268.
[18] Tlusty, J and Andrews, GC. Critical review of sensors for unmanned machining. CIRP Annals, 1983. 32(2): p. 563–572.
[19] Soliman, E and Ismail, F. Chatter detection by monitoring spindle drive current. International Journal of Advanced Manufacturing Technology, 1997. 13(1): p. 27–34.
[20] McBride, R, Carolan, TA, Barton, JS, Wilcox, SJ, Borthwick, WKD, and Jones, JDC. Detection of acoustic emission in cutting processes by fibre optic interferometry. Measurement Science and Technology. 1993. 4(10): p. 1122–1128.
[21] Delio, T, Tlusty, J, and Smith, S. Use of audio signals for chatter detection and control. Journal of Engineering for Industry, Transactions of the ASME, 1992. 114(2): p. 146–157.
[22] Bayly, PV, Lamar, MT, and Calvert, SG. Low–frequency regenerative vibration and the formation of lobed holes in drilling. Journal of Manufacturing Science and Engineering, Transactions of the ASME, 2002. 124(2): p. 275–285.
[23] Mann, BP, Bartow, MJ, Young, KA, Bayly, PV, and Schmitz, TL. Machining accuracy due to tool or workpiece vibrations. in ASME Mechanical Engineering Congress Exposition. 2003. Washington, DC., United States: American Society of Mechanical Engineers, New York, NY 10016–5990, United States. 14: p. 55.

[24] Dohner, JL, et al., Mitigation of chatter instabilities in milling by active structural control. Journal of Sound and Vibration, 2004. 269(1–2): p. 197–211.
[25] Weck, M, Verhaag, E, and Gather, M. Adaptive control for face–milling operations with strategies for avoiding chatter vibrations and for automatic cut distribution. CIRP Annals, 1975. 24–1: p. 405–409.
[26] Bartow, MJ, Bayly, PV, and Mann, BP. Fiber Bragg grating sensors for measurement of tool displacement. in ASME International Mechanical Engineering Congress Exposition. 2004. Anaheim, CA, United States: American Society of Mechanical Engineers, New York, NY 10016–5990, United States. 15: p. 941.
[27] Sims, ND. The self–excitation damping ratio: a chatter criterion for time–domain milling simulations. Journal of Manufacturing Science and Engineering. 2005. 127(3): p. 433–445.
[28] Zhang, Y and Sims, ND. Milling workpiece chatter avoidance using piezoelectric active damping: a feasibility study. Smart Materials and Structures, 2005. 14: p. N65–N70.
[29] Schmitz, TL, Medicus, K, and Dutterer, B. Exploring once–per–revolution audio signal variance as a chatter indicator. Machining Science and Technology, 2002. 6(2): p. 215–233.
[30] Nayfeh, A and Balachandran, B, Applied Nonlinear Dynamics: Analytical, Computational and Experimental Methods. 1994: Wiley.
[31] Welch, PD. The use of fast Fourier transform for the estimation of power spectra: a method based on time averaging over short, modified periodograms. IEEE Trans. Audio Electroacoust., 1967. AU–15: p. 70–73.
[32] Mann, BP, Young, KA, Schmitz, TL, and Dilley, DD. Simultaneous stability and surface location error in milling. Journal of Manufacturing Science and Engineering, 2005. 127(3): p. 446–453.
[33] MetalMax Harmonizer. Manufacturing Laboratories, Inc, Las Vegas, 2004.
[34] Li, XQ, Wong, YS, and Nee, AYC. Tool wear and chatter detection using the coherence function of two crossed accelerations. International Journal of Machine Tools and Manufacture. 1997. 37(4): p. 425–435.
[35] Kondo, E, Ota, H, and Kawai, T. A new method to detect regenerative chatter using spectral analysis, Part 1: basic study on criteria for detection of chatter. Journal of Manufacturing Science and Engineering, 1997. 119(4(A)): p. 461–466.
[36] Smith, S and Tlusty, J. Stabilizing chatter by automatic spindle speed regulation. CIRP Annals, 1992. 41: p. 433–436.
[37] El Hafsi, M and Yeralan, S. Computer control system for a metal cutting machine. Computers & Industrial Engineering, 1992. 23(1–4): p. 345–348.
[38] Smith, S and Delio, T. Sensor–based chatter detection and avoidance by spindle speed. Journal of Dynamic Systems, Measurement and Control, Transactions of the ASME, 1992. 114(3): p. 486–492.
[39] Altintas, Y and Munasinghe, WK. Modular CNC design for intelligent machining, Part 2: Modular integration of sensor based milling process monitoring and control tasks. Journal of Manufacturing Science and Engineering, Transactions of the ASME, 1996. 118(4): p. 514–521.
[40] Balachandran, B. Nonlinear dynamics of milling processes. Philosophical Transactions of the Royal Society of London, Part A, 2001. 359: p. 793–819.
[41] Pratt, JR and Nayfeh, AH. Chatter control and stability analysis of a cantilever boring bar under regenerative cutting conditions. Philosophical Transactions of the Royal Society of London, Part A, 2001. 359: p. 759–792.
[42] Schmitz, TL and Davies, MA, Method and device for avoiding chatter during machine tool operation, US Patent and Trademark Office. 2002, 20020146296

[43] Bayly, PV, Mann, BP, Schmitz, TL, Peters, DA, Stepan, G, and Insperger, T. Effects of radial immersion and cutting direction on chatter stability in end milling. in ASME International Congress and Exposition. 2002. 13: p. 351–363.
[44] Campomanes, ML and Altintas, Y. An improved time domain simulation for dynamic milling at small radial immersions. Journal of Manufacturing Science and Engineering, 2003. 125: p. 416–422.
[45] CutPro. Manufacturing Automation Laboratories, Inc, Vancouver, 2004.
[46] Rioul, O and Vetterli, M. Wavelets and signal processing. IEEE Signal Processing Magazine, 1991. 8(4).
[47] Mallat, S, A Wavelet Tour of Signal Processing. 1998, London: Academic Press.
[48] Khraisheh, MK, Pezeshki, C, and Bayoumi, AE. Time series based analysis for primary chatter in metal cutting. Journal of Sound and Vibration, 1995. 180(1): p. 67–87.
[49] Berger, BS, Minis, I, Harley, J, Rokni, M, and Papadopoulos, M. Wavelet based cutting state identification. Journal of Sound and Vibration, 1998. 213(5): p. 813–827.
[50] Choi, T and Shin, YC. On–line chatter detection using wavelet–based parameter estimation. Journal of Manufacturing Science and Engineering, Transactions of the ASME, 2003. 125(1): p. 21–28.
[51] Yoon, MC and Chin, DH. Cutting force monitoring in the endmilling operation for chatter detection. Proceedings of the Institution of Mechanical Engineers, Part B: Journal of Engineering Manufacture, 2005. 219(6): p. 455–465.
[52] Tansel, IN, Wagiman, A, and Tziranis, A. Recognition of chatter with neural networks. International Journal of Machine Tools and Manufacture, 1991. 31(4): p. 539–552.
[53] Tarng, YS and Chen, MC. Intelligent sensor for detection of milling chatter. Journal of Intelligent Manufacturing, 1994. 5(3): p. 193–200.
[54] Li, XQ, Wong, YS, and Nee, AYC. Comprehensive identification of tool failure and chatter using a parallel multi–ART2 neural network. Journal of Manufacturing Science and Engineering, Transactions of the ASME, 1998. 120(2): p. 433–442.
[55] Fansen, K, Junyi, Y, and Xiaoqin, Z. Analysis of fuzzy dynamic characteristics of machine cutting process: fuzzy stability analysis in regenerative–type–chatter. International Journal of Machine Tools and Manufacture, 1999. 39(8): p. 1299–1309.
[56] Sim, WM, Dewes, RC, and Aspinwall, DK. An integrated approach to the high–speed machining of moulds and dies involving both a knowledge–based system and a chatter detection and control system. Proceedings of the Institution of Mechanical Engineers, Part B: Journal of Engineering Manufacture, 2002. 216(12): p. 1635–1646.
[57] Liu, Q, Chen, X, and Gindy, N. Fuzzy pattern recognition of AE signals for grinding burn. International Journal of Machine Tools and Manufacture, 2005. 45(7–8): p. 811–818.
[58] Gradisek, J, Govekar, E, and Grabec, I. Using coarse–grained entropy rate to detect chatter in cutting. Journal of Sound and Vibration, 1998. 214(5): p. 941–952.
[59] Gradisek, J, Baus, A, Govekar, E, Klocke, F, and Grabec, I. Automatic chatter detection in grinding. International Journal of Machine Tools and Manufacture, 2003. 43(14): p. 1397–1403.
[60] Berger, B, Belai, C, and Anand, D. Chatter identification with mutual information. Journal of Sound and Vibration, 2003. 267(1): p. 178–186.

5

Tool Design, Tool Wear and Tool Life

Jiwang Yan[a], Yoshihiko Murakami[b] and J. Paulo Davim[c]

[a]Department of Nanomechanics
Tohoku University
Aoba 6-6-01, Aramaki, Aoba-ku, Sendai 980-8579, Japan

[b]R&D Centre
OSG Corporation
Honnogahara 1-15, Toyokawa, Aichi Prefecture 442-8544, Japan

[c]Department of Mechanical Engineering
University of Aveiro
Campus Santiago, 3810-193 Aveiro, Portugal

Notation

a	Exponent of feed in tool life equation
α	Major cutting edge angle
b	Exponent of depth of cut in tool life equation
β	Minor cutting edge angle
C	Constant based on tool and workpiece
d	Depth of cut
f	Tool feed
γ	Tool rake angle
h	Undeformed chip thickness
KT	Crater wear depth
n	Tool-material dependant constant
θ	Tool relief angle
R_{max}	Maximum height (peak to valley) of a surface
R	Tool nose radius
r	Edge radius
T	Tool life
VB_B	Flank wear land width
VB_{Bmax}	Maximum flank wear land width
V_c	Cutting velocity
AE	Acoustic emission
Al_2O_3	Aluminium oxide
BUE	Built-up edge

CaF$_2$	Calcium fluoride
CBN	Cubic boron nitride
CrN	Chromium nitride
CVD	Chemical vapour deposition
DLC	Diamond-like carbon
DM	Ductile machining
DRM	Ductile regime machining
EFM	Environmentally friendly machining
HSM	High speed machining
HSS	High speed steel
MMC	Metal matrix composite
MQL	Minimum quantity lubricant
PCD	Poly-crystalline diamond
Ps	Cutting edge plane
PVD	Physical vapour deposition
SEM	Scanning electron microscope
SCD	Single-crystalline diamond
Si$_3$N$_4$	Silicon nitride
SKD	Dies tool steel
SKS	Special tool steel
TiAlN	Titanium aluminium nitride
TiC	Titanium carbon
TiCN	Titanium carbon nitride
TiN	Titanium nitride
UPM	Ultraprecision machining
WC	Tungsten carbide

5.1 Tool Design

5.1.1 The Tool-workpiece Replication Model

Metal cutting is a process of removing material from a workpiece in the form of chips using single- or multi-point cutting tools with a clearly defined geometry. To some extent, the performance of a cutting tool determines the cutting behaviour and the process capability. In order to design high-performance cutting tools, it is important to clearly understand the tool-workpiece interfaces and the mechanism of surface generation in machining.

Surface roughness is one of the major parameters in describing the quality of a produced surface by cutting processes. Surface roughness is a result of tool-workpiece replication based on the movement provided by the machine tool. Therefore, theoretically, surface roughness can be predicted if tool geometry and cutting parameters are given.

If depth of cut is sufficiently large, theoretical surface roughness is a function of tool feed and tool geometry. It represents the best possible finish which can be obtained for a given tool shape and feed. Here, we consider cutting models with two kinds of simple tool shapes: straight-nosed tools and round-nosed tools. The tool-workpiece replication models of straight- and round-nosed tools are shown in Figure 5.1 and Figure 5.2, respectively.

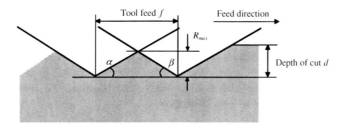

Figure 5.1. Tool-workpiece replication model for a straight-nosed tool

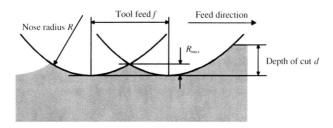

Figure 5.2. Tool-workpiece replication model for a round-nosed tool

As shown in Figure 5.1, for a sharp straight-nosed tool without a nose radius, surface roughness is the maximum height (peak to valley, Rmax) of the scallops, and can be given by:

$$R_{max} = \frac{f}{\cot\alpha + \cot\beta} \tag{5.1}$$

where d is the depth of cut, f is the tool feed, α is the major cutting edge angle, and β is the minor cutting edge angle. For a round-nosed tool with a nose radius R, as shown in Figure 5.2, the maximum height of the scallops is given by:

$$R_{max} = R - \left(R^2 - \frac{f^2}{4}\right)^{\frac{1}{2}} \approx \frac{f^2}{8R} \tag{5.2}$$

For example, when cutting with a straight-nosed tool with major/minor cutting edge angles of $\alpha = \beta = 45°$ at a tool feed of 10μm/rev, Rmax will be 5 μm according to Equation 5.1. In the case of cutting with a round-nosed tool with a nose radius of 1 mm at a tool feed of 10 μm/rev, Rmax will be 0.0125 μm or 12.5 nm, according to Equation 5.2. Generally speaking, round-nosed tools are beneficial in terms of low surface roughness compared to straight-nosed tools. To obtain a smooth surface with straight-nosed tools, it is important to use sufficiently small cutting edge angles [1].

Theoretical surface roughness can be achieved only if all irregularities, such as built-up edges (BUEs), chatters and inaccuracies in the machine tool movements are eliminated completely. However, this cannot happen in practice. It is impossible to achieve the above-mentioned perfect conditions. One of the main factors contributing to actual surface roughness is the occurrence of built-up edges, which deteriorates the theoretical tool-workpiece replication. The larger the built-up edge, the rougher is the surface produced.

Other factors affecting the surface finish include: A) Dynamic errors in the machining variables, including cutting speed, tool feed, and depth of cut, which are mostly caused by the machine movement errors; B) Cutting tool geometry errors, including initial geometrical errors and errors due to tool wear; C) Workpiece material heterogeneity, such as grain boundaries, impurity, and existing defects *etc.*; D) Tool-workpiece interface conditions, including cooling, lubricant and chemical effects; E) Vibrations of the workpiece and the cutting tool, and machine tool chatters. Due to these factors, the actual surface roughness for a certain cutting process is always higher than the theoretical surface roughness.

Among the above factors, the geometrical error of cutting tools has the most direct effects on surface roughness and plays an important role in determining the quality of the surface. Some geometric factors which affect achieved surface finish include nose radius, rake angle, relief angle, cutting edge angle, and cutting edge sharpness and evenness. To design and fabricate high quality and wear-resistance cutting tools is the fist step for high quality machining.

5.1.2 Tool Design Principles

Tool design is one of the most important aspects of the machining process design. In tool design, it is important to comprehensively consider tool geometry, tool material, surface treatment, and their combination to improve the total machining performance of a tool. Basic factors one should consider in tool design are shown in Figure 5.3, and the necessary items for a cutting tool are listed in Table 5.1. Tool material and surface treatment will be discussed in Sections 5.2 and 5.3, respectively. In this section, we emphatically discuss some basic aspects of tool geometry design.

As shown in Figure 5.4, when we consider two-dimensional cutting with a single-point tool, tool rake angle γ, relief angle θ and edge radius (sharpness) r are three major geometrical parameters, of which tool rake angle is the most important one. For tools that have 3-dimensional helical flutes such as drills, end mills and taps, the tool rake angle is not easily to define. For example, a drill has no nominal rake angle. The effective rake angle depends on the helical angle at each cutting point.

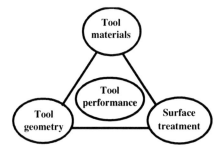

Figure 5.3. Fundamental factors of cutting tool design

Table 5.1. Basic requirements of a cutting tool

Geometrical requirements	Form accuracy
	Edge sharpness
	Edge uniformity
Material requirements	High strength (hardness and toughness)
	High thermo-chemical stability
	Low affinity with workpiece materials
	High thermal conductivity

Generally speaking, the tool rake angle should be small or negative when cutting hard materials in order to strengthen the cutting edge. On the contrary, the rake angle should be large when cutting soft materials for the purpose of improving the edge sharpness of the tool and decreasing the cutting forces. For example, recently, pre-hardened mould steel has become possible to be directly machined with end mills and drills instead of electric discharge machining. In this case, the cutting depth is very thin and the rotation speed of spindle is very high; thus, the size of the chip is very small. Therefore, the flutes of end mills or drills should be very narrow and shallow in order to maintain the strength of cutting edges. Moreover, the cutting edge section of the tool is designed relatively short, so that the tool can have a higher rigidity. A ball end mill designed based on this concept is shown in Figure 5.5. Additionally, in the case of cutting hard materials, because the temperature of the cutting edge tends to be very high, usually the tool is treated by surface coating technology with a material of high thermal strength.

The other case is the machining of relatively soft and sticky materials, such as stainless steel and aluminium alloys. These materials are also difficult-to-cut materials in a different meaning from pre-hardened steel. Because the cutting chips of these materials are soft and sticky, the chips easily adhere to the flute surface of the tool. In this case, one should design wide flutes on end mills or drills to ease the chip removal. Another solution is to make the cutting edges very sharp to improve the fluent formation of chips. Also, a large helical angle of the flute is often used to increase the effective sharpness of cutting edges. A drill designed for drilling stainless steel based on this concept is shown in Figure 5.6, with a comparison with a conventional drill.

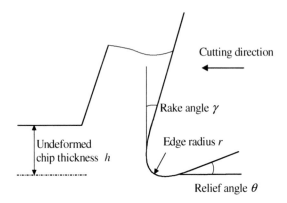

Figure 5.4. Two-dimensional cutting model showing edge geometry

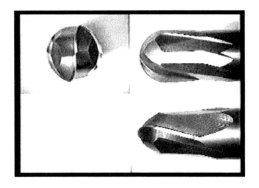

Figure 5.5. Ball end mill designed for cutting hardened steel

(a) Conventional drill

(b) Newly designed drill

Figure 5.6. Drills designed for cutting stainless steel and aluminium alloys

5.1.3 The Tool Design for New Machining Technologies

Tool design is a key technology for the manufacturing industry and it must meet new requirements and match new trends of machining technology. Recently, machining technology is developing toward three main directions. They are:

- High speed machining;
- High precision machining; and
- Low pollution machining.

Machining speed is a key factor for manufacturing economics. Provided other conditions are the same, a high machining speed leads to a high machining efficiency, and in turn, a low production cost. On the other hand, a high cutting speed leads to a high cutting temperature, which accelerates tool wear and shortens tool life. Therefore, the selection of tool material and surface treatment process is the most important factor for high speed machining technology. Machining with traditional steel tools has been performed at a speed of only a few tens of m/min. However, recent machining using ultra-hard materials or high performance coated tools has enabled cutting speeds up to a few thousand m/min.

The second important trend of machining technology is high precision machining, or, alternatively termed, ultraprecision machining (UPM). Machining precision is a deterministic factor for product quality and reliability. The conception of "precision" changes with date and products thus cannot be dealt with impartially. However, generally speaking, high precision machining means that workpiece material is removed at an extremely small scale, in the micron/nano level, to produce surfaces with a form accuracy at the micron level or better and surface roughness at the submicron level or better.

To perform high precision machining, we must have high precision, high rigidity and environmentally controlled machine tools, and extremely precise cutting tools. To realize precise tool-workpiece replication for smooth surfaces, the tool itself must be fabricated precisely, with sharp and strong cutting edges. Edge sharpness is one of the most important parameters for precision cutting tools. As indicated in Figure 5.4, when the machining scale (undeformed chip thickness h) becomes small enough, it approaches the same level as the edge radius r of commercially available cutting tools. In this case, the edge radius effects become very significant and cannot be neglected as in traditional machining. Edge radius effects include many aspects. One of the effects is that the effective rake angle induced by the edge radius becomes a large negative value; thus, material in front of the cutting edge is downward suppressed, and after the tool has passed, recovers due to elastic deformation. This recovering deformation will cause machining errors in terms of the change in effective cutting depth.

Another recent trend of machining is environmentally friendly machining (EFM), namely dry machining and semi-dry machining technology. According to the quantity of used coolant, machining can be divided into wet machining, dry machining and semi-dry machining. Semi-dry machining includes minimum quantity lubricant (MQL) machining and hybrid mist machining. The cost of coolant is approximately 15 percent of the life-cycle operational cost of a machining process. It includes the costs associated with procurement, filtration, separation, disposal and record keeping for the environmental protection. On most occasions, the costs for disposal of coolant are higher than the initial cost of the coolant. Recently, very strict regulations are under construction for coolant usage, disposal and worker protection. As a result, machining with MQL is gaining acceptance as a cost-saving and environmentally friendly option in place of some wet machining processes. MQL machining not only requires the precise control of lubricant mixing systems to maintain the thermal stability of the machining process, but also needs appropriate cutting tool materials and tool coating technologies to improve the high-temperature strength, wear resistance and friction efficient of the cutting tool to assist the chip removal under dry or semi-dry conditions.

5.2 Tool Materials

Machining is a high-speed material removal process that involves extremely high temperatures, high pressures and high stress conditions. It is known that the cutting temperature during machining metal materials can be higher than 1000 °C and the pressure acted on the tool can reach a few GPa. For this reason, tool materials are required to have not only a high hardness and a high toughness but also excellent thermal stability to withstand high temperatures and high pressures, but also severe corrosion. Practically, the hardness of a tool should be at least 3 times higher than that of the workpiece materials. Steel is a typical commercial tool material, which includes high speed steel (HSS), alloyed tool steel (alternatively termed special tool steel, or SKS), die tool steel (SKD) and so on. Other new tool materials are cemented carbide, cermet, cubic boron nitride (CBN), ceramics, and diamond, *etc*. A comparison of tool material characteristics in terms of wear resistance and toughness is shown in Figure 5.7. The features of some of these materials are simply summarized below.

5.2.1 High Speed Steel

High speed steel is a steel alloy with tungsten, as well as cobalt, molybdenum, chromium and vanadium *etc*. Appropriate heat treating can improve the tool properties significantly. The cobalt component gives the material a hot hardness value much greater than carbon steels, and its hardness is in the range of 1000 Hv. It can cut materials with high strengths at a speed up to a few tens of m/min. It is used as all type of cutters, single- and multi-point tools, and rotary tools.

5.2.2 Cemented Carbide

Cemented carbide is a complex alloy by sintering the carbide particles of IV, V or VI group metals (Ti, Zr, Hf, V, Nb, Ta, Cr, Mo, W) using the VIII group metals (Fe, Co, Ni) as binders. These carbide particles have a high oxidization resistance, high fusion point and high hardness. It is technologically possible to obtain many kinds of cemented carbide by combining 9 kinds of carbides and 3 binder metals. In practice, tungsten carbide alloys such as WC-Co, WC-TiC-Co and WC-TiC-Ta are often used as cutting tool materials, among which WC-Co is the most important and widely-used one.

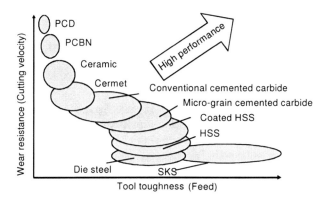

Figure 5.7. Comparison of tool materials in terms of wear resistance and toughness

WC-Co is produced by sintering grains of tungsten carbide in a cobalt matrix. The WC tool bits are often brazed to steel shanks, or used as inserts in holders. The high temperature hardness properties are very good, and thus a cutting speed up to 100 m/min is common on mild steels. Coolants and lubricants can be used to increase tool life, but are not necessary.

5.2.3 Cermet

"Cermet" is a word derived from two words "ceramics" and "metal". It can be classified into the cemented carbide categories, but generally, TiC-Ni-based alloy is independently called cermet. It is generally considered that the cutting performance of TiC based cermet is slightly lower in comparison with WC-Co based cemented carbide, but cermet has excellent corrosion resistibility in comparison with cemented carbide.

5.2.4 Ceramics

Ceramic tools are often made of sintered or cemented ceramic oxides. One of the most widely used ceramic tool materials is aluminium oxide (Al_2O_3). It can be used for turning and facing most metals. The tools are often used as inserts in special holders. The high temperature hardness properties are excellent, and thus generally coolants are not needed. Mild steels can be cut by Al_2O_3 tools at a speed up to 450 m/min. There is also no occurrence of chip welding or built-up edges. What should be noted is that the Al_2O_3 tools are best used in continuous cutting operations, while in interrupted cutting, micro chippings may occur on the cutting edge due to the brittleness of Al_2O_3.

Another ceramic tool material widely used in the industry is silicon nitride (Si_3N_4). Because Si_3N_4 particles cannot be bonded to each other during sintering, usually it is necessary to add sintering additives such as alumina and magnesium oxide. Silicon nitride cutting tool has excellent mechanical and thermal properties. Especially, the bending strength and hardness at high temperatures are excellent in comparison with cemented carbide and cermet. Thus, it is applied to high speed cutting where the edge of a tool is under a high temperature state. It should be noted that silicon nitride is easy to react with steel during cutting steel materials; therefore, Si_3N_4 cutting tools are often limited to cast iron machining.

5.2.5 Diamond

Diamond has the highest hardness and highest heat conductivity among the substances and is excellent in chemical stability. Because of these excellent characteristics, diamond is used as a tool material for high performance machining. It is applicable to turning and boring, producing a very good surface finish. Diamond is also used as diamond dust in a metal matrix for grinding and lapping, which is not discussed here. On the other hand, diamond is a brittle material having low fracture toughness; therefore, it is easily fractured due to shocks and impacts. For this reason, cutting operations must be optimised to minimize vibrations and impacts in order to avoid edge chipping and prolong tool life [2-5].

According to microstructures, diamond can be divided into single crystals and poly crystals. Single-crystal diamond (SCD) is the only tool material that can be used for ultraprecision machining, because an extremely sharp cutting edge can only be obtained on SCD. The estimated edge sharpness (edge radius r_0 in Figure 5.4) is in the level of a few nanometers to a few tens of nanometers. According to the source of production, SCD can also be divided into two categories: natural diamond and composite diamond. Recently, the composite diamond, *i.e.*, the artificially made SCD, has been commercialised as the cutting tool material for the ultraprecision cutting process in place of natural diamond.

SCD tools can be used to machine various nonferrous metals such as copper, aluminium, and electroless plated nickel, *etc.*, which are used as laser mirrors, magneto-optical discs, optic lenses and their moulds. Under specific conditions, SCD tools can also be used to machine brittle crystalline materials such as germanium (Ge), silicon (Si), calcium fluoride (CaF_2), and a few kinds of optical glasses in a ductile mode. This technology is termed ductile regime machining (DRM) or simply ductile machining (DM) [6, 7]. High quality surfaces obtained on these materials provide key parts for the advanced optoelectronic and optical technologies [8-10].

Poly-crystal diamond (PCD) is sintered by diamond powders with some binders such as cobalt powders under high pressure (~ 5000 atmospheric pressure) and high temperature (~ several thousand degrees of Centigrade). The diamond sintered body shows excellent abrasion resistance when cutting nonferrous metals and other new materials and is taking an active part as high accuracy, high efficiency cutting tool.

However, it must be mentioned that a diamond tool is not suitable for machining ferrous materials such as iron and steels. This is due to the fact that the element carbon (C) in diamond will react with the element Fe in the ferrous materials, which accelerates the wear of the tools.

5.2.6 Cubic Boron Nitride

Cubic boron nitride (CBN) has the second highest hardness and thermal conductivity next to diamond. It has better chemical and thermal stability than those of diamond. Due to these properties, CBN has been used as an important tool material. Because CBN has less reactivity with the iron system materials, it has become an especially important tool material for machining ferrous materials which cannot be machined by diamond tools.

CBN is synthesized artificially under the conditions of high pressure and high temperature similar to the production process of diamond. The mechanical and thermal properties of CBN are influenced largely by the kinds and quantities of binder materials. According to the content of the binder, CBN can be generally divided into two categories: the poly-crystalline type (with low binder content) and the composite type (with a high binder content). The poly-crystalline type CBN has an excellent property for cutting cast iron materials and heat resisting alloys; while the composite type CBN is suitable for machining quenched steels. By making full uses of the characteristics of CBN, the high efficiency cutting of hard ferrous components such as automobile parts, engine blocks, gears, shafts, and bearings is possible.

5.3 High-performance Coated Tools

A coated tool is a new type of tool where long tool life is achieved by hard coatings. As indicted by the arrow in Figure 5.7, tool manufacturers are currently trying to shift the tool cutting performance toward the upper right direction, *i.e.*, toward high wear resistance and high toughness, by advanced tool coating technology. By coating a tool, the cutting performance of the tool can be upgraded significantly in many aspects, such as tool life and machining dynamics, *etc*.

Recently, powder-type high speed steel and micro-grain cemented carbide are usually used as major tool body materials for coated tools in order to achieve both high cutting performance and low production cost. The scanning electron microscope (SEM) photographs of these two materials are shown in Figure 5.8.

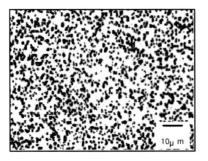 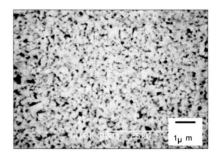

(a) Powder-type high speed steel   (b) Micro-grain cemented carbide

Figure 5.8. SEM photographs of two tool body materials. **a** powder-type high speed steel **b** micro-grain cemented carbide

5.3.1 Tool Coating Methods

Tool coating methods can be classified roughly into two categories: chemical vapour deposition (CVD) and physical vapour deposition (PVD). CVD and PVD can be further divided into various kinds, as shown in Table 5.2.

Table 5.2. Classification of tool coating methods

Chemical vapour deposition (CVD)	Heat CVD
	Moderate temperature (MT) CVD
	Plasma CVD
Physical vapour deposition (PVD)	Vacuum evaporation
	Sputtering
	Ion plating

CVD is a high-temperature process (~1000 °C), and thus on some occasions, micro-structural transformation and thermal deformation of the tools by the high temperature becomes a problem. For this reason, CVD is only used for simple-shape tools such as corner radius tools, insert chips and punches etc., which do not need subsequent edge sharpening processes. It is difficult to apply CVD to the tools having very sharp edges that need grinding for edge sharpening, such as drills, end mills and taps. For these tools, PVD is preferable. Among different kinds of CVD methods, plasma CVD involves the lowest temperature and is applied for diamond-like carbon (DLC) coating and diamond coating.

PVD technology was initially developed by the National Aeronautics and Space Administration (NASA) of the USA in the late 1980s for the purpose of solid lubrication. At the beginning, only simple metal composites such as titanium nitride (TiN) and titanium carbon (TiC) were used as coating materials. However, at present, various materials, such as titanium carbon nitride (TiCN), titanium aluminium nitride (TiAlN), and chromium nitride (CrN) have been developed and used for coating cutting tools to increase cutting performance.

Table 5.3 shows popular coating materials that are widely applied to cutting tools. It is evident that all the coatings have higher hardness than those of high speed steel (900Hv) and cemented carbide (1500Hv). From this aspect, it is easy to suppose that these coatings have a superior wear resistance. Additionally, low friction coefficient and high oxidation temperature of these coatings are beneficial for high speed machining by preventing heat generation and oxidation wear at extremely high temperatures.

Table 5.3. Characteristics of typical tool coatings

Coating	Colour	Hardness [HV]	Friction coefficient	Temperature oxidation [°C]
TiN	Golden	2000	0.4	500
TiCN	Violet	2700	0.3	400
TiAlN	Black	2800	0.3	850
CrN	Grey	1800	0.25	700
DLC	Black	3000	0.1	300
Diamond	Black	9000	0.15	600

5.3.2 The Cutting Performance of PVD Coated Tools

Figure 5.9 a is a comparison of tool performance in terms of the number of drilled holes before tool failure between an uncoated drill and a TiN coated drill. It is evident that the coated drill has a remarkably longer life than the uncoated drill. Figure 5.9 b is a comparison of the effective cutting length between an uncoated end mill and a TiN coated end mill. It is clear that the coated end mill have a significantly longer life than the uncoated one.

Figure 5.10 is a comparison of the cutting performance in terms of tapping torque between an uncoated tap and a TiN coated tap. It is evident that the torque of the coated tap is fluctuating violently compared to the uncoated tap. In this case, the chips generated by the coated tap were very long due to the low friction coefficient on tool rake face, and the chips tend to be winded to the tap body and cause the increase in the tapping torque. From this aspect, we can say that the shape of the tool body should be designed with considering the property of coating material. In other words, to obtain the highest performances of coated tools, the combination of tool geometry design and coating design becomes very important.

Figure 5.11 shows cutting test results of 3 kinds of coated end mills and uncoated ones. The coating materials are TiN, TiCN and TiAlN, respectively. The work material is a cold-type die steel after annealing. We can see that the coated tools by TiN, TiCN and TiAlN have effective cutting lengths of 2, 3 and 5 times, respectively, that of the uncoated tool.

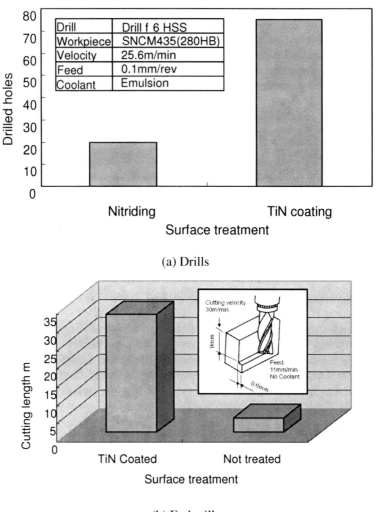

Figure 5.9. Comparisons of tool performance between TiN coated tools and uncoated tools. **a** drills **b** end mills

In recent PVD tool coating technologies, multi-layer coating is of concentrated interests. Compared to single layer coatings, multi-layer coatings can improve residue stress conditions of the coated layers and prevent coatings from separating from tool bodies. Figure 5.12 shows a cross-sectional SEM photograph of the TiAlN multi-layer coating on a micro-grain cemented carbide tool. Such multi-layer coated tools show a remarkably longer life than conventionally coated tools.

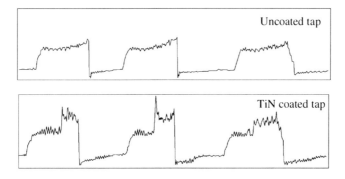

Figure 5.10. Comparison of the tapping torque between uncoated and TiN coated taps

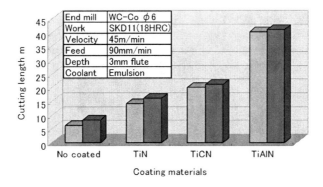

Figure 5.11. Comparison of the cutting performance of end mills coated with different materials

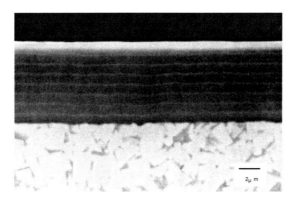

Figure 5.12. Cross-sectional SEM photograph of a TiAlN multi-layer coating

5.3.3 The Cutting Performance of CVD Coated Tools

As mentioned in Section 5.1.3, recently, for the purposes of environment protection, dry machining and MQL machining are required. In the case of steel machining, dry cutting or MQL machining has been achieved by PVD coated tools. However, for aluminium alloys, it is difficult to drill or mill under dry or MQL conditions. This is because the chips of aluminium alloys are often adhered to the tool and degrade the tool life due to the low hardness and high affinity of aluminium.

One solution for the above problem is to cut aluminium alloys with DLC or diamond coated tools. DLC and diamond coatings have an extremely low friction coefficient and a low affinity for aluminium alloys. DLC coatings are especially effective for aluminium alloys of duralumin series such as A5052 and A7075.

Figure 5.13 shows photographs of a DLC coated end mill and an uncoated one after machining super duralumin A7075, where machining conditions are the same. It is clear that there is a distinct difference in the chip adhesion state. It was impossible for the uncoated tool to cut any longer because the flutes were filled up with aluminium chips and the edges were damaged. However, when cutting with the DLC coated end mill, no chip adhesion and damages were observed.

(a) Uncoated   (b) DLC coated

Figure 5.13. Comparison of uncoated and DLC coated end mills after machining an aluminium alloy

Diamond coated tools are effective for high-silicon content materials such as cast iron and metal matrix composite (MMC). Figure 5.14 shows a comparison of cutting test results of diamond coated drills and uncoated ones under air blowing and MQL conditions. Uncoated drills could drill only a few tens of holes under air blowing and a few hundred holes under MQL conditions. As a contrast, diamond coated drills could drill 3,000 holes under air blowing and over 9,000 holes under MQL conditions. Figure 5.15 shows the photographs of these drills after machining. Almost no chip adhesion was observed for the diamond coated drills.

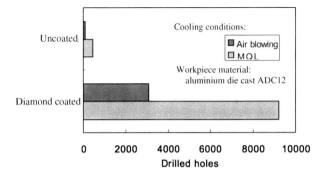

Figure 5.14. Cutting test results of diamond coated and uncoated drills

Figure 5.15. Appearances of diamond coated and uncoated drills after machining

5.3.4 Recoating of Worn Tools

After a coated tool has been used and worn, it can be ground and coated again for reuse. Usually, both the rake face and the flank face need recoating. However, in case of tools such as hobs and roughing end mills, the tools can be reused directly after grinding the rake faces without recoating. This is because the remaining coatings on the flank face are still effective.

On most occasions, the remaining coating needs to be removed by a chemical reaction before grinding and recoating. For example, the reagent involving hydrogen peroxide is used for removing coatings from high speed steel tools. However, we should note that this reagent is not suitable for cemented carbide tools because of the chemical reactions between the reagent and tungsten carbide.

5.4 Tool Wear

Tool wear and tool failure are critical problems in machining, which not only raises the production cost but also degrades the product quality. Tool failure interrupts the machining process intermittently and substantially increases the preparation time. Tool wear/failure can be directly indicated by two facts: one is that the work surface finish is inadequate, and the other is that the work dimension becomes outside tolerance. Tool wear also affects the machining dynamics, and thus can be monitored by measuring cutting forces and torques and so on.

Tool wear/failure is a complicated issue, which has not been well clarified. Generally, tool wear/failure depends on tool material and geometry, workpiece material, cutting parameters (cutting velocity, feed rate and depth of cut), cutting fluids and machine tool characteristics.

5.4.1 Tool Wear Classification

Tool failure/wear can be typically grouped into two categories: premature tool failure and progressive tool wear. Figure 5.16 shows some types of failures and wear usually occurring on cutting tools.

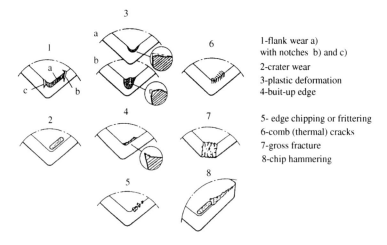

Figure 5.16. Basic types of failures and wear of cutting tools (source: Sandvik®)

Premature tool failure takes place on various occasions and mostly occurs as the breakage of the cutting edge. The reason for edge breakage includes the rapid growth of the crater wear, pre-existing potential cracks in the tool, and impacts during cutting. Tool edge breakage will lead to complete tool failure and the tool thus (or then) becomes unusable.

The useful life of a tool can be defined in terms of the progressive wear. Progressive tool wear mainly includes the wear on the tool rake face (crater wear) and that on the clearance face (flank wear). Of these two, flank wear is often used to define the end of effective tool life. As the flank wear land width (VB_B) has grown to a certain level, it will influence the dimensional accuracy and surface finish of the component as well as the stability of the machining process. The tool failure due to flank wear can be estimated by the maximum value of VB_B and predicted by a function of time. Flank wear is also commonly used for tool wear monitoring.

An established industrial standard on tool wear is ISO 3685:1993. Figure 5.17 shows the typical progressive tool wear geometries according to this standard. In the figure, the wear of the major cutting edges of the tool can be divided into four regions:

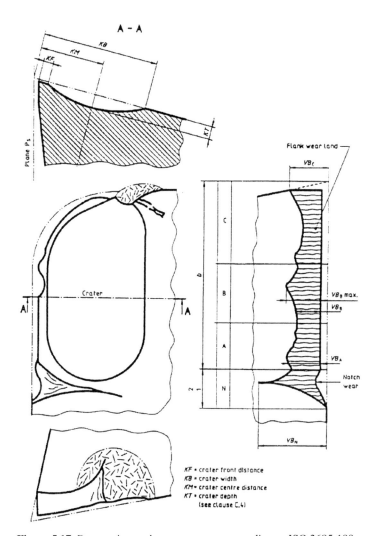

Figure 5.17. Progressive tool wear geometry according to ISO 3685:199

- Region C is the curved part of the cutting edge at the tool corner;
- Region B is the remaining straight part of the cutting edge between region C;

- Region A is the quarter of the worn cutting edge length farthest away from the tool corner;
- Region N extends beyond the area of mutual contact between the tool workpiece for approximately 1 to 2 mm along the major cutting edge. The wear in this region is of the notch type.

The width of the flank wear land VB_B should be measured within region B in the cutting edge plane Ps1 perpendicular to the major cutting edge. The crater depth (KT) should be measured as the maximum distance between the crater bottom and the original face in region B. Tool wear is most commonly measured using microscopes with video imaging systems with a resolution less than 0.01 mm or a stylus instrument similar to a profilometer with precise diamond styluses.

5.4.2 Tool Wear Evolution

A typical curve of the evolution of flank wear land VB_B with cutting time for different cutting velocities is shown in Figure 5.18. The curve can be divided into three zones:

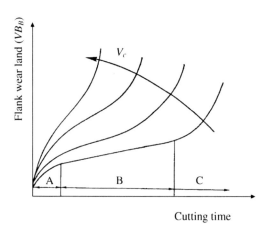

Figure 5.18. Evolution of flank wear with cutting time at various cutting

- Zone A, where the initial flank wear land is established (initial wear);
- Zone B, where wear progresses at a uniform rate (steady wear);
- Zone C, where wear occurs at a gradually increasing rate (several wear).

Also, from Figure 5.18, the influence of the cutting velocity on flank wear land can be seen. An increase in cutting velocity causes an increase in flank wear land VB_B

---

[1] Tool cutting edge plane Ps is the plane containing the major cutting edge and the assumed direction of the primary motion (ISO3685:1993).

during the same cutting time; while a PCD insert is used in turning the metal matrix composite (MMC) (f = 0.1 mm/rev; d=1 mm with cutting fluid) [11].

Here, we give an example of tool wear in turning the MMC with PCD inserts [11]. A photograph of the wear of the insert is presented in Figure 5.19. The dominant wear mechanism in turning MMC is identified as the abrasive wear on the flank face of the turning insert. The wear curves in Figure 5.20 were obtained in turning the MMC with PCD inserts at different cutting velocities. At a cutting speed of 250 m/min, it was possible to perform cutting for 45 minutes with an average flank wear VB_B=0.24 mm. However, at a cutting velocity of 700 m/min, it was only possible to cut for 2 minutes with an average flank wear VB_B=0.42 mm.

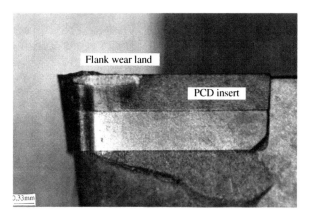

Figure 5.19. Flank wear land of a PCD insert in turning MMC (Vc= 250 m/min; f = 0.1 mm/rev; d=1 mm with cutting fluid) [11]

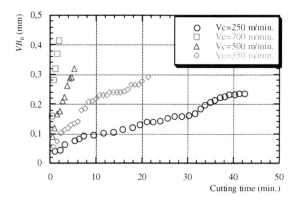

Figure 5.20. Wear curves of P

As the criterion to define the effective tool life, ISO3685:1993 recommended some limit values for the tool wear of cemented carbide tools, HSS tools and ceramic tools, as described below:

For cemented carbide tools:

- VB_B=0.3mm, or
- VB_{Bmax}=0.6mm, if the flank is irregularly worn, or
- KT=0.06 + 0.3 f, where f is the feed.

For HSS and ceramic tools:

- Catastrophic failure, or
- VB_B=0.3 mm, if the flank is regular in region B; or
- VB_{Bmax}=0.6mm, if the flank is irregular in region B.

In the manufacturing industries, general recommendations used in practice for the limit of flank wear VB_B for several cutting tool materials are given in Table 5.4.

Table 5.4. Practical recommendations of flank wear limits for various tool materials

Tool material		HSS	Cemented carbides	Cemented carbide coated	Ceramics	
Operation	(mm)				Al_2O_3	Si_3N_4
Roughing	VB_B	0.35-1.0	0.3-0.5	0.3-0.5	0.25-0.3	0.25-0.5
Finishing	VB_B	0.2-0.3	0.1-0.25	0.1-0.25	0.1-0.2	0.1-0.2

5.4.3 The Material-dependence of Wear

On most occasions, several tool wear mechanisms such as abrasion, adhesion, oxidation, diffusion, etc. can operate simultaneously in a machining process. The dominant wear mechanism will depend on the cutting conditions and tool/workpiece materials. It need hardly be said that tool wear characteristics are dependent on the materials of both tool and workpiece. Different combinations of tool/workpiece materials give rise to different machining performance and tool wear characteristics.

For example, during machining steels with WC tools, the most likely dominant wear mechanisms and the corresponding cutting speeds/temperatures are: abrasion at low speeds/temperatures, followed by adhesion at moderate speeds/temperatures and then diffusion at high speeds/temperatures. The wear of CBN tools is mostly due to temperature controlled processes such as chemical wear and diffusion [12].

5.4.4 The Wear of Diamond Tools

As mentioned in Section 5.2.5, diamond is an important tool material for the precision machining of optical and optoelectronic components or their moulds. Thus, the wear of diamond tools is a critical problem for precision manufacturing processes. Several different mechanisms such as mechanical, thermal, chemical and possible electrical effects can contribute to diamond wear and multiple mechanisms may be at work under some circumstances. However, the predominant wear mechanism will depend on machining conditions and the workpiece material.

Various workers in the past have made contributions to the understanding of the processes that take place when a diamond tool wears during machining metal materials [2-4, 13-14]. In the case of machining aluminium alloys, the dominant tool wear mechanism is likely to be abrasion and chipping caused by impurities on the grain boundaries. As a result, severe edge rounding and flank wear are significant. However, when cutting copper the predominant tool wear is the crater wear, which is related to an increase in the temperature of the rake face. In the cutting of electroless plated nickel, micro chipping of the cutting edge is significant. In steels, the chemical effect is predominant.

Recently, precision machining of hard brittle materials is a research area of concentrated interests. Here, we use silicon as an example. Single-crystal silicon is not only a dominant substrate material for the fabrication of micro-electro and micro-mechanical components, but also an important infrared optical material. Although silicon is a nominally hard and brittle material, it can be deformed plastically in ultraprecision cutting, yielding ductile chips and smooth surfaces [1, 7]. This requires the use of an extremely rigid, environmentally controlled ultraprecision machine tool and a SCD cutting tool with a negative rake angle. Also, the machining scale (undeformed chip thickness) must be controlled to be extremely small, down to the range of a few tens of nanometres, so that a high pressure will be yielded in the cutting region, which facilitates a brittle material to deform plastically. This approach can machine silicon in a ductile mode without the need for subsequent polishing.

However, in practice the ductile machining of silicon is often limited by the rapid wear of diamond tools. As the geometry of the contact zone between the tool tip and the workpiece is extremely important for the attainment of the plastic flow rather than fracture in cutting, the wear of the diamond tool becomes a limiting factor in developing a machining process for silicon. Tool wear in silicon machining can be classified into two types: micro-chipping and gradual wear, the predominant wear mechanism depending on undeformed chip thickness [15]. Brittle mode cutting at an undeformed chip thickness of the micrometer level leads to edge micro-chippings, whereas ductile mode cutting at an undeformed chip thickness of a few tens of nanometres causes gradual wear involving crater wear and flank wear land. Some results of SCD tool wear in silicon machining are given below [15].

Figure 5.21 shows an SEM photograph of the main cutting edge of a new diamond tool. The edge appears to be extremely smooth and sharp without any visible defects. Edge radius is estimated to be 50 nm. Figure 5.22 a is an SEM photograph of the cutting edge after face-turning silicon for 1.27 kilometres at an undeformed chip thickness of 900 nm, *i.e.*, brittle mode cutting. In the figure no obvious wear can be observed on the rake face; whereas a few micro-chippings have occurred to the edge. As the cutting distance increased, both the number and the size of the micro chippings increased. Figure 5.22 b is the same cutting edge after cutting for 7.62 kilometres. The entire edge has been covered with micro-chippings that are a few micrometers in size.

In the brittle mode cutting, cleavage fracturing occurs to the work material, forming micro craters in front of the cutting edge. The micro craters make the cutting process intermittent and cause micro impacts. These micro impacts will take place at a very high frequency, leading to edge chippings. This mechanism may be similar to the fracture of other brittle material tools in traditionally interrupted cutting or when cutting materials containing hard particles and inclusions.

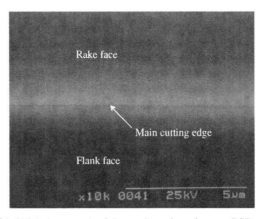

Figure 5.21. SEM photograph of the cutting edge of a new SCD tool [15]

However, in ductile mode cutting, progressive wear becomes dominant. Figure 5.23 a is an SEM photograph of the cutting edge after cutting for 1.27 kilometres at an undeformed chip thickness of 90 nm. A crater wear has been formed on the rake face. Immediately near the crater wear, a flank wear land is formed. Both the crater wear and the wear land are smooth and uniform along the entire cutting edge, indicating that the tool wear has been a stable and gradual process. Figure 5.23 b is the cutting edge after cutting for 7.62 kilometres. The flank wear land has increased up to 2 μm wide. The flank wear land is covered with numerous micro grooves oriented along the cutting direction and a step structure can be observed on the wear land. These features indicate that the tool wear has been accelerated and has changed from a steady process to an unsteady process.

Tool Design, Tool Wear and Tool Life 141

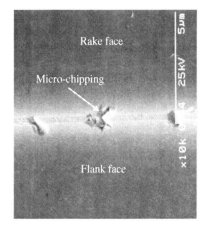 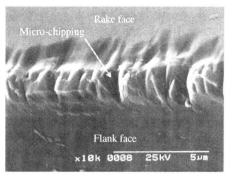

(a) After cutting for 1.27 kilometres   (b) After cutting for 7.62 kilometres

Figure 5.22. SEM photographs of the cutting edge after brittle mode cutting showing the occurrence of micro chippings [15]

 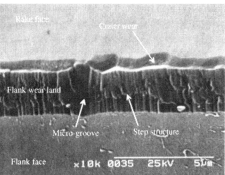

(a) After cutting for 1.27 kilometres   (b) After cutting for 7.62 kilometres

Figure 5.23. SEM photographs of the cutting edge after ductile mode cutting, showing crater wear and flank wear [15]

In ductile mode cutting, the initial tool wear tends to be a gradual process and can be considered as the presence of a combination of adhesive wear, abrasive wear and possible diffusion wear. That is, when a diamond tool is new and sharp, conditions on the rake face are much more severe than those on the flank face. Temperature and pressure are very high and the rake face constitutes a heavily loaded slider [16]. Under such conditions, diffusion wear will be predominant and the characteristic wear pattern is a crater. As the cutting distance increases, the cutting edge recedes and the flank wear land becomes predominant. The wear land results in a loss of relief angle, which gives rise to increased friction resistance (cutting forces). The wear land is normally a loaded slider with the maximum temperature at the trailing edge which increases with increase in wear land length [16]. The wear rate rises abruptly when the temperature at the trailing edge of the wear land reaches the thermal deterioration point of diamond [17]. The thermal deterioration under these circumstances may involve the diamond-graphite transformation at a high temperature [8] and the chemical reaction between carbon, silicon and oxygen. Subsequently the deteriorated layer at the trailing edge drops off due to the rubbing force between the tool and the workpiece, forming step structures on the wear land. This kind of wear is catastrophic and will lead to the total destruction of the tool. The flank wear land also causes unacceptable figure error in the part and induces surface micro fractures.

5.5 Tool Life

5.5.1 The Definition of Tool Life

Tool life is the time a tool can be reliably used for cutting before it must be discarded or repaired. In other words, tool life is the length of time in minutes between two neighbouring changes of the cutting tools. Tool life is important in machining since considerable time is lost whenever a tool is replaced and reset. Tool life depends on a number of factors that include machine tool, tool material and geometry, work material, coolant and lubricant conditions and other cutting conditions. Recently, numerous new tool materials have been especially, developed to machine numerous new workpiece materials, and thus the situation of tool wear and tool life becomes further complex.

The accurate prediction of the tool life during machining is important for the design of cutting tools and machining processes. The extensive research in this area has contributed greatly to the understanding of tool wear mechanisms. However, there is not yet a comprehensive technique which can be used to precisely predict tool life for all combinations of tool/workpiece materials and machining conditions.

5.5.2 Taylor's Tool Life Model

Conventionally, tool life is predicted by experience. In the last century, F.W. Taylor proposed an equation to describe tool life and this equation has been widely used until now [18]. The equation developed by Taylor is outlined as:

$$V_c \times T^n = C \qquad (5.3)$$

where:
    V_c - cutting velocity in m/min;

T - tool life in minutes, time taken to have a certain flank wear VB_B;
n - a tool-material dependant constant;
C - a constant based on the tool and the workpiece.

Note that C is the cutting velocity at $T=1$ min. To use the C value, one needs to know the cutting conditions which include the workpiece material, the cutting tool material, and the recommended cutting speed (V_c). Generally, the V_c value can be found in a machinery handbook or from charts supplied by the cutting tool manufacturer. However, one can increase the value of C by reducing the cutting speed, and vice versa. The n constant is a value based on experimentation with various cutting tool materials and cutting parameters. The n value can also be found in the charts supplied by the cutting tool manufacturer.

Tool wear, especially the flank wear land VB_B is always used as a tool life criterion because it is easy to determine quantitatively and has significant influences on workpiece surface roughness and accuracy. Figure 5.24 a shows a simplified presentation of the typical wear curves (VB_B versus cutting time) for several cutting velocities (1, 2 and 3) and Figure 5.24 b shows the corresponding tool life curve (cutting velocity versus tool life).

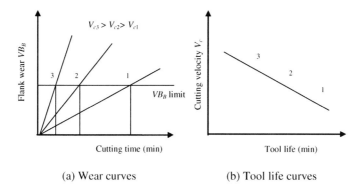

(a) Wear curves (b) Tool life curves

Figure 5.24. Schematic presentations of wear curves and tool life curves for different cutting velocities

For example, choosing two extreme points, points 1 and 3 in Figure 5.24 b, where V_c=100 m/min, T=30 min, and V_c =300 m/min and T=5 min, respectively. Then we obtain Equations 5.4 and 5.5, as follows:

$$100 \times 30^n = C \tag{5.4}$$
$$300 \times 5^n = C \tag{5.5}$$

By applying the natural logarithms, the above equations equal to:

$$\ln 100 + n \ln 30 = \ln 300 + n \ln 5 \tag{5.6}$$

After the calculation, we have:

$$4.605 + 3.401n = 5.704 + 1.609n \tag{5.7}$$

$$n = 0.613 \tag{5.8}$$

Substituting the value of n in Equations 5.4 and 5.5, we obtain the value of C, as follows:

$$C = 100 \times 30^{0.613} = 804 \text{ or } C = 300 \times 5^{0.613} = 804 \tag{5.9}$$

Finally, the Taylor equation obtained from the data of Figure 5.24 b is:

$$V_c \times T^{0.613} = 804 \tag{5.10}$$

Table 5.5. Practical values of n for some tool materials

Tool material	HSS	Cemented carbides	Ceramics
n	0.1-0.2	0.2-0.5	0.5-0.7

Table 5.5 shows the ranges of n values determined in industrial practice.

5.5.3 The Extended Taylor's Model

Note that in Equation 5.3, only one cutting parameter, the cutting velocity V_c, is considered. Although the cutting velocity is the most important cutting parameter in the tool life, the tool feed and the depth of cut are also significant factors. Generally speaking, as the feed and depth of cut decrease, tool life increases, and usually feed has greater effects than depth of cut. In this case, an extended equation of Taylor's model can be expressed as follows:

$$V_c T^n f^a d^b = C \tag{5.11}$$

where d is the depth of cut (mm), and f is the feed (mm/rev). The exponents a and b are determined experimentally for each cutting parameter. In practice, the values for HSS tools are $n=0.17$, $a=0.77$ and $b=0.37$ [19]. According to these data, we can consider that the importance of the cutting parameters in tool life decreases in an order of cutting velocity, feed and depth of cut.

Equation 5.11 can also be rewritten as:

$$T = C^{\frac{1}{n}} V_c^{\frac{-1}{n}} f^{\frac{-a}{n}} d^{\frac{-b}{n}} \tag{5.12}$$

Using the above practical data, the Taylor's model will become:

$$T = C^{5.88} V^{-5.88} f^{-4.53} d^{-2.18} \tag{5.13}$$

The Taylor tool life model is generally used to calculate the length of time a cutting tool will last before a given amount of flank wear will occur. Alternatively, if given the cutting time, the model will also help to calculate the proper feeds and speeds and depth of cuts. Thus, the Taylor tool life model is useful to find the most economical tool life.

Generally speaking, a reasonable agreement of the Taylor's tool life model has been shown over a wide range of cutting processes normally used in practice. On the other hand, it has also been shown that, for some given tool/work material combinations, the above equation does not agree well with experimental results. This is attributed to the changes in the dominant tool wear mechanism with changes in tool/work material and machining conditions. Tool life depends on the tool (material and geometry), the cutting parameters (cutting velocity, feed, depth of cut, cutting fluids), the workpiece material properties (chemical composition, hardness, strength), the machining processes (turning, drilling, and milling etc.), the machine tool stiffness and state of maintenance, and other conditions. For this reason, to make universal and precise tool life criteria seems impossible at the present stage. This area remains to be a future research topic.

5.5.4 Tool Life and Machining Dynamics

Tool wear/tool life is also affected by machining dynamics. Machining dynamics involves multiple aspects as discussed in other chapters of this book. In this section, we only take up the problem of dynamic instability, *i.e.*, vibrations and chatters, and consider their correlations with tool wear and tool life.

Dynamic instability where a cyclic relative motion exists between tool and workpiece, is of two types: forced vibration and self-exited vibration, where the latter is also termed machining chatter. The forced vibration occurs when the external force varies at a frequency close to one of the natural frequencies of the machine system. It can be reduced by eliminating the exiting force, changing the frequency of the external force, increasing machine stiffness and increasing damping.

Machining chatter can be divided into a few categories, the causes of which are very different. One of the primary causes of machining chatter is regeneration. During machining, cutting forces cause time-varying tool vibrations that are imprinted onto the workpiece surface. Subsequent cutting encounters this variable surface, which results in time-varying forces and deflections that depend on the phase relationship between the current tool position and previous cuts. If it has been not dealt with appropriately, this cyclic feedback mechanism can produce regenerative machining chatter.

Chatter has been a common problem on the shop floor. It is undesirable because it not only affects the surface quality and dimensional accuracy of a workpiece, but also leads to higher tool wear rates and possibly tool breakage. There have been a number of studies conducted to reveal the interrelationship or interdependence between machining chatter instability and tool wear/tool life [20-24]. To solve the problem of machining chatter is of great importance for machining operations.

Traditionally, the solution to chatter problems is to decrease surface speed and increase the feed rate. However, taking this approach may result in very unproductive cutting parameters. This will be especially a process limit for recent high speed machining (HSM) technology. In order to improve process efficiency, it is vital to identify chatter free cutting conditions without the sacrifice of material removal rates. Typically, the stability behaviour of a particular machining operation can be characterized by a stability chart. On such a chart, the regions of stable (non-chattering) and unstable (chattering) behaviour are plotted as a function of practical machining parameters, such as spindle speed and depth of cut. Stable regions (lobes) become more pronounced at higher spindle speeds, and the ability to take advantage of these regions is a major potential benefit of machining. With the help of suitable predictive models, optimal chatter free cutting conditions can be identified for high speed machining.

On the other hand, tool wear may inversely affect machining dynamics by causing instability to cutting processes. The process instability caused by tool wear involves variations in cutting forces and torques, and machining chatter. It has been known that the onset of chatter instability changes as tool wear progresses. By comparing the chatter stability for a fresh cutting tool and a worn cutting tool and analyzing the chatter behaviour for a cutting tool with wear flats on the tool flank, explicit analytical expressions describing the onset of chatter instability and tool wear have been derived for some machining operations [20, 21]. Therefore, when selecting cutting conditions and cutting tools to improve process stability, one should always consider the tool wear characteristics.

Here, we also use the silicon machining experiments [15] mentioned in Section 5.4.4 to demonstrate the effect of wear on machining instability. Figure 5.25 a shows the variations in principal force F_c, thrust force F_t per unit of cutting edge length (mm) and their ratio r with the cutting distance when face turning silicon with an SCD tool. Until the cutting distance of 5.08 kilometres, both F_c and F_t tended to increase linearly, indicating a steady tool wear. After that, the forces increased with significant fluctuations. Figures 5.25 b and 5.25 c show the force waviness at 1.27 and 7.62 kilometres, respectively. In **b**, the force waviness is very steady, whereas in **c** sudden fluctuations can be observed frequently, indicating the unsteadiness of the cutting process caused by severe tool wear. From **a**, it can also be seen that the force ratio keeps increasing linearly as the tool wears. These results demonstrated the possibility of monitoring the tool wear using cutting force signals.

Acoustic emission (AE) is another useful process signal which can be used for the on-line monitoring of tool wear in metal cutting [24]. The high sensitivity of AE to friction and wear phenomena makes it possible to detect the contact and deformation mechanism in material removal processes. Recently, the AE signal has become an effective and practical tool for the monitoring and analysis of tool wear and failure in cutting. The wear of a cutting tool is expected to affect the AE signal through the change in the tool geometry and the change in the chip form. Thus, the dynamic variation of the AE signal often carries information regarding the state of the tool wear. The analysis of AE can also be a relevant tool for the monitoring of machining chatters as well [24]. In the case of chatter, AE signals a change in response to the variation in the basic mechanics of the cutting process accompanying the tool-workpiece vibration. This includes the change in strain rate due to the change of the tool tip velocity and varying undeformed chip thickness.

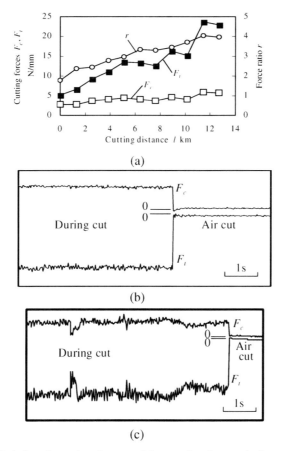

Figure 5.25. a Variations in cutting forces and force ratio; changes in force waviness with cutting distance from **b** 1.27 kilometers to **c** 7.62 kilometers in ductile regime cutting (face turning) of silicon [15]

From these points of view, it is considerable that a clear understanding of the interrelationships and interdependence between the tool wear/tool life and process parameters such as cutting forces, torques, and AE signals provides a possibility for monitoring the tool wear and assists the judgment of tool life on line based on recent sensing and signal processing technology.

References

[1] J. Yan, K. Syoji, T. Kuriyagawa and H. Suzuki. (2002). Ductile regime turning at large tool feed. Journal of Materials Processing Technology, 121, 363–372.
[2] D. Keen. (1971). Some observations on the wear of diamond tools used in piston machining. Wear, 17, 195–208.
[3] R. Wada, H. Kodama, K. Nakamura, Y. Mizutani, Y. Shimura and N. Takenaka. (1980). Wear characteristics of single crystal diamond tool. Annals of the CIRP, 29 (1), 47–52.
[4] C.J. Wong. (1981). Fracture and wear of diamond cutting tools. Transactions of the ASME: Journal of Engineering Materials and Technology, 103, 341–345.
[5] N. Ikawa and S. Shimada. (1982). Microfracture of diamond as fine tool material. Annals of the CIRP, 31 (1), 71–74.
[6] P.N. Blake and R. O. Scattergood. (1990). Ductile regime machining of germanium and silicon. Journal of American Ceramics Society, 73 (4), 949–957.
[7] J. Yan, M. Yoshino, T. Kuriyagawa, T. Shirakashi, K. Syoji and R. Komanduri. (2001). On the Ductile Machining of Silicon for Micro Electro–mechanical Systems (MEMS), Opto–electronic and Optical Applications, Materials Science and Engineering, A, 297, 1–2, 230–234.
[8] N. Ikawa, R. R. Donaldson, R. Komanduri, W. Konig, P. A. McKeown, T. Moriwaki and I. F. Stowers. (1991). Ultraprecision metal cutting – the past, the present and the future, Annals of the CIRP, 40 (2), 587–600.
[9] J. Yan, J. Tamaki, K. Syoji, T. Kuriyagawa. (2004). Single–point diamond turning of caf_2 for nanometric surface. International Journal of Advanced Manufacturing Technology, 24 (9–10), 640–646.
[10] J. Yan, K. Maekawa, J. Tamaki and T. Kuriyagawa. (2005). Micro grooving on single–crystal germanium for infrared Fresnel lenses. Journal of Micromechanics and Microengineering, 15, 1925–1931.
[11] J. P. Davim and A. M. Baptista. (2000). Relationship between cutting force and PCD cutting tool wear in machining silicon carbide reinforced aluminium. Journal of Materials Processing Technology, 103 (3), 417–423.
[12] J. A. Arsecularatne, L.C. Zhang and C. Montross. (2006). Wear and tool life of tungsten carbide, PCBN and PCD cutting tools. International Journal of Machine Tools and Manufacturing, 46 (5), 482–491.
[13] J. M. Oomen and J. Eisses. (1992). Wear of monocrystalline diamond tools during ultraprecision machining of nonferrous metals. Precision Engineering, 14 (4), 206–218.
[14] E. Paul, C. J. Evans, A. Mangamelli, M.L. McGlauflin and R.S. Polvani. (1996). Chemical aspects of tool wear in single point diamond turning. Precision Engineering, 18 (1), 4–19.
[15] J. Yan, K. Syoji and J. Tamaki. (2003). Some observations on the wear of diamond tools in ultra–precision cutting of single–crystal silicon, Wear, 255 (7–12), 1380–1387.
[16] M.C. Shaw. (1986). Metal Cutting Principles, Clarendon Press, Oxford.
[17] K. Cheng, X. Luo, R. Ward and R. Holt. (2003). Modeling and simulation of the tool wear in nanometric cutting. Wear, 255 (7–12), 1427–1432.
[18] F. W. Taylor. (1907). On the art of cutting metals, Transactions of the ASME, 28, 31–58.

[19] O. W. Boston: Metal Processing, Wiley, New York, 1941.
[20] R. Y. Chiou and S. Y. Liang. (1998). Chatter stability of a slender cutting tool in turning with tool wear effect. International Journal of Machine Tools and Manufacturing, 38 (4), 315–327.
[21] M. S. Fofana, K. C. Ee and I. S. Jawahir. (2003). Machining stability in turning operation when cutting with a progressively worn tool insert. Wear, 255 (7–12), 1395–1403.
[22] B. E. Clancy and Y. C. Shin. (2002). A comprehensive chatter prediction model for face turning operation including tool wear effect. International Journal of Machine Tools and Manufacturing, 42 (9), 1035–1044.
[23] Y. S. Chiou, E. S. Chung and S. Y. Liang. (1995). Analysis of tool wear effect on chatter stability in turning. International Journal of Mechanical Science, 37 (4), 391–404.
[24] R. Y. Chiou and S. Y. Liang. (2000). Analysis of acoustic emission in chatter vibration with tool wear effect in turning. International Journal of Machine Tools and Manufacturing, 40 (7), 927–941.

6

Machining Dynamics in Turning Processes

E. O. Ezugwu[a], W. F. Sales[b] and J. Landre Jr.[b]

[a] Machining Research Centre
Department of Engineering Systems
London South Bank University
London, SE1 0AA, UK

[b] Manufacturing Research Centre
Mechanical and Mechatronics Engineering
Pontifical Catholic University of Minas Gerais
PUC Minas, Av. Dom José Gaspar, 500, Belo Horizonte, MG, Brazil

6.1 Introduction

The dynamic stability of a machine tool in the turning process depends essentially on the compliance of the lathe turning structure, as well as on the properties of the cutting process [1]. However, the design of the machine tool, the material(s) employed for its manufacture and their mechanical properties are extremely important for the dynamic behaviour of the machining system (comprising the entire lathe and the work material) [1-13]. Theoretical details of dynamic stability and how to quantify, measure and monitor them as well as other phenomena such as chatter (self-excited vibration) and forced vibration have been covered in previous chapters. This chapter will, therefore, focus on the illustration and the discussion of practical details regarding the turning process. The influence of the input on the output parameters will be evaluated with regards to the dynamic stability in a turning process. The main input parameters affecting the machining system vibration are: work material, work material geometry, tool material, tool geometry, lathe rigidity, cutting conditions (cutting speed, vc, feed rate, f and depth of cut, doc) and tool wear. The behaviour of the machining system during vibration is a major output parameter.

6.2 Principles

Dynamic loads generated due to seizure phenomena during machining processes such as turning, milling, drilling, etc. can directly affect the machine tool and work piece interaction, and consequently alter the machining quality [9]. This can be accentuated by irregularities in the work material (trapped air bubbles, hard particles, etc) and/or disturbances in the machine tool components (unbalanced rotational parts, motor misalignments, etc.). The entire machining system and the resulting dynamic loads can generate high instability and therefore directly affect the process efficiency, *i.e.*, compromising the dimensional accuracy of the workpiece, the acceleration of tool wear and/or damaging some machine tool components. The instability can occur due to two specific conditions:

- The resonance caused by the coincidence of the rotational speed and the natural frequencies of the workpiece-machine tool system;
- The periodic excitation generated by dynamic loads, which occur due to external applied forces such as those from gear drives, unbalanced parts of the machine tool or the work material.

Exact identification of the main factors responsible for instability of the machining process can minimise or completely eliminate some inherent problems. For this reason, numerical and experimental evaluations are useful (and available) tools to better understand the dynamic behaviour associated with each machining process [5]. Recently, numerical methods have been proposed for this study. The most useful are Finite Differences, Finite Volume, Boundary Elements, Finite Elements, etc. The Finite Element method will be extensively discussed in this chapter due to its powerful application potentials as well as its reliability - mainly when the evaluation is applied to materials in the elastic state. The numerical evaluations applied in the study of dynamic behaviour are basically divided into modal analysis and frequency-domain response.

When experimental evaluation is required, vibration levels are commonly measured using accelerometer transducers [14-21]. This technique will ensure that measurements can be taken while the machine tool is in full operation. It is therefore necessary to specify compatible accelerometers that can be plugged in predetermined points in the machine tool. When machine tools are being evaluated, it is essential that they should possess high mass and rigidity, which, from the dynamics standpoint, exhibit low natural frequency values. This should be considered before specifying the correct transducer types and their work range for those applications.

The use of numerical and experimental methods for a complex investigation of the machine tool performance was recently achieved [5, 12, 13]. Baker and Rouch [10] employed the finite element method for the stability evaluation of the machine tool and reported two principal parameters of machine stability: the machine tool structure and cutting parameters used in the process. Sturesson *et al.* [11] developed their work using finite element modelling. A model was developed using solid elements to verify the vibrational behaviour during continuous turning operation.

It can be shown, using numerical methods (mainly the finite element method modelling) that it is possible to study the dynamic stability of the machine tool and work material system during the turning process. However, the numerical results are necessary and essential for eventual corrections and/or adjustments on the virtual models. Since these models are based on mathematical studies, it is necessary to make some simplifications, in order to make the study possible [5, 10-13]. The resultant model is therefore susceptible, in most cases, to predictable errors that can be minimised after introducing the experimental techniques and their results including the vibration measurements.

Experimental techniques can be illustrated from many studies [14, 16, 20, 21-26], some of which need to be highlighted. For instance, the work of Lee and Kim [6] used the modal analysis technique to evaluate the dynamic condition of the tool holder. Continuing this investigation, other researchers [8, 13, 17, 19, 26] measured vibrations when studying the dynamic stability for the machine tool and reported that examples of both applications are powerful techniques for studying stability.

Rezende [12, 13] evaluated the machined workpiece by employing mainly previous knowledge of the natural dynamic behaviour of the machine tool and its response when receiving external excitations using both techniques. A model generated using a finite element method was developed and later adjusted using results from vibrations measured in some locations of the lathe during the turning process.

6.2.1 The Turning Process

During turning, the work material is fixed in the jaw of a lathe machine that has a rotational movement [9]. The tool is fixed in a tool holder that moves along the longitudinal axis, parallel to the work material axis. The combined and simultaneous movements result in a seizure (cutting) of a portion of the work material. The surface produced can be a cylinder or a complex profile. Figure 6.1 shows schematically the turning process, its components and movements.

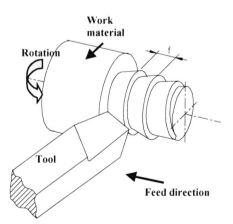

Figure 6.1. Schematic representation for the turning process

6.3 Methodology and Tools for the Dynamic Analysis and Control

The Finite Element Method (FEM) is a powerful mathematical and computational tool employed for the dynamic analysis and control of turning processes. It is largely used for process simulation and also estimates the machine dynamic behaviour during the machining operation [5, 10, 12].

Experimental measurements applied as monitoring techniques such as vibration, acoustic emission, cutting forces and power consumption, etc. are intensively used as a support to evaluate the status of the system composed by the machine/toolholder/tool/workpiece [13, 20]. Recent mathematical and experimental techniques have been jointly applied to improve the understanding of the dynamic behaviour of the machine system and the effect(s) on the machined components.

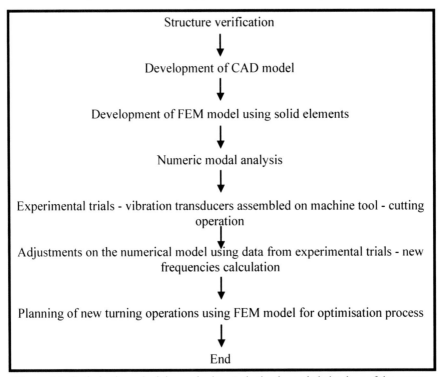

Figure 6.2. Optimisation schedule required to study the dynamic behaviour of the system

An overall understanding of the activities and interactions between them is extremely important. The evaluation of the dynamic behaviour and the adjustment of the work rotations involved are essential to prevent problems associated with resonance during machining operations. Modal analyses are useful since this technique can identify the natural frequencies of the machine tool. The present investigation can therefore be performed out of those work frequencies [13]. On the other hand, if the main objective is to determine the dynamic response resulting in the optimized work condition, some further analyses will be necessary such as determining the natural frequencies using the modal analyses numeric method, followed by a vibration analysis through experimental measurements. Finally, a study to determine the response on frequency-domain ensures the evaluation of the machine tool, and tool holder/tool performance at a wide range of work frequencies [5, 6, 10, 13]. If the turning process optimisation is required using the dynamic knowledge of the system, the study can be carried out using the sequence illustrated in Figure 6.2.

If the natural frequencies and the resulting optimised work condition will be determined (in a reduce resonance problems), the study has to be conducted using the steps shown in Figure 6.3.

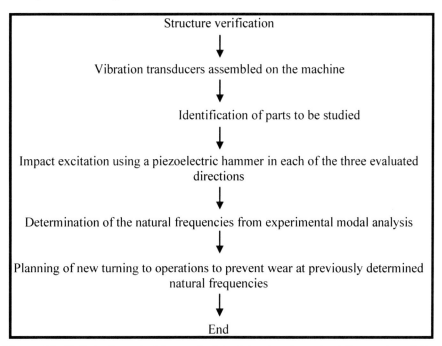

Figure 6.3. Optimisation schedule required to determine the natural frequencies of the system

6.4 Implementation Perspectives

Evaluation of the dynamic stability and its control for turning processes is currently used in many cases worldwide, mainly using the experimental approach with piezoelectric accelerometers as sensors. In most cases, technicians' expertise in vibration control will just measure some vibration parameters, like displacement, velocity and/or acceleration. They do not possess sufficient knowledge to attempt to obtain relationships among the vibration signal, the cutting parameters and the dynamic stability for the entire system. Therefore, analyses through just using the signal from sensors cannot explain with reasonable accuracy what is taking place with the process since a systemic approach could be applied. In other words, other components of the system, like the machine tool, the tool holder and the tool fixture system need to be carefully evaluated.

On the other hand, powerful techniques based on numerical methods are widely used for computational simulations. The FEM, for instance, needs to be better studied and consequently the real cases applied for dynamic evaluation need to be exposed. It seems that most of these numerical techniques are mainly used in an academic environment. Their utilisation for product manufacture should therefore be stimulated. There are many training programs for technicians and engineers to enhance the optimization of the turning processes in order to improve the tool life and product quality, and to reduce associated turning costs.

6.5 Applications

6.5.1 The Rigidity of the Machine Tool, the Tool Fixture and the Work Material

This primarily involves a dynamic evaluation using modelling and finite element techniques to modal analysis and natural frequency identification [5]. Rezende [12, 13] recently conducted a study on the dynamic behaviour of the steel turning process using modelling and experimental techniques. The lathe's geometric model and the workpieces studied were developed using commercial design software. The initial study involved an extensive investigation of the lathe machine including all parts and their connections, followed by a consideration of each part's geometry and material composition. This information, together with the boundary conditions, would be employed in the FE model.

A CAD 3D model using commercial software would be developed (Figure 6.4) to well represent the real machine. Since this geometric model will be used to perform the FE analysis, the reliability of the numerical results depends on the accuracy of the virtual model relative to the machine tool.

The structure in Figure 6.4 was later divided into a finite element mesh using an automatic mesher in the software. The element shape was a solid tetragonal element with 10 nodes which best adjusts to complex solid structures such as the lathe. Some adjustments were performed to evaluate the mesh refinement that has the best cost-benefit ratio, *i.e.*, the mesh should have a good precision with a reasonable processing time. Therefore, it was determined that the mesh's global edge length would be 22 mm, which has an insignificance error of about 0.31% and a 39% shorter processing time compared to a more refined mesh with a global edge length equivalent to 17 mm. The mesh generated from the turning machine-workpiece system of Figure 6.4 is shown in Figure 6.5. This consists of 443,648 elements, 714,811 nodes and 4,288,866 degrees of freedom.

The lathe does not have any special fixture system with the floor; hence, it is just supported by the floor. Therefore, the boundary conditions applied to the model where the restriction of the nodes' degrees of freedom (3 translations) is in contact with the floor. After this, the material properties of the turning machine (grey cast iron) and work material (steel AISI 4140) were assumed, according to Table 6.1.

Figure 6.4. CAD 3D-model for the lathe machine

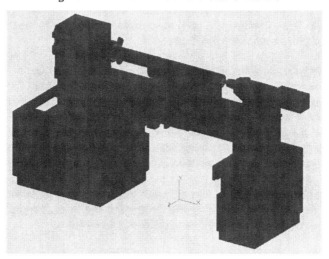

Figure 6.5. Lathe and workpiece model after the "meshed" step

After the previous assignments were concluded, with the machine tool divided into small elements and the appropriate boundary conditions defined, the model was ready to be analysed. The dynamic analysis was used to fully understand the influence of each part of the lathe and the work piece dimensions under the modal responses, *i.e.*, the natural frequencies and vibration modes.

Table 6.1. Typical properties for grey cast iron and steel

	Grey cast iron	Steel AISI 4140
Modulus of elasticity (GPa)	90 to 113	207
Poisson ratio	0.26	0.30
Density (kg/m^3)	7,300	7,850
Crystallographic orientation	Isotropic	Isotropic

Another point that should be considered concerns the position of the tool holder and tailstock, which in real conditions must be properly adjusted and also needs to be reproduced in the model. In addition, for each of their positions, different natural frequency values can be obtained. This is the reason that the natural frequencies must be calculated for each condition. A mobile part can be assumed during the real turning operation.

Figures 6.6 - 6.9 show the calculated results for the first natural frequencies under specific conditions of the lathe, with different dimensions of the work material. Also, the CAD model with the deformation isolayers for that frequency is shown, for different positions of the tool holder and tailstock. The work material was removed from the lathe in order to evaluate the machine tool structure isolated from the workpiece as illustrated in Figure 6.6.

Figure 6.6 and Figure 6.7 illustrate a different dynamic behaviour. In both cases, the tool-holder magazine and tailstock were in the same position. The difference between them is the presence of the work material in Figure 6.7. When observing the right side of the work material (Figure 6.7), it tends to be removed from the tailstock. Certainly this operation could be dangerous, because if the cutting occurs at excitation frequencies next to some natural frequencies, the system can operate in resonance range and the work material can be expulsed from the lathe.

Analyses of Figures 6.8 and Figure 6.9 show different dynamic behaviours. The tool-holder magazine and tailstock were in different positions. Another difference is the presence of the work material with different dimensions.

In Figure 6.9, the work material is long (L=1,000 mm), with a shorter diameter (ϕ=40 mm), and promoting a high fluctuation in the entire system. This can have a negative effect on the turning process, in terms of the tool life and the surface finish generated.

Machining Dynamics in Turning Processes 159

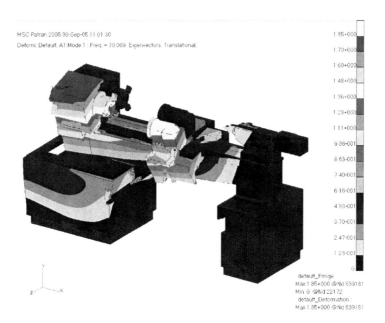

Figure 6.6. First natural frequency of the turning machine structure (without the work material)

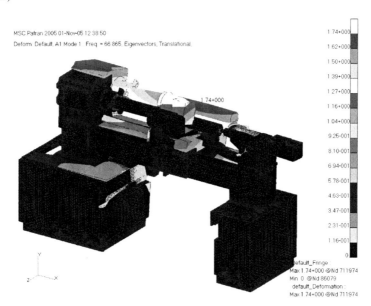

Figure 6.7. First natural frequency of the turning machine and the work material with a diameter of ϕ=100 mm and length of L=1,000 mm

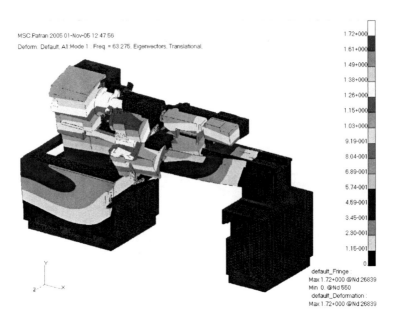

Figure 6.8. First natural frequency of the turning machine and the work material with a diameter of φ=100 mm and a length of L=350 mm

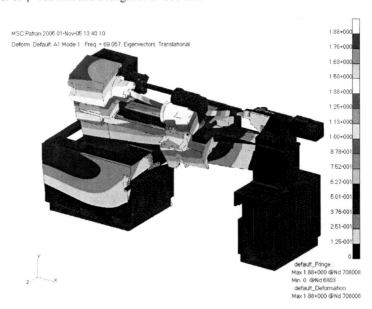

Figure 6.9. First natural frequency of the turning machine and the work material with a diameter of φ=40 mm and a length of L=1,000 mm

Table 6.2 shows the natural frequencies obtained from the finite element analysis for the particular configurations shown in Figures 6.6 - 6.9. The table shows from the first to tenth natural frequencies for each case and that the first natural frequency (mode shape 1) for the lathe without the workpiece, and the tool holder magazine in a centralised position on the machine, the material was 70.1 Hz; 66.7 Hz for the work material with a diameter of ϕ=40 mm and a length of L=1,000 mm assembled on the lathe, 63.3 for the work material with the same diameter, but with reduced length of 350 mm, and 69.1 for the work material with a reduced diameter of 40 mm and a length of 100 mm. Analyses of Table 6.2 clearly show that the influence of the workpiece dimensions on the natural frequency (when there are variations in dimensions) together with the position of the tool-holder magazine can drastically alter the dynamic behaviour during the turning operation.

Table 6.2. Natural frequencies obtained from the finite element analysis (Hz)

	Work material dimensions (mm)	ϕ=100 and L=1,000	ϕ=100 and L=350	ϕ=40 and L=1,000
Mode shapes	Lathe	Lathe and workpiece 1	Lathe and workpiece 2	Lathe and workpiece 3
1	70.1	66.7	63.3	69.1
2	108.2	87.3	112.1	98.6
3	122.4	97.5	124.2	101.5
4	147.5	109.1	134.2	109.8
5	154.2	117.7	163.5	122.5
6	167.4	145.8	166.6	147.8
7	172.2	154.6	172.0	155.5
8	203.4	170.4	178.0	171.2
9	229.2	170.5	205.6	171.9
10	306.7	204.8	218.4	206.0

Rezende [12, 13] also reported that all natural frequencies determined numerically were also identified in the modal tests within a frequency range. Therefore, the lathe-workpiece model generated corresponds to the real structure with good accuracy, considering that the maximum error was 16.7% for the second mode shape. Relating these results to other models investigated show that all of them or any other model required could be used as a preliminary knowledge about the vibration behaviour of the machine-tool-workpiece system. These findings are in agreement with Mahdavinejad's results [5], obtained in a different structure.

The materials used to construct the lathe structure have a great high influence on the resulting machine dynamic stability. As it is well known, grey cast iron has good damping properties relative to steel. This, coupled with their low manufacturing costs, make them ideal material for the structure of machine tools. In order to improve the damping effect, other structural materials such as synthetic granite were evaluated. A new generation of high-speed machine tools has been manufactured using synthetic granite for their structure. Because of their low density, the resultant structure has reduced weight. Rahman et al. [23, 27-28] conducted detailed studies on lathe machines consisting of ferrocement bed and grey cast iron bed. The improved damping ability for the ferrocement material when compared with grey cast iron was responsible for the better results obtained, such as lower vibration levels and lower tool flank wear.

In order to improve the dynamic stability forced and self-excited vibrations can be suppressed if the magnitude of the frequency response function of the cutting tool is reduced. This is why, in practical applications, a dynamic vibration absorber should be designed and assembled together with the tool-holder system [8].

6.5.2 The Influence of the Input Parameters

It is well known that the dynamic response of machine tools depends largely on the process input parameters such as work and tool materials, tool geometry, tool wear and cutting parameters (cutting speed, feed rate and depth of cut). The individual effects of each of these parameters are discussed below.

6.5.2.1 The Effect of the Work and Tool Materials
As discussed previously, the dynamic stability is highly influenced by output parameters from the cutting process such as cutting forces. In addition, phenomena that occur at the chip-tool interface such as resistance on the secondary rake face is highly responsible for the high cutting forces generated. On the other hand, dimensions of those phenomena depend on the tool geometry, mechanical property of the work material (tensile strength, toughness, hardness, etc.) and their chemical affinity with the tool material [9]. Therefore, the correct selection of the tool material must consider all those cited factors as well as the chemical affinity.

6.5.2.2 The Effect of the Tool Geometry
Mei *et al.* [29] developed a study using a mathematical model followed by experimental trials with an objective of improving the active chatter suppression by on-line variation of the rake and clearance angles. They concluded that the variation of those angles had influences on the dynamic stability of the cutting system. In addition, they found that the simultaneous increase of the rake angle and the decrease of the clearance angle by the same amount tends to increase the stability of the cutting system.

Generally the use of tools with a negative rake angle, a large nose radius or with a chamfered edge, tend to work with a large chip-tool interface area. This usually tends to increase the cutting forces with a consequent negligible influence on the stability of the machining system. In applications that require the use of tool geometry to prevent vibrations, the machine tool as well as the tool fixture system and cutting parameters should be properly selected.

6.5.2.3 The Effect of Tool Wear
Tool wear is a real phenomenon that unfortunately occurs during a cutting process [9]. The quality of machined components is usually affected by conditions at the chip-tool interface. Changes in tool geometry, promoted by wear during machining, can drastically change the dynamics of the cutting process. Evaluated

parameters such as cutting forces, power consumption and chip-tool interface temperature normally encounter significant alterations, increasing or decreasing depending on the extent of tool wear. Dimensional tolerances and surface roughness of the machined workpiece are frequently measured and monitored in-process or post-process, depending on the level of automation employed. It has been shown that the rigidity of the machining system (the entire lathe and work material) has a large influence on the dynamic stability. Additionally, vibrations encountered during machining can drastically alter the quality of the workpiece. This has, therefore, prompted numerous studies on how the dynamics of the lathe, evaluated mainly by vibration measurements, can alter the geometry of the machined components.

Diniz and Bonifácio [14], reported that the vibration signal showed a rapid increase after cutting for a while, for all cutting conditions evaluated, thereby indicating the end of tool life when finishing turning AISI 4340 steel with coated carbide tools. The results confirm the suitability of using the vibration signal to establish tool life in finish turning. Lim [21] monitored tool wear through vibration measurements when turning AISI 1045 steel with cemented carbide tools. He concluded that the vibration signature analysis, evaluated through vibration acceleration amplitudes during in-process monitoring appears to be a very promising method for the detection of tool flank-wear.

6.5.2.4 The Effect of Cutting Parameters
Cutting parameters such as cutting speed, feed rate and depth of cut frequently have influence on the dynamic behaviour of the machining system. The cutting speed is directly associated with the external excitation frequency while the other cutting parameters influence the cutting forces. The cutting speed influences the excitation frequency and, at the same time, the chip-tool interface temperature. When the cutting speed increases, there is a tendency for the cutting temperature to rise and for the cutting force to drop. Both phenomena can drastically alter the dynamic stability of the machining system.

The feed rate and depth of cut influences the chip-tool contact area, *i.e.*, when one or both of them increase, the cutting forces increase and in addition acts as excitation loads on the machining system. Therefore, the mechanical parts assembled in the lathe can suffer some deformation and thus the machining conditions can be altered due to changes in the dynamic behaviour - mainly if the lathe was not properly designed and dimensioned for machining under those conditions.

In cases where there are maintenance problems and the lathe machine has old parts with a high level of wear, it is possible that there may be wide gaps between those parts and consequently, the entire machine tool will be encountering that load during machining. This tends to promote excessive vibration, thus compromising the cutting process and the workpiece quality.

6.6 Conclusions

Some applications of numeric and experimental methods of machining dynamics in turning operation have been reviewed. It is evident that a numeric method, such as the FEM, could be used as a useful tool and methodology for understanding the dynamic performance of any lathe-workpiece system. A good understanding of this area can help the process operator in choosing optimum acceptable cutting conditions that will prevent the instability of the machining system and the consequent deterioration of the quality of the manufactured components. The use of established information on machine tool dynamics coupled with experimental techniques can more precisely be used to monitor tool life and the surface finish on-line. Knowledge of the natural structural frequencies and the amplitude changes for the frequencies related to tool wear or surface finish would assist comparisons with the natural dynamic behaviour.

Therefore, the previous knowledge obtained by the use of numeric (virtual) and experimental methods, followed by intense analyses of the dynamic behaviour of the system, can be used to better specify the cutting parameters for a process in order to establish the workpiece quality during the design process.

References

[1] Ahn, T.Y., Eman, K.F. and Wu, S.M., 1984, Identification of the transfer function of dynamic cutting processes – a comparative assessment, International Journal of Machine Tools and Manufacture, Vol. 25, pp. 75–90.

[2] Mackerle, J., 1999, Finite–element analysis and simulation of machining: a bibliography (1976–1996), Journal of Materials Processing Technology, 86, pp. 17–44.

[3] Glavonjic, M. and Milancic, V.R., 1988, A practical procedure for conceptual design and testing of machine tool structure, Robotics and Computer–Integrated Manufacturing, Vol. 4, No. ¾, pp. 317–333.

[4] Fofana, M.S., Ee, K.C. and Jawahir, I.S., 2003, Machining stability in turning operation when cutting with a progressively worn tool insert, Wear, 255, pp. 1395–1403.

[5] Mahdavinejad, R., 2005, Finite element analysis of machine and workpiece instability in turning, International Journal of Machine Tools and Manufacture, 45, pp. 753–760.

[6] Lee, J. and Kim, D.H., 1995, Experimental modal analysis and vibration monitoring of cutting tool support structure, International Journal of Mechanical Sciences, Vol. 37, pp. 1133–1146.

[7] Rahman, M. and Matin, M.A., 1991, Effect of tool nose radius on the stability of turning processes, Journal of Materials Processing Technology, 26, pp. 13–21.

[8] Lee, E.C., Nian, C.Y. and Tarng, Y.S., 2001, Design of a dynamic vibration absorber against vibrations in turning operations, Journal of Materials Processing Technology, 108, pp. 278–285.

[9] Trent, E.M. and Wright, P.K., 2000, Metal Cutting, 4th. Ed., ISBN 0–7506–7069–X, Butterworth–Heinemann, London.

[10] Baker, J.R. and Rouch, K.E., 2002, Use of finite element structural models in analyzing machine tool chatter, Finite Elements in Analysis and Design, Volume 38, Issue 11, pp. 1029–1046.
[11] Sturesson, P.O.H., Håkansson, L. and Claesson, I., 1997, Identification of statistical properties of cutting tool vibrations in a continuous turning operation–correlation to structural properties, Mechanical Systems and Signal Processing, Vol. 11, Issue 3, pp. 459–489.
[12] Rezende, C.R., Landre, Jr. and Sales, W.F., 2008, Finite element analysis applied to the turning machine dynamic behaviour, International Journal of Advanced Manufacturing Technology (in press).
[13] Rezende, C.R., 2005, Evaluation of the machining dynamics in turning process using modelling and experimental techniques [In Portuguese], Master Science Dissertation, PUC Minas, Belo Horizonte, MG, Brazil.
[14] Diniz, A.E. and Bonifácio, M.E.R., 1994, Correlating tool wear, tool life, surface roughness and tool vibration in finish turning with coated carbide tool, Wear, 173, pp. 137–144.
[15] Skelton, R.C., 1969, Surface finish produced by a vibrating tool during turning, International Journal of Machine Tools and Manufacture, Vol. 9, pp. 375–389.
[16] Dimla Sr., D.E., 2004, The impact of cutting conditions on cutting forces and vibration signals in turning with plane face geometry inserts, Journal of Materials Processing Technology, 155–156, pp. 1708–1715.
[17] Chiou, Y.S., Chung, E.S. and Liang, S.Y., 1995, Analysis of tool wear effect on chatter stability in turning, International Journal of Mechanical Sciences, Vol. 37, No. 4, pp. 391–404.
[18] Selvam, M.S., 1975, Tool vibration and its influence on surface roughness in turning, Wear, 35, pp. 149–157.
[19] Lasota, A. and Rusek, P., 1983, Influence of random vibrations on the roughness of turned surfaces, Journal of Mechanical Working Technology, 7, pp. 277–284.
[20] Jang, D.Y., Choi, Y.G., Kim, H.G. and Hsiao, A., 1996, Study of the correlation between surface roughness and cutting vibrations to develop an on–line roughness measuring technique in hard turning, International Journal of Machine Tools and Manufacture, Vol. 36, No. 4, pp. 453–464.
[21] Lim, G.H., 1995, Tool–wear monitoring in machining turning, Journal of Materials Processing Technology, 51, pp. 25–36.
[22] Thomas, M., Beauchamp, Y. Youssef, A.Y. and Masounave, J., 1996, Effect of tool vibrations on surface roughness during lathe dry turning process, Computers and Industrial Engineering, Vol. 31, No. 3–4 pp. 637–644.
[23] Rahman, M., Mansur, M.A. and Lau, S.H., 2001, Tool wear study in a lathe made of cementitious material, Journal of Materials Processing Technology, 113, pp. 317–321.
[24] Chiou, R.Y. and Liang, S.Y., 2000, Analysis of acoustic emission in chatter vibration with tool wear effect in turning, International Journal of Machine Tools and Manufacture, 40, pp. 927–941.
[25] Risbood, K.A., Dixit, U.S. and Sahasrabudhe, A.D., 2003, Prediction of surface roughness and dimensional deviation by measuring cutting forces and vibrations in turning process, Journal of Materials Processing Technology, 132, pp. 203–214.
[26] Liao, C.L. and Tarng, Y.S., 1992, Dynamic response of a workpiece in turning with continuously varying speed, Computer and Structures, 1992, Vol. 45, No. 5/6, pp. 901–909.
[27] Rahman, M., Mansur, M.A. and Chua, K.H., 1988, Evaluation of advanced cementitious composites for machine tool structures, Annals of the CIRP, 37(1), pp. 373–376.

[28] Rahman, M., Mansur, M.A. and Chua, K.H., 1993, Evaluation of a lathe with ferrocement bed, Annals of the CIRP, 42(1), pp. 437–440.
[29] Mei, Z., Yang, S., Shi, H., Chang, S. and Ehmann, K.F., 1994, Active chatter suppression by on–line variation of the rake and clearance angles in turning – principles and experimental investigations, International Journal of Machine Tools and Manufacture, Vol. 34, No. 7, pp. 981–990.

7

Machining Dynamics in Milling Processes

Xiongwei Liu

School of Aerospace, Automotive and Design Engineering
University of Hertfordshire
Hatfield AL10 9AB, UK

7.1 Introduction

The milling process is distinguished by a rotating tool with one or more teeth that removes material while travelling along various axes with respect to the workpiece. The engagement of each tooth with the workpiece is discontinuous in the milling process different to the turning process. As a tooth engages the workpiece, it receives a shock followed by a varying cutting force. The cyclic shock and variation of the cutting force induces vibrations between the tool and workpiece. It can also provide the necessary energy to excite a natural mode of vibration in any part of the machining system. These vibrations should be minimized because they can, and frequently do, degrade machining accuracy and surface finish. Moreover, under unfavourable conditions they may become unstable, leading to chatter, which can cause accelerated tool wear and breakage, accelerated machine wear, and even damage to the machine and part. For high speed machining and especially high speed milling, unstable vibration or chatter is a major factor limiting production rate.

So, what is chatter?

In practice, chatter can be recognized by a characteristic noise, chatter marks, and the undulated or dissected chip [1]. However it is essential to understand the nature of chatter so that it can be effectively detected and controlled.

There are two main sources of chatter vibration in milling operation: (a) forced vibration, (b) self-excited vibration.

7.1.1 Forced Vibration

A forced vibration occurs when a cyclic exciting force applies to an elastic structure. For a machining system of a milling operation, the tool, tool holder and spindle together form the tool assembly, and the workpiece and its fixture form the work assembly. When the cyclic shock and varying cutting force apply to the tool and work assemblies, they induce vibrations in these elastic structures.

Forced vibrations have a great impact on the machining process when one or more of the frequencies of the cyclic shock and varying cutting force are equal or close to one or more natural frequencies of the machining system. For instance, if the tooth passing frequency is equal or very close to a natural frequency of the tool assembly, the cyclic varying cutting force will excite a system resonance and the tool will vibrate excessively.

For high speed milling, if the spindle speed is high enough for the tooth passing frequency to approach a dominant natural frequency of the machining system, forced vibrations can lead to machining instability or chatter, *i.e.*, an unstable machining process. In this case, the problem can generally be eliminated by carefully choosing the spindle speed so that the tooth passing frequency is not equal or close to any of the dominant natural frequencies of the machining system once they are known or identified. Therefore, it is vital to have the natural frequencies of the machining system identified before a cutting process so that an appropriate spindle speed can be chosen to avoid spindle speed related frequencies identical or close to any of the natural frequencies of the machining system.

However there are uncertainties about the natural frequencies of a machining system which may challenge the efforts to reduce or eliminate this kind of chatter. The natural frequencies of a machining system may not be consistent for different machining processes due to different machining conditions, and even during a specific machining process due to continuous feed motions, dynamic cutting conditions, etc. This makes it difficult to identify the natural frequencies of a machining system. It also indicates that for high precision and high speed machining, the machining conditions should be maintained as consistently as possible to obtain a steady and stable machining process.

7.1.2 Self-excited Vibration

In machining, chatter is commonly referred to the vibration that feeds on itself as the tool moves across the workpiece.

What does "feed on itself" mean?

During the milling process, the tool and work assemblies will vibrate due to variations in the cutting forces. As a matter of fact, the machining system tends to vibrate at its natural frequencies during the machining process. At the tool tip, the vibration leaves waves on the machined surface. The waviness can cause the next cutting edge to experience a variable undeformed chip thickness. The variable undeformed chip thickness due to vibrations from the previous tooth path causes additional cutting force fluctuations, which feeds the vibration that already exists, such as a closed-loop system, and can make the vibration worse. When the dynamic cutting force is out of phase with the instantaneous relative movement between the tool and workpiece, this leads to the development of self-excited vibration [2]. This type of instability (an unstable condition) is called regenerative chatter because the vibration reproduces itself in a subsequent tooth pass through the generation of the waviness. It is the most practically significant form of self-excited vibration. There is another self-excited vibration, non-regenerative chatter, which occurs without undulation, and is relatively uncommon.

The fundamental features of self-excited vibration are: (a) its amplitude increases with time until a limit is attained due to nonlinear effect; (b) its frequency is equal to one of the natural frequencies of the machining system.

Generally speaking, self-excited vibrations can be distinguished from forced vibrations, because the frequencies of forced vibrations are equal to some multiple of the spindle speed and tooth passing frequency, and the frequencies of self-excited vibrations may not be equal to some multiple of the spindle speed. For different machining processes with different cutting speeds, if vibration persists at a specific frequency, the vibration is self-excited.

However there are some factors which may challenge the efforts to detect and reduce self-excited vibrations. For example, the natural frequencies of a machining system may spread in a wide frequency band; no matter whatever efforts you make, there will be one or more frequencies of the multiple of the spindle speed within the natural frequency band. This means not only forced vibrations will be excited, but also very likely self-excited vibrations within the natural frequency band will be generated once enough energy supporting the vibration is obtained from the machining process. This means the elimination of chatter in a particular machining process can be a laborious exercise and frequently can be accomplished only by reducing the production rate, *i.e.*, to reduce the energy supporting the chatter vibration.

7.1.3 The Scope of this Chapter

There are many types of milling operations; very basic ones include peripheral milling, face milling and ball-end milling. This chapter will mainly address machining dynamics in peripheral milling. Because ball-end milling (high speed ball-end milling in particular) is getting more and more popular in mould manufacture, the dynamic cutting force model is addressed as well in this chapter.

Cutting tools for peripheral milling and ball-end milling may have straight or helical teeth. Helical teeth generally are preferred over straight teeth because each tooth is gradually engaging and partially engaged with the workpiece as the tool rotates. Consequently it results in a smoother machining process and reduces vibration. In this chapter, it is assumed that a helical angle applies to all the tools.

Many researchers have studied machining dynamics and machining stability, and several theories [1, 2, 3] have been published to explain, identify and reduce chatter vibration. This chapter, based on previous research efforts, will provide some new means to address chatter vibration rather than repeating those published theories.

The study of machining dynamics has two basic objectives: (1) to identify rules for designing a stable machine system, a tool assembly system and a work assembly system; and (2) to identify rules for choosing dynamically stable cutting conditions. This chapter will focus on the second objective. To achieve this objective, it is vital to model the dynamic cutting forces and machining system vibrations, *i.e.*, the vibrations of the tool and work assemblies. This chapter begins with the dynamic cutting force modelling for both peripheral milling and ball-end milling, which lays the foundation for the following mathematical modelling of machining dynamics. The machining dynamics model can be used for the prediction of the machining system vibration and chatter, the simulation of machining process, the prediction of the machined surface roughness and waviness, the prediction of machining accuracy, and the optimization of the machining process for a maximum production rate. In practice, however, it is essential to have the natural frequency of the machining system been identified by either calculation/estimation using finite element analysis (FEA) or experimental measurement before applying the machining dynamics model. All these topics will be covered in this chapter.

7.1.4 Nomenclature in This Chapter

Ω	Radial immersion angle
α_e	Effective rake angle
α_n	Normal rake angle
α_r	Radial rake angle
β	Helix angle of tool, inclination angle of oblique cutting
α_j, β_j	Modal parameters corresponding to mode j
φ	Helix lag angle
φ_i	Position angle of a point on the cutting edge of the ith helical flute
ψ	Axial immersion angle of a tooth within b_a
δ_e	Cutter run-out
$\varepsilon(t)$	Error signal
ω	Angular spindle speed
ω_d	Modal frequency or damped natural frequency
ω_{nx}, ω_{ny}	Natural frequencies
ζ_x, ζ_y	Damping ratios
F_{ix}, F_{iy}	Total cutting force components of the ith helical flute in x and y directions
F_x, F_y	Total cutting force components in x and y directions

G	Transfer function
K'_s	Cutting force coefficient
R	Tool radius
SF	Spindle frequency
T	Tooth passing period
TPF	Tooth passing frequency
b	Width of cut
b_a	Axial depth of cut
c_1	Radial force ratio
c_2	Axial force ratio
c_x, c_y	Damping coefficients
d	Radial depth of cut
$e(t)$	Disturbance signal
f_t	Feedrate per tooth per revolution
m	Flute number of tool
m_x, m_y	Mass
n	Spindle rotation speed (rpm)
k_x, k_y	Stiffness
$p_k = -\zeta_k \omega_{n,k} + j\omega_{d,k}$	Pole location in the s-plane for the kth mode
$r_k = \sigma_k + jv_k$	Complex residue of the kth mode
s	Laplace factor
t	Undeformed chip thickness
$t_i(\varphi_i)$	Undeformed chip thickness of the cutting point on the ith helical flute at position angle φ_i
u, u_0	Total specific cutting energy per unit volume
$u(t)$	Input signal
v_i	Dynamic displacement of the ith tooth in angle position φ_i
x, y	Dynamic displacements in X, Y directions respectively
$y(t)$	Output signal
c	In superscript and subscript stand for tool
w	In superscript and subscript stand for workpiece

7.2 The Dynamic Cutting Force Model for Peripheral Milling

To model the cutting force in peripheral milling, we take the common assumption that the helical flute of an end mill is treated as a combination of a series of oblique cutting edge segments. Therefore, modelling the dynamic cutting force can be based on the oblique cutting model, which is a direct extension of orthogonal cutting [4].

7.2.1 Oblique Cutting

For oblique cutting, as shown in Figure 7.1, the principal cutting forces F_P (in the cutting direction VC) and F_Q (in the undeformed chip thickness t direction) are [4]:

$$F_P = uA_v = ubt = ult \cos \beta \qquad (7.1)$$
$$F_Q = cF_P \qquad (7.2)$$

where b is the width of cut, l is the engaged length of cutting edge, β is the inclination angle, c is the cutting force ratio, and u is the total specific cutting energy per unit volume.

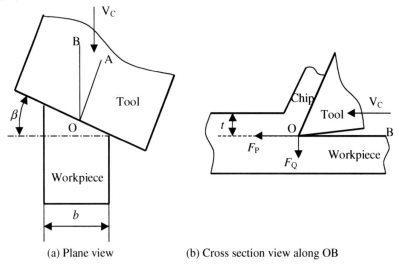

(a) Plane view (b) Cross section view along OB

Figure 7.1. Oblique cutting

The specific cutting energy is essentially independent of the cutting speed over a wide range of values, provided a large built-up edge (BUE) is not generated. However it is influenced by the workpiece material, the effective rake angle α_e of tool (u decreases about 1% per degree increase in α_e) and undeformed chip thickness t, and takes the form:

$$u \sim 1/t^{0.2} \qquad (7.3)$$

This inverse relationship is sometimes referred to as the size effect [4].

Table 7.1 gives approximately average values for a few common classes of work materials cut at representative hardness levels.

Table 7.1. Approximate values of specific energy for different materials cut with $a_c = 0°$ and t =0.25 mm for cutting with continuous chip and no BUE [4]

Materials	u_0 (J/m³)
Aluminium alloy	7.02×10⁸
Grey cast iron	10.53×10⁸
Free-machining steel (AISI 1213)	17.55×10⁸
Mild steel (AISI 1018)	21.06×10⁸
Titanium alloy	35.10×10⁸
Stainless steel (18-8)	49.14×10⁸
High-temperature alloy (Ni or Co base materials)	49.14×10⁸

7.2.2 The Geometric Model of a Helical End Mill

The end mill can be divided into a number of slices along its z-direction, as shown in Figure 7.2. Within each slice, the cutting action for an individual tooth can be modelled as oblique cutting, the tangential and normal cutting forces on the rake face can be obtained from the oblique cutting model. It is worth noting that the tangential and normal directions together with chip thickness vary during the formation of a single chip in the milling process. Accordingly, a dynamic model must account for these variations in the magnitude and the direction of the cutting forces.

To relate the helical flute geometry of the end mill to the tangential cutting force, it is essential to determine the position angle φ_i of point P on the ith flute at any time in the coordinate system:

$$\varphi_i = \varphi + \theta + (i-1)\frac{2\pi}{m} \quad (1 \leq i \leq m, 0 \leq \varphi \leq \psi) \tag{7.4}$$

Where:

$$\theta = -\omega t_{ime} \tag{7.5}$$

is the instantaneous rotational angle of the flute tip from the X-axis, φ is the helix lag angle of point P relating to the flute tip, m is the flute number of the tool, and ω is the spindle angular speed:

$$\psi = \frac{b_a \tan \beta}{R} \tag{7.6}$$

is the axial immersion angle of a flute within the axial depth of cut b_a, R is the tool's radius, and β is the tool's helix angle, which has the same meaning as the tool's inclination angle in oblique cutting.

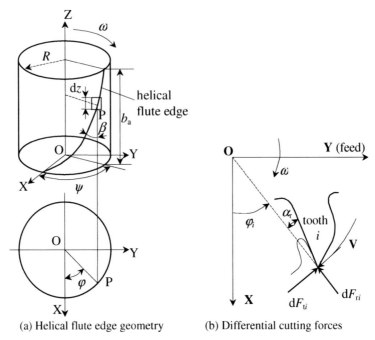

(a) Helical flute edge geometry   (b) Differential cutting forces

Figure 7.2. A diagrammatic illustration of the differential cutting force model of peripheral milling

7.2.3 Differential Tangential and Normal Cutting Forces

As shown in Figure 7.2, within each slice of the ith flute, the differential tangential, normal and axial cutting forces at any point on the rake face can be obtained from the oblique cutting model:

$$dF_{ti}(\varphi_i) = ut_i(\varphi_i)dz = ut_i(\varphi_i)R\cot\beta d\varphi \tag{7.7}$$

$$dF_{ri}(\varphi_i) = c_1 dF_{ti}(\varphi_i) \tag{7.8}$$

$$dF_{ai}(\varphi_i) = c_2 dF_{ti}(\varphi_i) \tag{7.9}$$

where the radial force ratio c_1 varies depending on the tool's material and geometry, work material, and cutting conditions, and the axial force ratio can be determined using [6]:

$$c_2 = \frac{\sin\beta(1-\sin\alpha_n) - c_1\cos\alpha_n\tan\beta}{\sin\beta\sin\alpha_n\tan\beta + \cos\beta} \tag{7.10}$$

where α_n is the normal rake angle of the helical flute [4], which can be expressed in term of radial rake angle α_r as follows:

$$\tan\alpha_n = \tan\alpha_r \cos\beta \tag{7.11}$$

Considering the size effect of the undeformed chip thickness and the influence of the effective rake angle, the total specific cutting energy takes the form [4]:

$$u = u_0(1 - \frac{\alpha_e - \alpha_{e0}}{100})(\frac{t_0}{t_i(\varphi_i)})^{0.2} \tag{7.12}$$

where u_0 is the initial total cutting energy per unit volume, α_e is the effective rake angle of the tool, $\alpha_{e0} = 0$ is its initial value, $t_i(\varphi_i)$ is the undeformed chip thickness of the cutting point on the ith helical flute at position angle φ_i, and $t_0 = 0.25$ mm is the initial undeformed chip thickness (refer to Table 7.1). u_0, c_1 and c_2 are supposed to be variable, depending on the materials of the tool and the workpiece and the cutting conditions, and should be experimentally calibrated [5, 6].

The effective rake angle α_e, which has an influence on the total specific cutting energy u, is determined by [4]:

$$\sin\alpha_e = \sin^2\beta + \cos^2\beta\sin\alpha_n \tag{7.13}$$

7.2.4 Undeformed Chip Thickness $t_i(\varphi_i)$

During the milling process, both the tool and the workpiece can be considered as a two-degree-of-freedom spring-damper vibratory system in the two mutually perpendicular directions of X and Y, as shown in Figure 7.3.

The cutting forces excite vibrations of the tool and workpiece simultaneously in the normal (X) and feed (Y) directions, causing dynamic displacements x_c, y_c and x_w, y_w of the tool and workpiece, respectively. The dynamic displacements are carried to the undeformed chip thickness direction v with the coordinate transformation as following:

$$v_{ic} = -x_c \cos(\varphi_i) - y_c \sin(\varphi_i) \qquad (7.14)$$
$$v_{iw} = -x_w \cos(\varphi_i) - y_w \sin(\varphi_i) \qquad (7.15)$$

where v_{ic} and v_{iw} are the dynamic displacements of the tool and workpiece in the v direction at the position φ_i of the ith tooth.

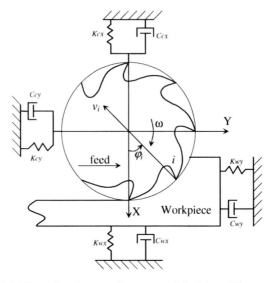

Figure 7.3. The spring-damper vibratory model of the milling operation

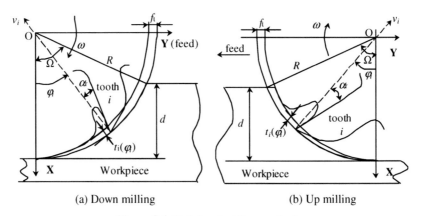

(a) Down milling (b) Up milling

Figure 7.4. Peripheral milling method

Considering the vibrations of the tool/workpiece and the influence of the cutter run-out, the undeformed chip thickness removed by a point on the ith helical flute can be calculated as the following:

For down milling, as shown in Figure 7.4 a:

$$t_i(\varphi_i) = \begin{cases} f_t \sin(\varphi_i) + (v_{ic(-T)} - v_{ic}) - (v_{iw(-T)} - v_{iw}) + \delta_i(\varphi_i) & \text{if } 0 \leq \varphi_i \leq \Omega \\ 0 & \text{else} \end{cases} \quad (7.16)$$

For up milling, as shown in Figure 7.4 b:

$$t_i(\varphi_i) = \begin{cases} f_t \sin(-\varphi_i) + (v_{ic(-T)} - v_{ic}) - (v_{iw(-T)} - v_{iw}) + \delta_i(\varphi_i) & \text{if } -\Omega \leq \varphi_i \leq 0 \\ 0 & \text{else} \end{cases} \quad (7.17)$$

where Ω is the tool's radial immersion angle within the radial depth of cut:

$$\Omega = \arccos(1 - \frac{d}{R}) \quad (7.18)$$

$v_{ic(-T)}$ and $v_{iw(-T)}$ are the dynamic displacements of the tool and workpiece in the v direction from the previous tooth pass at the position φ_i of the ith tooth, and $\delta_i(\varphi_i)$ is the contribution of the cutter run-out to the undeformed chip thickness [6], which is given by:

$$\delta_i(\varphi_i) \approx R_i'(\varphi_i') - R_{i-1}'(\varphi_i') \quad (7.19)$$

where $R_i'(\varphi_i')$ and $R_{i-1}'(\varphi_i')$ are the actual cutting radius of the ith tooth and the preceding tooth respectively, as shown in Figure 7.5.

$$R_i'(\varphi_i') = \sqrt{R^2 + \delta_e^2 - 2R\delta_e \cos(\angle O'OP)} \quad (7.20)$$

where $\angle O'OP = \pi - |\varphi_i - \varphi_e|$, $\delta_e = \overline{OO'}$ is the cutter run-out value, O' is the spindle centre, and O is the tool centre, as shown in Figure 7.5.

In the initial position, the spindle rotational angle $\theta = -\omega t_{ime} = 0$, let the initial location angle of the tool centre be φ_{e0} ($0 \leq \varphi_{e0} \leq 2\pi$). Assuming the initial location angle of the first tooth ($m = 1$, $\varphi = 0$) is 0, so for the ith tooth of the tool, when the helix lag angle $\varphi \neq 0$ and the spindle rotational angle $\theta \neq 0$, the location angle of the tool centre $\varphi_e = \varphi_{e0} + \theta$. By calling Equation 7.4, we can rewrite Equation 7.20 as:

$$R_i'(\varphi_i') = \sqrt{R^2 + \delta_e^2 - 2R\delta_e \cos(\pi - |\varphi - \varphi_{e0} + 2(i-1)\pi/m|)} \quad (7.21)$$

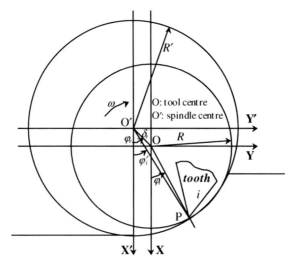

Figure 7.5. The cutter run-out model

7.2.5 Differential Cutting Forces in X and Y Directions

As shown in Figure 7.2 b, resolving the differential cutting forces of Equations 7.7-7.9 into X, Y and Z components yields:

$$\begin{cases} dF_{tix} = -dF_{ti}(\varphi_i)\sin\varphi_i = -ut_i(\varphi_i)R\cot\beta\sin\varphi_i d\varphi \\ dF_{tiy} = dF_{ti}(\varphi_i)\cos\varphi_i = ut_i(\varphi_i)R\cot\beta\cos\varphi_i d\varphi \end{cases} \quad (7.22)$$

$$\begin{cases} dF_{rix} = -dF_{ri}(\varphi_i)\cos\varphi_i = -c_1 ut_i(\varphi_i)R\cot\beta\cos\varphi_i d\varphi \\ dF_{riy} = -dF_{ri}(\varphi_i)\sin\varphi_i = -c_1 ut_i(\varphi_i)R\cot\beta\sin\varphi_i d\varphi \end{cases} \quad (7.23)$$

$$dF_{iz} = -c_2 ut_i(\varphi_i)R\cot\beta d\varphi \quad (7.24)$$

Considering Equations 7.16-7.17 and summing up each force component gives the differential cutting forces dF_{ix}, dF_{iy} and dF_{iz}.

(a) For down milling:

$$\begin{cases} dF_{ix} = -u[f_t\sin\varphi_i + (v_{ic(-T)} - v_{ic}) - (v_{iw(-T)} - v_{iw}) + \delta_i(\varphi_i)]R\cot\beta(\sin\varphi_i + c_1\cos\varphi_i)d\varphi \\ dF_{iy} = u[f_t\sin\varphi_i + (v_{ic(-T)} - v_{ic}) - (v_{iw(-T)} - v_{iw}) + \delta_i(\varphi_i)]R\cot\beta(\cos\varphi_i - c_1\sin\varphi_i)d\varphi \\ dF_{iz} = -u[f_t\sin\varphi_i + (v_{ic(-T)} - v_{ic}) - (v_{iw(-T)} - v_{iw}) + \delta_i(\varphi_i)]c_2 R\cot\beta d\varphi \end{cases}$$

Substituting Equations 7.14-7.15 into the equation gives:

$$\begin{cases} dF_{ix} = -u[f_t \sin\varphi_i + (x_c \cos\varphi_i + y_c \sin\varphi_i - x_{c(-T)} \cos\varphi_i - y_{c(-T)} \sin\varphi_i) \\ \quad - (x_w \cos\varphi_i + y_w \sin\varphi_i - x_{w(-T)} \cos\varphi_i - y_{w(-T)} \sin\varphi_i) + \delta_i(\varphi_i)]R \cot\beta(\sin\varphi_i + c_1\cos\varphi_i)d\varphi \\ dF_{iy} = u[f_t \sin\varphi_i + (x_c \cos\varphi_i + y_c \sin\varphi_i - x_{c(-T)} \cos\varphi_i - y_{c(-T)} \sin\varphi_i) \\ \quad - (x_w \cos\varphi_i + y_w \sin\varphi_i - x_{w(-T)} \cos\varphi_i - y_{w(-T)} \sin\varphi_i) + \delta_i(\varphi_i)]R \cot\beta(\cos\varphi_i - c_1\sin\varphi_i)d\varphi \\ dF_{iz} = -u[f_t \sin\varphi_i + (x_c \cos\varphi_i + y_c \sin\varphi_i - x_{c(-T)} \cos\varphi_i - y_{c(-T)} \sin\varphi_i) \\ \quad - (x_w \cos\varphi_i + y_w \sin\varphi_i - x_{w(-T)} \cos\varphi_i - y_{w(-T)} \sin\varphi_i) + \delta_i(\varphi_i)]c_2 R \cot\beta d\varphi \end{cases}$$

Considering the nature of the total specific cutting energy u in Equation 7.12 and the undeformed chip thickness $t_i(\varphi_i)$ in Equations 7.16-7.17, for the sake of simplicity, assume $t_i(\varphi_i) \approx f_t \sin\varphi_i$, then:

$$uR \cot\beta = u_0 R \cot\beta (1 - \frac{\alpha_e - \alpha_{e0}}{100})(\frac{t_0}{f_t})^{0.2} \sin^{-0.2}\varphi_i = u' \sin^{-0.2}\varphi_i$$

where $u' = u_0 R \cot\beta (1 - \frac{\alpha_e - \alpha_{e0}}{100})(\frac{t_0}{f_t})^{0.2}$. Then the differential cutting forces can be rewritten as:

$$\begin{cases} dF_{ix} = -u'[(f_t + y_c - y_{c(-T)} - y_w + y_{w(-T)})(\sin^{1.8}\varphi_i + c_1 \sin^{0.8}\varphi_i \cos\varphi_i) \\ \quad + (x_c - x_{c(-T)} - x_w + x_{w(-T)})(\sin^{0.8}\varphi_i \cos\varphi_i + c_1 \sin^{-0.2}\varphi_i \cos^2\varphi_i) \\ \quad + \delta_i(\varphi_i)(\sin^{0.8}\varphi_i + c_1 \sin^{-0.2}\varphi_i \cos\varphi_i)]d\varphi_i \\ dF_{iy} = u'[(f_t + y_c - y_{(-T)} - y_w + y_{w(-T)})(\sin^{0.8}\varphi_i \cos\varphi_i - c_1 \sin^{1.8}\varphi_i) \\ \quad + (x_c - x_{(-T)} - x_w + x_{w(-T)})(\sin^{-0.2}\varphi_i \cos^2\varphi_i - c_1 \sin^{1.8}\varphi_i) \\ \quad + \delta_i(\varphi_i)(\sin^{-0.2}\varphi_i \cos\varphi_i - c_1 \sin^{0.8}\varphi_i)]d\varphi_i \\ dF_{iz} = -u'[(f_t + y_c - y_{(-T)} - y_w + y_{w(-T)})\sin^{0.8}\varphi_i \\ \quad + (x_c - x_{(-T)} - x_w + x_{w(-T)})\sin^{-0.2}\varphi_i \cos\varphi_i \\ \quad + \delta_i(\varphi_i)\sin^{-0.2}\varphi_i]c_2 d\varphi_i \end{cases}$$

$$\begin{pmatrix} \varphi_i = \varphi - \omega t_{ime} + (i-1)\frac{2\pi}{m} \\ 0 \leq \varphi_i \leq \Omega \end{pmatrix} \quad (7.25)$$

(b) For up milling, similarly, we have:

$$\begin{cases} dF_{ix} = u'[(f_t + y_{c(-T)} - y_c - y_{w(-T)} + y_w)(\sin^{1.8}(-\varphi_i) - c_1 \sin^{0.8}(-\varphi_i)\cos\varphi_i) \\ \qquad + (x_{c(-T)} - x_c - x_{w(-T)} + x_w)(-\sin^{0.8}(-\varphi_i)\cos\varphi_i + c_1 \sin^{-0.2}(-\varphi_i)\cos^2\varphi_i) \\ \qquad + \delta_i(\varphi_i)(\sin^{0.8}(-\varphi_i) - c_1 \sin^{-0.2}(-\varphi_i)\cos\varphi_i)]d\varphi_i \\ dF_{iy} = u'[(f_t + y_{c(-T)} - y_c - y_{w(-T)} + y_w)(\sin^{0.8}(-\varphi_i)\cos\varphi_i + c_1 \sin^{1.8}(-\varphi_i)) \\ \qquad + (x_{c(-T)} - x_c - x_{w(-T)} + x_w)(-\sin^{-0.2}(-\varphi_i)\cos^2\varphi_i - c_1 \sin^{0.8}(-\varphi_i)\cos\varphi_i) \\ \qquad + \delta_i(\varphi_i)(\sin^{-0.2}(-\varphi_i)\cos\varphi_i + c_1\sin^{0.8}(-\varphi_i))]d\varphi_i \\ dF_{iz} = u'[(f_t + y_{c(-T)} - y_c - y_{w(-T)} + y_w)\sin^{0.8}(-\varphi_i) \\ \qquad + (x_{c(-T)} - x_c - x_{w(-T)} + x_w)(-\sin^{-0.2}(-\varphi_i)\cos\varphi_i) \\ \qquad + \delta_i(\varphi_i)\sin^{-0.2}(-\varphi_i)]c_2 d\varphi_i \end{cases}$$

$$\begin{pmatrix} \varphi_i = \varphi - \omega t_{ime} + (i-1)\dfrac{2\pi}{m} \\ -\Omega \le \varphi_i \le 0 \end{pmatrix} \quad (7.26)$$

where $1 \le i \le m$, $0 \le \varphi \le \psi$.

7.2.6 Total Cutting Forces in X and Y Directions

The total cutting force applied on the whole cutting edge is given by:

$$\begin{cases} F_{ix} = \int_{\varphi_s}^{\varphi_e} dF_{ix} d\varphi_i \\ F_{iy} = \int_{\varphi_s}^{\varphi_e} dF_{iy} d\varphi_i \\ F_{iz} = \int_{\varphi_s}^{\varphi_e} dF_{iz} d\varphi_i \end{cases} \quad (7.27)$$

where φ_s and φ_e are the lag angular locations of the start and end points of the contact of the cutting edge, and are defined in the following kinematics analysis.

Summing up the cutting forces acting on all the m helical flutes gives the total force applied on the whole tool:

$$\begin{cases} F_x(t_{ime}) = \sum_{i=1}^{m} F_{ix} \\ F_y(t_{ime}) = \sum_{i=1}^{m} F_{iy} \\ F_z(t_{ime}) = \sum_{i=1}^{m} F_{iz} \end{cases} \quad (7.28)$$

7.2.7 The Calibration of the Cutting Force Coefficients

In the cutting force model, the following coefficients have to be calibrated by experiment:
- u_0, the initial total cutting energy per unit volume, under the initial cutting condition $\alpha_{e0} = 0°$ and $t_0 = 0.25$ mm (refer to Table 7.1) [4],
- c_1, the radial force ratio, and
- c_2, the axial force ratio.

u_0 depends on the workpiece material, the tool material, the cutting edge radius, and the friction characteristics between the workpiece and the tool (no built-up edge is assumed), whereas the ratios c_1 and c_2 rely mainly on the tool geometry. Although c_2 can be obtained from c_1 and the tool geometry using Equation 7.10, it needs to be calibrated by experiments, due to the inconsistency between the calculated value and the experimental data.

The Reference [6] recorded the whole procedure for the calibration of the cutting force coefficients; however, for its application, a brief summarization is described below.

A. In order to avoid the interference of the cutting force generated by adjacent teeth, the best way is to use a one-tooth helical end mill in the cutting trials for the calibration of cutting force coefficients. The reason is that a small cutter run-out (no matter how much the value and its initial location angle φ_{e0} are) theoretically has no influence on the cutting force (refer to Equation 7.19).
B. In order to obtain a maximum radial immersion angle, the best way is to use slot milling for the calibration.
C. The vibrations of the tool and workpiece should be minimised so as to ignore their influence and simplify the formula.
D. For the estimation of the cutting force coefficients, the easiest and probably the best way is to use the average cutting forces rather than the instant cutting forces. The theoretical average cutting force components can be asserted from Equations 7.25-7.26 as:

$$\begin{cases} \overline{F}_x = u_0 \sum_{i=1}^{m} \overline{f}_{ix1} + u_0 c_1 \sum_{i=1}^{m} \overline{f}_{ix2} \\ \overline{F}_y = u_0 \sum_{i=1}^{m} \overline{f}_{iy1} + u_0 c_1 \sum_{i=1}^{m} \overline{f}_{iy2} \\ \overline{F}_z = u_0 c_2 \sum_{i=1}^{m} \overline{f}_{iz} \end{cases} \qquad (7.29)$$

where (a) for down milling:

$$f_{ix1} = \int_{\varphi_s}^{\varphi_e} -c_s \sin^{0.8} \varphi_i d\varphi_i$$

$$f_{ix2} = \int_{\varphi_s}^{\varphi_e} -c_s \sin^{-0.2} \varphi_i \cos\varphi_i d\varphi_i$$

$$f_{iy1} = \int_{\varphi_s}^{\varphi_e} c_s \sin^{-0.2} \varphi_i \cos\varphi_i d\varphi_i$$

$$f_{iy2} = \int_{\varphi_s}^{\varphi_e} -c_s \sin^{0.8} \varphi_i d\varphi_i$$

$$f_{iz} = \int_{\varphi_s}^{\varphi_e} -c_s \sin^{-0.2} \varphi_i d\varphi_i$$

$$c_s = (1 - \frac{\alpha_e - \alpha_{e0}}{100})(\frac{t_0}{f_t})^{0.2}(f_t \sin(\varphi_i) + \delta_i(\varphi_i'))R \cot \beta$$

(b) For up milling:

$$f_{ix1} = \int_{\varphi_s}^{\varphi_e} c_s \sin^{0.8}(-\varphi_i) d\varphi_i$$

$$f_{ix2} = \int_{\varphi_s}^{\varphi_e} -c_s \sin^{-0.2}(-\varphi_i) \cos\varphi_i d\varphi_i$$

$$f_{iy1} = \int_{\varphi_s}^{\varphi_e} c_s \sin^{-0.2}(-\varphi_i) \cos\varphi_i d\varphi_i$$

$$f_{iy2} = \int_{\varphi_s}^{\varphi_e} c_s \sin^{0.8}(-\varphi_i) d\varphi_i$$

$$f_{iz} = \int_{\varphi_s}^{\varphi_e} -c_s \sin^{-0.2}(-\varphi_i) d\varphi_i$$

$$c_s = (1 - \frac{\alpha_e - \alpha_{e0}}{100})(\frac{t_0}{f_t})^{0.2}(f_t \sin(-\varphi_i) + \delta_i(\varphi_i'))R \cot \beta$$

Averaging the cutting forces from the measured dynamic ones and substituting them in Equation 7.29 can yield the cutting force coefficients u_0, c_1 and c_2.

Table 7.2 and Figure 7.6 show the calibrated cutting force coefficients for carbon steel EN8 using a single fluted solid carbide end-mill.

Table 7.2. Experimentally calibrated cutting force coefficients (EN8)

f_t (mm/tooth)	0.0197	0.0296	0.0395	0.0494	0.0592	0.0691	0.0790	0.0889
u_0 (GJ/m^3)	2.4067	2.4433	2.3851	2.4532	2.3891	2.5146	2.4429	2.5184
c_1	0.5114	0.4832	0.4776	0.3889	0.3457	0.4052	0.3992	0.4171
c_2	0.2617	0.2871	0.3191	0.3529	0.3525	0.3596	0.3570	0.3599
Calculated c_2	0.2038	0.2413	0.2448	0.3667	0.4241	0.3540	0.3530	0.3292

where:

Cutter parameters: $R = 10$ mm, $\beta = 45°$, $\alpha_r = 5°$.

Cutting parameters: $b_a = 10.5$ mm, $v = 70$ m/min.

Cutting condition: with cutting fluid.

It is worth mentioning that using Equation 7.10 with the calibrated c_1 gives the calculated c_2, as illustrated in Figure 7.6 c. There is a small difference between the experimental calibration and the estimated evaluation for the axial force ratio c_2.

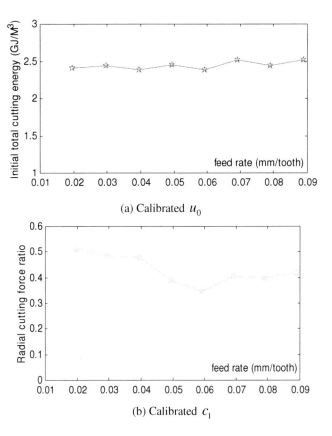

(a) Calibrated u_0

(b) Calibrated c_1

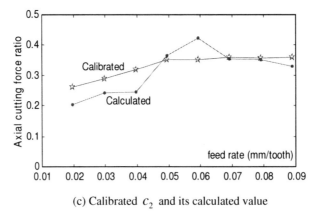

(c) Calibrated c_2 and its calculated value

Figure 7.6. Experimentally calibrated cutting force coefficients (EN8)

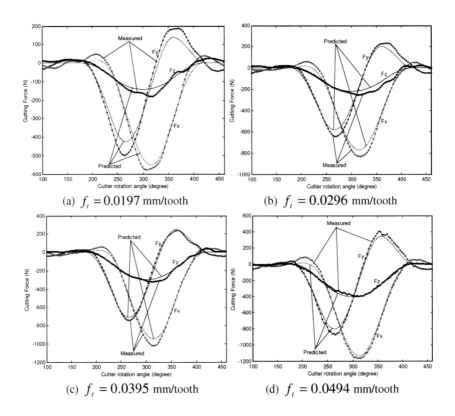

(a) $f_t = 0.0197$ mm/tooth

(b) $f_t = 0.0296$ mm/tooth

(c) $f_t = 0.0395$ mm/tooth

(d) $f_t = 0.0494$ mm/tooth

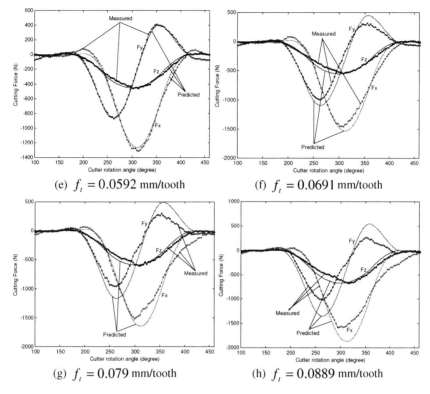

Figure 7.7. Measured and predicted cutting forces in slot milling

Figure 7.7 shows the measured and predicted dynamic cutting force components, in which the predicted ones are generated based on the proposed cutting force model with the calibrated force coefficients.

The results shown in Figure 7.7 reveal that there is a good agreement between the measured and predicted cutting forces when feedrate changes from $f_t = 0.0395$ mm/tooth to $f_t = 0.0592$ mm/tooth. When feedrate $f_t < 0.0395$ mm/tooth, the measured cutting force components F_x and F_y are greater than the predicted ones. That means the calibrated u_0 is smaller than its actual value. When feedrate $f_t > 0.0592$ mm/tooth, the measured cutting force components F_x and F_y are smaller than the predicted ones. That means the calibrated u_0 is greater than its actual value. This result indicates the size effect of the undeformed chip thickness and the influence of the effective rake angle may be more significant than the estimated value from Equation 7.12. For this case, the proposed model is valid only when the feedrate changes from $f_t = 0.0395$ mm/tooth to $f_t = 0.0592$ mm/tooth.

7.2.8 A Case Study: Verification

Table 7.3 lists the cutting conditions and parameters of a case study, and Figure 7.8 shows the measured and predicted dynamic cutting forces, in which the predicted values are obtained from the cutting force model using the evaluated cutting force coefficients.

Table 7.3. Cutting conditions and parameters (peripheral milling)

Tool: solid carbide end-mill, $R = 10$ mm, $\beta = 30°$, $\alpha_r = 5°$, $m = 2$, $\delta_e = 2\,\mu\text{m}$
Work material: carbon steel EN8
Cutting condition: spindle speed $n = 1592$ rpm, feedrate $f_t = 0.05$ mm/tooth, axial depth of cut $b_a = 5.105$ mm, radial depth of cut $d = 10.8$ mm, with fluid

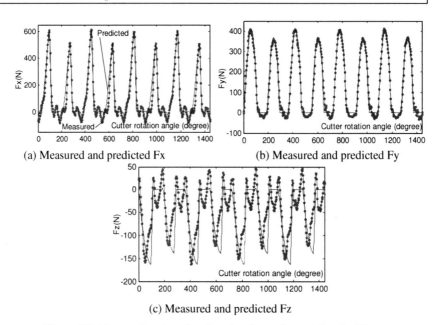

(a) Measured and predicted Fx
(b) Measured and predicted Fy
(c) Measured and predicted Fz

Figure 7.8. Measured and predicted cutting forces for peripheral milling

7.3 A Dynamic Cutting Force Model for Ball-end Milling

In a similar way to peripheral milling, we can model a dynamic cutting force for ball-end milling.

7.3.1 A Geometric Model of a Ball-end Mill

The geometry of a ball-end mill is shown in Figure 7.9. The flutes of the ball-end mill meet at the tip O of the sphere, and have a constant helix lead. The flutes have a helix angle of β_0 at the ball-shank meeting boundary. The ball part of the tool meets with the cylindrical shank which has a radius of R_0, which is the same as the ball radius. The z coordinate of a point P on the flute is:

$$z(\varphi) = R_0 \varphi \cot \beta_0 \quad (7.30)$$

for the constant helix-lead ball-end mill, where φ is the helix lag angle between the tip of the flute ($z = 0$) and the point P in the xy plane. It is measured clockwise from the $+x$ axis; see Figure 7.9.

The tool's radius in the xy plane at a point P can be expressed as:

$$R(\varphi) = \sqrt{x^2 + y^2} = R_0 \sqrt{1 - (1 - \varphi \cot \beta_0)^2} \quad (0 \leq \varphi \leq \tan \beta_0) \quad (7.31)$$

The local helix angle of the ball part of the tool is scaled by a radius factor and can be expressed as [7]:

$$\tan \beta(\varphi) = \frac{R(\varphi)}{R_0} \tan \beta_0 = \sqrt{1 - (1 - \varphi \cot \beta_0)^2} \, \tan \beta_0$$
$$(0 \leq \varphi \leq \tan \beta_0) \quad (7.32)$$

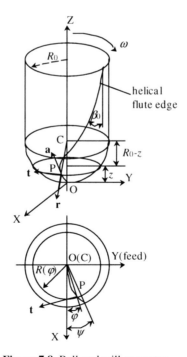

Figure 7.9. Ball-end mill geometry

To relate the helical flute geometry of the ball-end mill to the tangential cutting force, a curvilinear coordinate system (t, r, a) at point P on the flute is established, as shown in Figure 7.9, in which:

$$\begin{cases} \mathbf{t} = [\sin\varphi_i \quad -\cos\varphi_i \quad 0] \\ \mathbf{r} = [\sqrt{1-(1-\varphi\cot\beta_0)^2}\cos\varphi_i \quad \sqrt{1-(1-\varphi\cot\beta_0)^2}\sin\varphi_i \quad -(1-\varphi\cot\beta_0)] \\ \mathbf{a} = [(1-\varphi\cot\beta_0)\cos\varphi_i \quad (1-\varphi\cot\beta_0)\sin\varphi_i \quad \sqrt{1-(1-\varphi\cot\beta_0)^2}] \end{cases} \quad (7.33)$$

where $\varphi_i = \varphi + \theta + (i-1)\dfrac{2\pi}{m} \quad (1 \le i \le m, 0 \le \varphi \le \psi)$ (7.34)

is the position angle of a point P on the *i*th cutting edge, and $\theta = -\omega t_{ime}$ is the instantaneous rotation angle of the tool from the *x*-axis, φ is the helix lag angle, m is the flute number of the tool, ω is the angular velocity of the spindle, and:

$$\psi = \frac{b_a \tan\beta_0}{R_0} \quad (7.35)$$

is the axial immersion angle of a tooth within the axial depth of cut b_a. It is worth mentioning that b_a varies during the cutting of sculptured surfaces.

7.3.2 Dynamic Cutting Force Modelling

Similar to an end mill, a ball-end mill can be divided into a number of slices along its z-direction, as shown in Figure 7.10 a. Within each slice, the cutting action for an individual tooth can be modelled as a single point oblique cut, and the differential tangential, normal and axial cutting forces at any point on the rake face can be obtained from the oblique cutting model. It is important to note that the differential tangential, normal and axial directions, together with the chip thickness, vary during the formation of a single chip in milling. Accordingly, a dynamic model must account for these variations in the magnitude and direction of cutting forces.

The differential cutting forces acting on the helical flute's rake face are dependent on the undeformed chip thickness, and can be derived from the oblique cutting theory [4] as:

$$dF_{ti}(\varphi_i) = ut_i(\varphi_i)R_0 \cot\beta_0 d\varphi \quad (7.36)$$

$$dF_{ri}(\varphi_i) = c_1 dF_{ti}(\varphi_i) \quad (7.37)$$

$$dF_{ai}(\varphi_i) = c_2 dF_{ti}(\varphi_i) \quad (7.38)$$

as shown in Figure 7.11, where $t_i(\varphi_i)$ is the undeformed chip thickness in the xy plane, as shown in Figures 7.10 b and 7.10 c, u is the tangential cutting force coefficient, which is the same as the total energy per unit volume in Equation 7.12 and c_1 and c_2 are the ratios of radial and axial cutting force components, respectively.

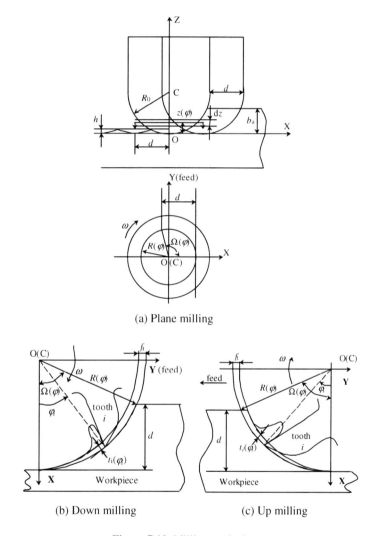

(a) Plane milling

(b) Down milling   (c) Up milling

Figure 7.10. Milling methods

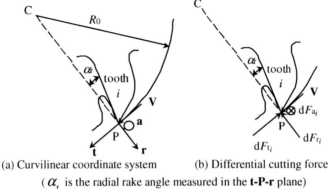

(a) Curvilinear coordinate system   (b) Differential cutting force
(α_r is the radial rake angle measured in the **t-P-r** plane)

Figure 7.11. Differential cutting forces acting on the element of the ith tooth

From the milling kinematics, the undeformed chip thickness removed by a point on the *i*th helical flute can be calculated as follows.

(a) For down milling, as shown in Figure 7.10 b:

$$t_i(\varphi_i) = \begin{cases} f_t \sin(\varphi_i) & \text{if} \quad 0 \leq \varphi_i \leq \Omega(\varphi) \\ 0 & \text{else} \end{cases} \quad (7.39)$$

(b) For up milling, as shown in Figure 7.10 c:

$$t_i(\varphi_i) = \begin{cases} f_t \sin(-\varphi_i) & \text{if} \quad -\Omega(\varphi) \leq \varphi_i \leq 0 \\ 0 & \text{else} \end{cases} \quad (7.40)$$

where Ω is the tool's radial immersion angle within the radial depth of cut, as shown in Figure 7.10, and

$$\Omega(\varphi) = \begin{cases} \arccos(1 - \dfrac{d}{R(\varphi)}) & \text{if} \quad \dfrac{h \tan \beta_0}{R_0} < \varphi < \dfrac{b_a \tan \beta_0}{R_0} \\ \pi & \text{if} \quad 0 \leq \varphi \leq \dfrac{h \tan \beta_0}{R_0} \end{cases} \quad (7.41)$$

where d is the step-over feedrate, and:

$$h = R_0 (1 - \sqrt{1 - \left(\dfrac{d}{2R_0}\right)^2}) \quad (7.42)$$

is the scallop height of the machined surface[1].

To develop the total force applied on the whole tool, the differential forces are resolved in the feed (y), normal (x) and axial (z) directions. As the differential cutting force components are just opposite to the corresponding directions of the curvilinear coordinate system (t, r, a), Equations 7.36-7.38 become:

$$\begin{cases} dF_{tix} = -ut_i(\varphi_i)R_0 \cot\beta_0 \sin\varphi_i d\varphi \\ dF_{tiy} = ut_i(\varphi_i)R_0 \cot\beta_0 \cos\varphi_i d\varphi \\ dF_{tiz} = 0 \end{cases} \quad (7.43)$$

$$\begin{cases} dF_{rix} = -c_1 ut_i(\varphi_i)R_0 \cot\beta_0 \sqrt{1-(1-\varphi\cot\beta_0)^2} \cos\varphi_i d\varphi \\ dF_{riy} = -c_1 ut_i(\varphi_i)R_0 \cot\beta_0 \sqrt{1-(1-\varphi\cot\beta_0)^2} \sin\varphi_i d\varphi \\ dF_{riz} = c_1 ut_i(\varphi_i)R_0 \cot\beta_0 (1-\varphi\cot\beta_0) d\varphi \end{cases} \quad (7.44)$$

$$\begin{cases} dF_{aix} = -c_2 ut_i(\varphi_i)R_0 \cot\beta_0 (1-\varphi\cot\beta_0)\cos\varphi_i d\varphi \\ dF_{aiy} = -c_2 ut_i(\varphi_i)R_0 \cot\beta_0 (1-\varphi\cot\beta_0)\sin\varphi_i d\varphi \\ dF_{aiz} = -c_2 ut_i(\varphi_i)R_0 \cot\beta_0 \sqrt{1-(1-\varphi\cot\beta_0)^2} d\varphi \end{cases} \quad (7.45)$$

Summing these three equations gives the differential forces:

$$\begin{cases} dF_{ix} = -ut_i(\varphi_i)R_0 \cot\beta_0 (\sin\varphi_i + c_1\sqrt{1-(1-\varphi\cot\beta_0)^2}\cos\varphi_i \\ \qquad + c_2(1-\varphi\cot\beta_0)\cos\varphi_i) d\varphi \\ dF_{iy} = ut_i(\varphi_i)R_0 \cot\beta_0 (\cos\varphi_i - c_1\sqrt{1-(1-\varphi\cot\beta_0)^2}\sin\varphi_i \\ \qquad - c_2(1-\varphi\cot\beta_0)\sin\varphi_i) d\varphi \\ dF_{iz} = ut_i(\varphi_i)R_0 \cot\beta_0 (c_1(1-\varphi\cot\beta_0) \\ \qquad - c_2\sqrt{1-(1-\varphi\cot\beta_0)^2}) d\varphi \end{cases} \quad (7.46)$$

Using the same notes used for peripheral milling, Equation 7.46 becomes:

(a) For down milling:

---

[1] Equations 7.41-7.42 are valid for plane milling only.

$$\begin{cases} dF_{ix} = -u't_i(\varphi_i)R_0\cot\beta_0(\sin^{0.8}\varphi_i + (c_1\sqrt{1-(1-\varphi\cot\beta_0)^2} \\ \qquad + c_2(1-\varphi\cot\beta_0))\sin^{-0.2}\varphi_i\cos\varphi_i)d\varphi \\ dF_{iy} = u't_i(\varphi_i)R_0\cot\beta_0(\sin^{-0.2}\varphi_i\cos\varphi_i \\ \qquad - (c_1\sqrt{1-(1-\varphi\cot\beta_0)^2} + c_2(1-\varphi\cot\beta_0))\sin^{0.8}\varphi_i)d\varphi \\ dF_{iz} = u't_i(\varphi_i)R_0\cot\beta_0(c_1(1-\varphi\cot\beta_0) \\ \qquad - c_2\sqrt{1-(1-\varphi\cot\beta_0)^2})\sin^{-0.2}\varphi_i d\varphi \end{cases} \quad (7.47)$$

By applying Equation 7.39, and noting that $d\varphi_i = d\varphi$, Equation 7.47 becomes:

$$\begin{cases} dF_{ix} = -u'f_tR_0\cot\beta_0(\sin^{1.8}\varphi_i + (c_1\sqrt{1-(1-\varphi\cot\beta_0)^2} \\ \qquad + c_2(1-\varphi\cot\beta_0))\sin^{0.8}\varphi_i\cos\varphi_i)d\varphi_i \\ dF_{iy} = u'f_tR_0\cot\beta_0(\sin^{0.8}\varphi_i\cos\varphi_i \\ \qquad - (c_1\sqrt{1-(1-\varphi\cot\beta_0)^2} + c_2(1-\varphi\cot\beta_0))\sin^{1.8}\varphi_i)d\varphi_i \\ dF_{iz} = u'f_tR_0\cot\beta_0(c_1(1-\varphi\cot\beta_0) \\ \qquad - c_2\sqrt{1-(1-\varphi\cot\beta_0)^2})\sin^{0.8}\varphi_i d\varphi_i \end{cases}$$

$$\begin{cases} \varphi_i = \varphi - \omega t_{ime} + (i-1)\dfrac{2\pi}{m} \\ 0 \le \varphi_i \le \Omega \end{cases} \quad (7.48)$$

(b) For up milling:

$$\begin{cases} dF_{ix} = -u't_i(\varphi_i)R_0\cot\beta_0(\sin\varphi_i\sin^{-0.2}(-\varphi_i) + (c_1\sqrt{1-(1-\varphi\cot\beta_0)^2} \\ \qquad + c_2(1-\varphi\cot\beta_0))\sin^{-0.2}(-\varphi_i)\cos\varphi_i)d\varphi \\ dF_{iy} = u't_i(\varphi_i)R_0\cot\beta_0(\sin^{-0.2}(-\varphi_i)\cos\varphi_i \\ \qquad - (c_1\sqrt{1-(1-\varphi\cot\beta_0)^2} + c_2(1-\varphi\cot\beta_0))\sin\varphi_i\sin^{-0.2}(-\varphi_i))d\varphi \\ dF_{iz} = u't_i(\varphi_i)R_0\cot\beta_0(c_1(1-\varphi\cot\beta_0) \\ \qquad - c_2\sqrt{1-(1-\varphi\cot\beta_0)^2})\sin^{-0.2}(-\varphi_i)d\varphi \end{cases} \quad (7.49)$$

By applying Equation 7.40, and noting that $\sin(-\varphi_i) = -\sin\varphi_i$, Equation 7.49 becomes:

$$\begin{cases} dF_{ix} = u' f_t R_0 \cot\beta_0 (\sin^{1.8}(-\varphi_i) - (c_1\sqrt{1-(1-\varphi\cot\beta_0)^2} \\ + c_2(1-\varphi\cot\beta_0))\sin^{0.8}(-\varphi_i))\cos\varphi_i)d\varphi_i \\ dF_{iy} = u' f_t R_0 \cot\beta_0 (\sin^{0.8}(-\varphi_i)\cos\varphi_i \\ + (c_1\sqrt{1-(1-\varphi\cot\beta_0)^2} + c_2(1-\varphi\cot\beta_0))\sin^{1.8}(-\varphi_i))d\varphi_i \\ dF_{iz} = u' f_t R_0 \cot\beta_0 (c_1(1-\varphi\cot\beta_0) \\ - c_2\sqrt{1-(1-\varphi\cot\beta_0)^2})\sin^{0.8}(-\varphi_i)d\varphi \end{cases}$$

$$\begin{pmatrix} \varphi_i = \varphi - \omega t_{ime} + (i-1)\dfrac{2\pi}{m} \\ -\Omega \leq \varphi_i \leq 0 \end{pmatrix} \quad (7.50)$$

where $(1 \leq i \leq m, 0 \leq \varphi \leq \psi)$.

The total cutting force applied to the whole cutting edge is given by:

$$\begin{cases} F_{ix} = \int_{\varphi_s}^{\varphi_e} dF_{ix} d\varphi_i \\ F_{iy} = \int_{\varphi_s}^{\varphi_e} dF_{iy} d\varphi_i \\ F_{iz} = \int_{\varphi_s}^{\varphi_e} dF_{iz} d\varphi_i \end{cases} \quad (7.51)$$

where φ_s and φ_e are the lag angular locations of the start and end points of the contact of the cutting edge, and are defined in the following kinematics analysis.

(a) For down milling:

Because $0 \leq \varphi \leq \psi$, $\varphi_i = \varphi - \omega t_{ime} + (i-1)\dfrac{2\pi}{m}$ and $0 \leq \varphi_i \leq \Omega$ gives the extreme values of the parametric angle φ_i as:

$$\varphi_s = \max(0, -\omega t_{ime} + (i-1)\dfrac{2\pi}{m}) \quad (7.52)$$

$$\varphi_e = \min(\Omega, \psi - \omega t_{ime} + (i-1)\frac{2\pi}{m}) \qquad (7.53)$$

(b) For up milling:

$0 \leq \varphi \leq \psi$, $\varphi_i = \varphi - \omega t_{ime} + (i-1)\frac{2\pi}{m}$ and $-\Omega \leq \varphi_i \leq 0$ gives the extreme values of the parametric angle φ_i as:

$$\varphi_s = \max(-\Omega, -\omega t_{ime} + (i-1)\frac{2\pi}{m}) \qquad (7.54)$$

$$\varphi_e = \min(0, \psi - \omega t_{ime} + (i-1)\frac{2\pi}{m}) \qquad (7.55)$$

Summing up the cutting forces acting on all the m helical flutes gives the total force applied to the whole tool:

$$\begin{cases} F_x = \sum_{i=1}^{m} F_{ix} \\ F_y = \sum_{i=1}^{m} F_{iy} \\ F_z = \sum_{i=1}^{m} F_{iz} \end{cases} \qquad (7.56)$$

7.3.3 The Experimental Calibration of the Cutting Force Coefficients

Similarly, a one-tooth helical ball-end mill should be used in slot milling for the calibration of cutting force coefficients. Table 7.4 lists the cutting conditions and parameters of a set of slot millings.

Similarly the average cutting forces are used for the estimation of the cutting force coefficients. The theoretical average cutting force components can be obtained from Equations 7.47-7.56 as:

$$\begin{cases} \overline{F}_x = u_0 \sum_{i=1}^{m} \overline{f}_{ix1} + u_0 c_1 \sum_{i=1}^{m} \overline{f}_{ix2} + u_0 c_2 \sum_{i=1}^{m} \overline{f}_{ix3} \\ \overline{F}_y = u_0 \sum_{i=1}^{m} \overline{f}_{iy1} + u_0 c_1 \sum_{i=1}^{m} \overline{f}_{iy2} + u_0 c_2 \sum_{i=1}^{m} \overline{f}_{iy3} \\ \overline{F}_z = u_0 c_1 \sum_{i=1}^{m} \overline{f}_{iz1} + u_0 c_2 \sum_{i=1}^{m} \overline{f}_{iz2} \end{cases} \qquad (7.57)$$

where :

(a) For down milling:

$$f_{ix1} = \int_{\varphi_s}^{\varphi_e} -c_s \sin^{1.8} \varphi_i d\varphi_i$$

$$f_{ix2} = \int_{\varphi_s}^{\varphi_e} -c_s \sqrt{1-(1-\varphi\cot\beta_0)^2} \sin^{0.8} \varphi_i \cos\varphi_i d\varphi_i$$

$$f_{ix3} = \int_{\varphi_s}^{\varphi_e} -c_s(1-\varphi\cot\beta_0)\sin^{0.8} \varphi_i \cos\varphi_i d\varphi_i$$

$$f_{iy1} = \int_{\varphi_s}^{\varphi_e} c_s \sin^{0.8} \varphi_i \cos\varphi_i d\varphi_i$$

$$f_{iy2} = \int_{\varphi_s}^{\varphi_e} -c_s \sqrt{1-(1-\varphi\cot\beta_0)^2} \sin^{1.8} \varphi_i d\varphi_i$$

$$f_{iy3} = \int_{\varphi_s}^{\varphi_e} -c_s(1-\varphi\cot\beta_0)\sin^{1.8} \varphi_i d\varphi_i$$

$$f_{iz1} = \int_{\varphi_s}^{\varphi_e} c_s(1-\varphi\cot\beta_0)\sin^{0.8} \varphi_i d\varphi_i$$

$$f_{iz2} = \int_{\varphi_s}^{\varphi_e} -c_s \sqrt{1-(1-\varphi\cot\beta_0)^2} \sin^{0.8} \varphi_i d\varphi_i$$

$$c_s = (1 - \frac{\alpha_e - \alpha_{e0}}{100})(\frac{t_0}{f_t})^{0.2} f_t R \cot\beta$$

(b) For up milling:

$$f_{ix1} = \int_{\varphi_s}^{\varphi_e} c_s \sin^{1.8}(-\varphi_i) d\varphi_i$$

$$f_{ix2} = \int_{\varphi_s}^{\varphi_e} -c_s \sqrt{1-(1-\varphi\cot\beta_0)^2} \sin^{0.8}(-\varphi_i) \cos\varphi_i d\varphi_i$$

$$f_{ix3} = \int_{\varphi_s}^{\varphi_e} -c_s(1-\varphi\cot\beta_0)\sin^{0.8}(-\varphi_i)\cos\varphi_i d\varphi_i$$

$$f_{iy1} = \int_{\varphi_s}^{\varphi_e} c_s \sin^{0.8}(-\varphi_i)\cos\varphi_i d\varphi_i$$

$$f_{iy2} = \int_{\varphi_s}^{\varphi_e} c_s \sqrt{1-(1-\varphi\cot\beta_0)^2} \sin^{1.8}(-\varphi_i) d\varphi_i$$

$$f_{iy3} = \int_{\varphi_s}^{\varphi_e} c_s (1-\varphi\cot\beta_0) \sin^{1.8}(-\varphi_i) d\varphi_i$$

$$f_{iz1} = \int_{\varphi_s}^{\varphi_e} c_s (1-\varphi\cot\beta_0) \sin^{0.8}(-\varphi_i) d\varphi_i$$

$$f_{iz2} = \int_{\varphi_s}^{\varphi_e} -c_s \sqrt{1-(1-\varphi\cot\beta_0)^2} \sin^{0.8}(-\varphi_i) d\varphi_i$$

$$c_s = (1 - \frac{\alpha_e - \alpha_{e0}}{100})(\frac{t_0}{f_t})^{0.2} f_t R \cot\beta$$

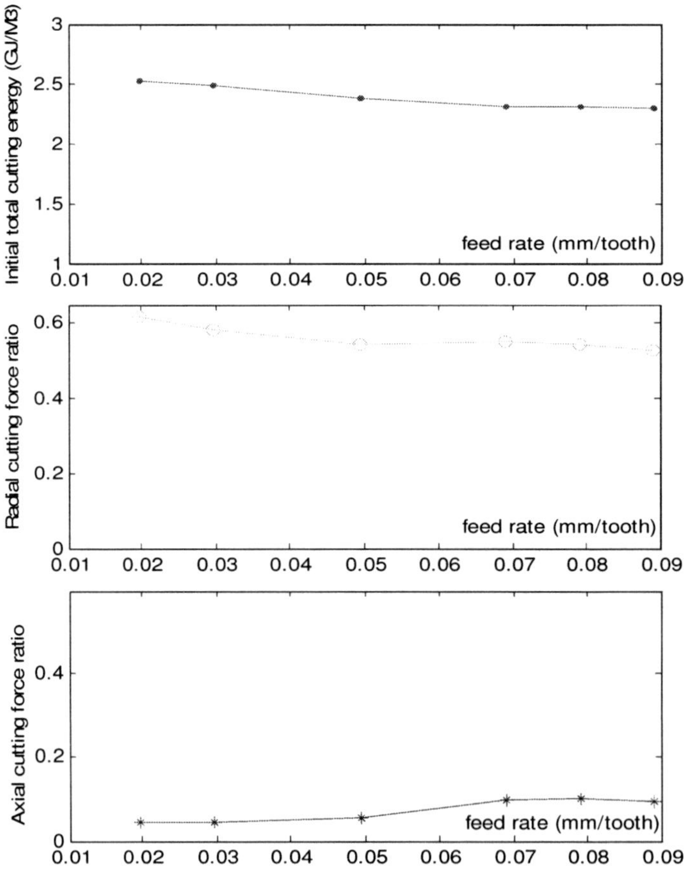

Figure 7.12. Experimental calibrated cutting force coefficients (by slot milling)

Table 7.4. Cutting conditions and parameters (slot milling)

Tool: a single fluted solid carbide ball End-mill, $R = 10$ mm, $\beta = 30°$, $\alpha_r = 5°$
Work material: carbon steel EN8
Cutting condition: with fluid
Cutting parameters: spindle rotation speed n =1,114 rpm (cutting speed $v = 70$ m/min), feed rate f_t in mm/tooth, axial depth of cut b_a in mm

$f_t = 0.0197$	$b_a = 9.95$
$f_t = 0.0296$	$b_a = 9.95$
$f_t = 0.0494$	$b_a = 9.95$
$f_t = 0.0691$	$b_a = 9.95$
$f_t = 0.0790$	$b_a = 9.95$
$f_t = 0.0889$	$b_a = 9.95$

Averaging the cutting forces from the measured dynamic ones and substituting them in Equation 7.57 can yield the cutting force coefficients u_0, c_1 and c_2. Table 7.5 and Figure 7.12 show the calibrated cutting force coefficients, in which u_0 varies between 2.2972×10^9 and 2.5215×10^9 J/m^3, c_1 between 0.5293 and 0.6155, and c_2 between 0.0453 and 0.1009.

Table 7.5. Experimental calibrated cutting force coefficients (by slot milling)

f_t (mm/toot)	0.0197	0.0296	0.0494	0.0691	0.0790	0.0889
u_0 (G J/m3)	2.5215	2.4860	2.3759	2.3075	2.3045	2.2972
c_1	0.6155	0.5842	0.5422	0.5515	0.5443	0.5293
c_2	0.0464	0.0453	0.0561	0.0999	0.1009	0.0957

Figure 7.13 shows the measured and predicted dynamic cutting force components in the x, y and z directions, in which the predicted values are obtained from the cutting force model using the calibrated force coefficients.

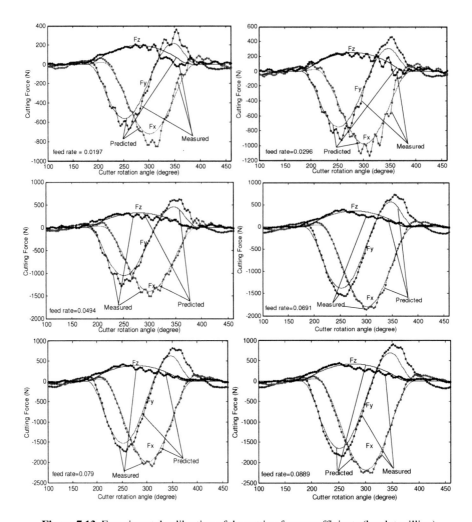

Figure 7.13. Experimental calibration of the cutting force coefficients (by slot milling)

7.3.4 A Case Study: Verification

A set of down-hill slope plane milling tests was carried out, as shown in Figure 7.14. Table 7.6 lists the corresponding cutting method and parameters, and the refined cutting force coefficients from the measured dynamic cutting forces. Figure 7.15 shows the measured and the calculated dynamic cutting forces.

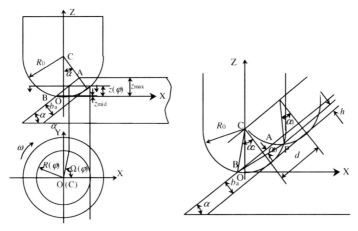

Figure 7.14. Down-hill slope plane milling

Table 7.6. Cutting methods and parameters (down-hill slope plane milling)

Spindle speed n = 2,228 rpm (cutting speed v = 70 m/min), d = 2 mm					
Test No.	1	2	3	4	5
f_t (mm/tooth)	0.035	0.0494	0.0494	0.0592	0.0592
b_a (mm)	2	2	2	2	2
Cutting method	up milling	down milling	up milling	down milling	up milling

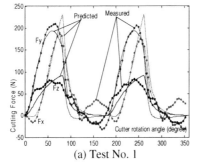

(a) Test No. 1

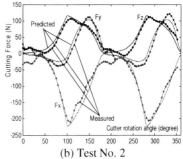

(b) Test No. 2

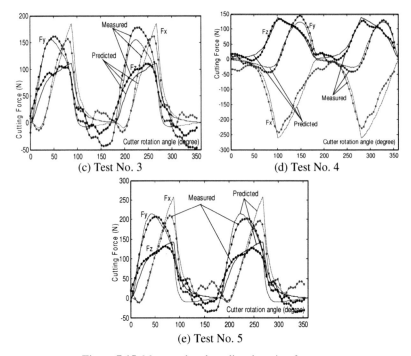

Figure 7.15. Measured and predicted cutting forces

7.4 A Machining Dynamics Model

Here we use peripheral milling as the case for machining dynamics modelling.

7.4.1 A Modularisation of the Cutting Force

For modelling machining dynamics, it is essential to simplify and modularise the cutting force model.

Integrating Equations 7.25-7.26 between the extreme values of the parametric angle φ_s and φ_e gives the total cutting force applied on the ith tooth below. Please note that we deal with only 2D machining dynamics, and therefore the cutting force in z direction is ignored.

(a) For down milling:

$$\begin{cases} F_{ix} \approx -u' f_t a_{ixy} - u' a_{ixx} (\Delta x_c - \Delta x_w) - u' a_{ixy} (\Delta y_c - \Delta y_w) - u' \delta_i(\varphi_i) a_{ix} \\ F_{iy} \approx u' f_t a_{iyy} + u' a_{iyx} (\Delta x_c - \Delta x_w) + u' a_{iyy} (\Delta y_c - \Delta y_w) + u' \delta_i(\varphi_i) a_{iy} \end{cases} \quad (7.58)$$

where $\Delta x_c = x_c - x_{c(-T)}$, $\Delta y_c = y_c - y_{c(-T)}$, $\Delta x_w = x_w - x_{w(-T)}$,

$$\Delta y_w = y_w - y_{w(-T)}$$

$$a_{ixx} = (0.5556\sin^{1.8}\varphi_i + 0.5c\varphi_i + 0.25c_1\sin 2\varphi_i)\Big|_{\varphi_s}^{\varphi_e}$$

$$a_{ixy} = (0.5\varphi_i - 0.25\sin 2\varphi_i + 0.5556c_1\sin^{1.8}\varphi_i)\Big|_{\varphi_s}^{\varphi_e}$$

$$a_{iyx} = (0.5\varphi_i + 0.25\sin 2\varphi_i - 0.5c_1\varphi_i + 0.25c_1\sin 2\varphi_i)\Big|_{\varphi_s}^{\varphi_e}$$

$$a_{iyy} = (0.5556\sin^{1.8}\varphi_i - 0.5c_1\varphi_i + 0.25c_1\sin 2\varphi_i)\Big|_{\varphi_s}^{\varphi_e}$$

$$a_{ix} = (-\cos\varphi_i + 1.25c_1\sin^{0.8}\varphi_i)\Big|_{\varphi_s}^{\varphi_e} \quad a_{iy} = (1.25\sin^{0.8}\varphi_i + c_1\cos\varphi_i)\Big|_{\varphi_s}^{\varphi_e}$$

Because $0 \leq \varphi \leq \psi$, $\varphi_i = \varphi - \omega t_{ime} + (i-1)\dfrac{2\pi}{m}$ and $0 \leq \varphi_i \leq \Omega$ gives the extreme values of the parametric angle φ_i as

$$\varphi_s = \max(0, -\omega t_{ime} + (i-1)\frac{2\pi}{m}) \tag{7.59}$$

$$\varphi_e = \min(\Omega, \psi - \omega t_{ime} + (i-1)\frac{2\pi}{m}) \tag{7.60}$$

(b) For up milling:

$$\begin{cases} F_{ix} \approx -u'f_t a_{ixy} + u'a_{ixx}(\Delta x_c - \Delta x_w) + u'a_{ixy}(\Delta y_c - \Delta y_w) + u'\delta_i(\varphi_i)a_{ix} \\ F_{iy} \approx -u'f_t a_{iyy} + u'a_{iyx}(\Delta x_c - \Delta x_w) + u'a_{iyy}(\Delta y_c - \Delta y_w) - u'\delta_i(\varphi_i)a_{iy} \end{cases} \tag{7.61}$$

where:

$$a_{ixx} = (-0.5556\sin^{1.8}\xi_i + 0.5c_1\xi_i + 0.25c_1\sin 2\xi_i)\Big|_{\xi_s}^{\xi_e}$$

$$a_{ixy} = (0.5\xi_i - 0.25\sin 2\xi_i - 0.5556c_1\sin^{1.8}\xi_i)\Big|_{\xi_s}^{\xi_e}$$

$$a_{iyx} = (-0.5\xi_i - 0.25\sin 2\xi_i - 0.5556c_1\sin^{1.8}\xi_i)\Big|_{\xi_s}^{\xi_e}$$

$$a_{iyy} = (0.5556\sin^{1.8}\xi_i + 0.5c\xi_i - 0.25c_1\sin 2\xi_i)\Big|_{\xi_s}^{\xi_e}$$

$$a_{ix} = (\cos\xi_i + 1.25c_1\sin^{0.8}\xi_i)\Big|_{\xi_s}^{\xi_e} \quad a_{iy} = (1.25\sin^{0.8}\xi_i - c_1\cos\xi_i)\Big|_{\xi_s}^{\xi_e},$$

And:
$$\xi_i = -\varphi_i.$$

Also, $\xi_i = -\varphi + \omega t_{ime} - (i-1)\dfrac{2\pi}{m}$ and $0 \le \xi_i \le \Omega$ gives the extreme values of the parametric angle ξ_i as

$$\xi_s = \max(0, -\psi + \omega t_{ime} - (i-1)\frac{2\pi}{m}) \tag{7.62}$$

$$\xi_e = \min(\Omega, \omega t_{ime} - (i-1)\frac{2\pi}{m}) \tag{7.63}$$

Summing up the cutting forces acting on all the m helical flutes gives the total force applied on the whole tool:

$$\begin{cases} F_x(t_{ime}) = \sum_{i=1}^{m} F_{ix} \\ F_y(t_{ime}) = \sum_{i=1}^{m} F_{iy} \end{cases} \tag{7.64}$$

(a) For down milling:

$$\begin{Bmatrix} F_x(t_{ime}) \\ F_y(t_{ime}) \end{Bmatrix} = \begin{bmatrix} -a_{xy}(t_{ime}) \\ a_{yy}(t_{ime}) \end{bmatrix} f_t + \begin{bmatrix} -a_{xx}(t_{ime}) & -a_{xy}(t_{ime}) \\ a_{yx}(t_{ime}) & a_{yy}(t_{ime}) \end{bmatrix} \begin{bmatrix} \Delta x(t_{ime}) \\ \Delta y(t_{ime}) \end{bmatrix}$$
$$+ \begin{bmatrix} -a_x(t_{ime}) \\ a_y(t_{ime}) \end{bmatrix} \delta_i(\varphi_i) \tag{7.65}$$

(b) For up milling:

$$\begin{Bmatrix} F_x(t_{ime}) \\ F_y(t_{ime}) \end{Bmatrix} = -\begin{bmatrix} a_{xy}(t_{ime}) \\ a_{yy}(t_{ime}) \end{bmatrix} f_t + \begin{bmatrix} a_{xx}(t_{ime}) & a_{xy}(t_{ime}) \\ a_{yx}(t_{ime}) & a_{yy}(t_{ime}) \end{bmatrix} \begin{bmatrix} \Delta x(t_{ime}) \\ \Delta y(t_{ime}) \end{bmatrix} \\ + \begin{bmatrix} a_x(t_{ime}) \\ -a_y(t_{ime}) \end{bmatrix} \delta_i(\varphi_i) \quad (7.66)$$

where $a_{xx} = u'\sum_{i=1}^{m} a_{ixx}$, $a_{xy} = u'\sum_{i=1}^{m} a_{ixy}$, $a_{yx} = u'\sum_{i=1}^{m} a_{iyx}$, $a_{yy} = u'\sum_{i=1}^{m} a_{iyy}$

$a_x = u'\sum_{i=1}^{m} a_{ix}$, $a_y = u'\sum_{i=1}^{m} a_{iy}$.

7.4.2 Machining Dynamics Modelling

The regenerative dynamic displacements of the tool and workpiece can be expressed in the frequency domain as a product of the transfer function of the structure and the dynamic cutting forces [9]:

$$\begin{Bmatrix} \Delta x_c(i\omega) \\ \Delta y_c(i\omega) \end{Bmatrix} = (1 - e^{-i\omega T}) \begin{bmatrix} G_{xx}^c(i\omega) & G_{xy}^c(i\omega) \\ G_{yx}^c(i\omega) & G_{yy}^c(i\omega) \end{bmatrix} \begin{Bmatrix} F_x(i\omega) \\ F_y(i\omega) \end{Bmatrix} \quad (7.67)$$

$$\begin{Bmatrix} \Delta x_w(i\omega) \\ \Delta y_w(i\omega) \end{Bmatrix} = (1 - e^{-i\omega T}) \begin{bmatrix} G_{xx}^w(i\omega) & G_{xy}^w(i\omega) \\ G_{yx}^w(i\omega) & G_{yy}^w(i\omega) \end{bmatrix} \begin{Bmatrix} F_x(i\omega) \\ F_y(i\omega) \end{Bmatrix} \quad (7.68)$$

where T is the tooth passing period, G_{xx}^c, G_{xy}^c, G_{yx}^c and G_{yy}^c are the direct and cross transfer functions of the tool in the X (normal) and Y (feed) directions, and G_{xx}^w, G_{xy}^w, G_{yx}^w and G_{yy}^w are the direct and cross transfer functions of the workpiece in the Y (normal) and Y (feed) directions. The regenerative displacements $[\Delta x_c, \Delta y_c]^T$, $[\Delta x_w, \Delta y_w]^T$ and the dynamic cutting forces $[F_x, F_y]$ have a closed loop interaction as illustrated in Figure 7.16. All of them grow exponentially when the operating conditions are unstable, which leads to chatter vibrations.

Experimental results reveal that the measured cross transfer functions were negligible, i.e., $G_{xy} = G_{yx} = 0$ [3]. So, the frequency domain response function in the Laplace domain can be expressed as:

$$\begin{cases} \Delta x_c(s) = (1-e^{-sT})G_{xx}^c(s)F_x(s) \\ \Delta y_c(s) = (1-e^{-sT})G_{yy}^c(s)F_y(s) \end{cases} \quad (7.69)$$

$$\begin{cases} \Delta x_w(s) = (1-e^{-sT})G_{xx}^w(s)F_x(s) \\ \Delta y_w(s) = (1-e^{-sT})G_{yy}^w(s)F_y(s) \end{cases} \quad (7.70)$$

$$\begin{Bmatrix} \Delta x(s) \\ \Delta y(s) \end{Bmatrix} = \begin{Bmatrix} \Delta x_c(s) - \Delta x_w(s) \\ \Delta y_c(s) - \Delta y_w(s) \end{Bmatrix}$$
$$= (1-e^{-sT}) \begin{bmatrix} G_{xx}^c(s) - G_{xx}^w(s) & 0 \\ 0 & G_{yy}^c(s) - G_{yy}^w(s) \end{bmatrix} \begin{Bmatrix} F_x(s) \\ F_y(s) \end{Bmatrix} \quad (7.71)$$

The spring-damper vibratory model in Figure 7.3 reveals that the transfer functions G_{xx}^c and G_{yy}^c of the tool can be expressed as:

$$G_{xx}^c(s) = \frac{x_c(s)}{F_x(s)} = \frac{\omega_{ncx}^2}{k_{cx}(s^2 + 2\zeta_{cx}\omega_{ncx}s + \omega_{ncx}^2)} \quad (7.72)$$

$$G_{yy}^c(s) = \frac{y_c(s)}{F_y(s)} = \frac{\omega_{ncy}^2}{k_{cy}(s^2 + 2\zeta_{cy}\omega_{ncy}s + \omega_{ncy}^2)} \quad (7.73)$$

where s is the Laplace operator, ω_n and ζ are the natural frequency and structural damping ratio respectively, and $\omega_{ncx} = \sqrt{k_{cx}/m_{cx}}$, $\zeta_{cx} = 0.5c_{cx}/\sqrt{k_{cx}/m_{cx}}$, $\omega_{ncy} = \sqrt{k_{cy}/m_{cy}}$, $\zeta_{cy} = 0.5c_{cy}/\sqrt{k_{cy}m_{cy}}$.

Similarly, the transfer functions G_{xx}^w and G_{yy}^w of the tool can be expressed as:

$$G_{xx}^w(s) = \frac{x_w(s)}{F_x(s)} = \frac{\omega_{nwx}^2}{k_{wx}(s^2 + 2\zeta_{wx}\omega_{nwx}s + \omega_{nwx}^2)} \quad (7.74)$$

$$G_{yy}^w(s) = \frac{y_w(s)}{F_y(s)} = \frac{\omega_{nwy}^2}{k_{wy}(s^2 + 2\zeta_{wy}\omega_{nwy}s + \omega_{nwy}^2)} \quad (7.75)$$

However, the practical cutting process usually has a multi-degree of freedom and the measured transfer functions can be represented in the form:

$$G_{xx}^c(s) = \sum_{j=1}^{l_{cx}} \frac{\alpha_{cxj}s + \beta_{cxj}}{s^2 + 2\zeta_{cxj}\omega_{ncxj}s + \omega_{ncxj}^2} \quad (7.76)$$

$$G_{yy}^{c}(s) = \sum_{j=1}^{l_{cy}} \frac{\alpha_{cyj} s + \beta_{cyj}}{s^2 + 2\zeta_{cyj}\omega_{ncyj} s + \omega_{ncyj}^2} \qquad (7.77)$$

$$G_{xx}^{w}(s) = \sum_{j=1}^{l_{wx}} \frac{\alpha_{wxj} s + \beta_{wxj}}{s^2 + 2\zeta_{wxj}\omega_{nwxj} s + \omega_{nwxj}^2} \qquad (7.78)$$

$$G_{yy}^{w}(s) = \sum_{j=1}^{l_{wy}} \frac{\alpha_{wyj} s + \beta_{wyj}}{s^2 + 2\zeta_{wyj}\omega_{nwyj} s + \omega_{nwyj}^2} \qquad (7.79)$$

where l is the mode number, and α_j and β_j are the corresponding modal parameters of mode j.

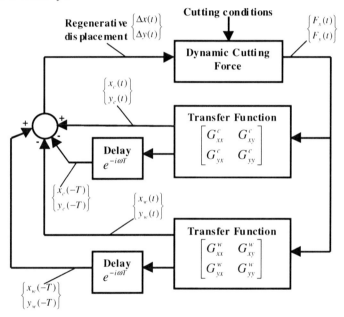

Figure 7.16. Block diagram of regenerative chatter in peripheral milling

7.4.3 The Surface Generation Model

The surface machined in the cutting process of peripheral milling is the envelop surface of all the traces of the moving teeth relative to the workpiece. Therefore, from the viewpoint of mechanistic movement, the generation of the envelop surface should take account of the geometric information of the tool (such as the diameter of the tool, the teeth number, the helix angle and the dimensional errors of each tooth), the cutter run-out (its value and initial location angle of the tool centre relative to the spindle centre), and the vibrations of the tool and workpiece. Figure 7.17 shows the schematic principle of the generation of the surface machined.

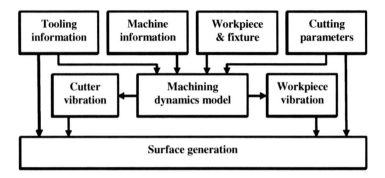

Figure 7.17. Schematic principle of surface generation

In order to generate the machined surface, a few section profiles of the surface, which are perpendicular to the tool axis, should be generated during the simulation. A surface profile is the envelope curve of the traces of all the teeth of the tool along the feed direction, as shown in the example in Figure 7.18, which has three teeth and a measure of cutter run-out.

The trace of tooth i of a specific section, which is defined by the helix lag angle φ, is a trochoidal curve and can be calculated using:

$$\begin{cases} X_i(t_{ime}) = R_i \cos(\varphi_i) + x_c - x_w \\ Y_i(t_{ime}) = R_i \sin(\varphi_i) + \dfrac{f_t m}{2\pi} \omega t_{ime} + y_c - y_w \end{cases} \quad \left(\begin{array}{c} i = 1, 2, \cdots, m \\ \varphi_i = \varphi - \omega t_{ime} + (i-1)\dfrac{2\pi}{m} \end{array} \right) \quad (7.80)$$

where f_t is the feedrate, m is the teeth number of the tool, ω is the spindle rotation angular velocity, x_c, y_c and x_w, y_w are the regenerative displacements of the tool and workpiece, respectively, and R_i is the actual cutting radius of the tooth i [6],

$$R_i = |\overline{O'P}| = \sqrt{R^2 + \delta_e^2 - 2R\delta_e \cos(\angle O'OP)} \quad (7.81)$$

The envelope curve:

$$X = X_{min}(Y) \quad (7.82)$$

of all the traces of the teeth can be calculated using the Z-Buffer scanning algorithm of computer graphics. The dimensional error of the corresponding section profile of the machined surface can be evaluated with:

$$\delta X(Y) = R - X_{\min}(Y) \tag{7.83}$$

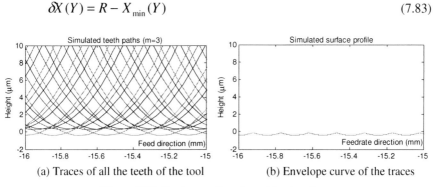

(a) Traces of all the teeth of the tool (b) Envelope curve of the traces

Figure 7.18. Section profile of machined surface

7.4.4 The Simulation Model

A simulation model is developed with Simulink based on the closed loop machining dynamics model as shown in Figure 7.16 integrated with the surface generation mechanism, and the simulation interface is shown in Figure 7.19.

7.5 The Modal Analysis of the Machining System

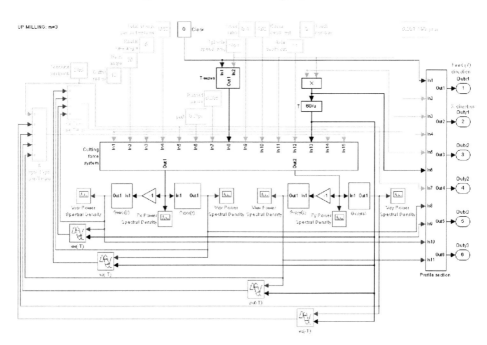

Figure 7.19. Simulation model (interface) of machining dynamics in peripheral milling

The reliability of the simulation results of the proposed machining dynamics model depends not only on an accurate dynamic cutting force model, but also on an accurate estimation of the transfer functions of the machining system. Basically there are two methods for determining the transfer functions of a mechanical system, such as the machining system. They are the analytical method, which is commonly practiced such as using finite element (FE) modelling, and the experimental method, namely experimental modal analysis. Here we are going to apply the experimental modal analysis to determine the modal parameters of the machining system.

7.5.1 The Mathematical Principle of Experimental Modal Analysis

The machining system of peripheral milling is considered to be a multi-degree-of-freedom vibration system, and its transfer functions can be identified by experimental modal analysis.

The most effective excitation instruments are electromagnetic or electro-hydraulic shakes. They are capable of delivering force over a wide frequency band containing the dominant natural modes of the tested structure. Alternatively, an impact hammer instrumented with a piezoelectric force transducer can be used. Figure 7.20 shows an experimental setup for the experimental modal analysis of a machining system. The setup includes an instrumented hammer, which is used to excite the tool and workpiece, and accelerometers, which measure the vibrations. The accelerometers and the force transducer are connected to the Data Physics signal processor, SignalCalc Mobilyzer, to calculate the transfer functions of the tool and workpiece. The transfer functions are then processed by the modal analysis software, STAR, to identify the modal parameters of the tool and workpiece [9].

(a) For the tool   (b) For the workpiece

Figure 7.20. Setup of experimental modal analysis

The modal analysis software, STAR, provides the modal parameters of the transfer functions with the following residue notation[2]:

$$h_k(s) = \frac{r_k}{2j(s-p_k)} + \frac{r_k^*}{2j(s-p_k^*)} \quad (7.84)$$

where n is the mode number, $p_k = -\zeta_k \omega_{n,k} + j\omega_{d,k}$ and $p_k^* = -\zeta_k \omega_{n,k} - j\omega_{d,k}$ are the pole locations in the s-plane for the kth mode, $r_k = \sigma_k + j\upsilon_k$ and $r_k^* = \sigma_k - j\upsilon_k$ are the complex residues of the kth mode.

Rewriting the transfer function in terms of its standard Laplace domain as:

$$h_k(s) = \frac{\alpha_k + \beta_k s}{s^2 + 2\zeta_k \omega_{n,k} s + \omega_{n,k}^2} \quad (7.85)$$

releases $\alpha_k = j(-\zeta_k \omega_{n,k} \sigma_k + \upsilon_k \omega_{d,k})$, $\beta_k = -j\sigma_k$. (Note: $\omega_{d,k} = \omega_{n,k}\sqrt{1-\zeta_k^2}$)

Note that the measurements are based on the frequency unit Hz, while the required frequency unit for the machining dynamics model is rad/s. Therefore, all the frequencies, ω_n and ω_d, must be scaled to rad/s by times 2π for each mode. The scaling operation must also be applied to the identified residues so that the transfer functions remain the same amplitude.

Attention must also be paid to the unit of the residues. The modal parameters are identified from the measurements made with accelerometers, and the residues are in accelerator unit m/s².N. For the machining dynamics model, the residues must be divided by $-\omega_d^2$ and scaled to displacement unit m/N for each mode [9].

7.5.2 A Case Study

For industrial applications, it is not practical to identify the transfer functions of all the tools, which have common characters with same diameter and length. It is reasonable to assume that they have the same transfer functions, because the determined modal parameters by the experimental modal analysis are only an approximation of the real values. Strictly speaking, the modal parameters will not remain the same for different tools. Even for the same tool, the modal parameters may vary if the tool is refitted to the chuck or the chuck remounted on the spindle head. Moreover, because the damping condition changes instantly during the cutting process due to the unstable friction conditions between the tool and the workpiece/chip, the modal parameters will change instantly and the difference may

---

[2] The STAR Users' Guide, Part 4 - Theory and Applications, Chapter 1 - Modal Analysis Theory of Operations.

be severe when chatter takes place. Therefore we will focus on the similarity of the main features of the cutting forces, the vibrations of the tool and workpiece, and the surface characters between the simulations and measurements.

The experimental work was performed on the three axis vertical CNC machine centre, Cincinnati Arrow2-500. A helical solid carbide end mill with 3 flutes, a 20 mm diameter, a 30° helix angle and 5° radial rake angle, which was fitted in a hydraulic chuck and mounted on the spindle head of the CNC machine centre, was used for peripheral milling of a workpiece of carbon steel EN8. The workpiece was fixed by the Microloc fixture system, Kit-75, which was mounted on the Kistler table dynamometer, 9257BA, which in turn was mounted on the worktable of the CNC machine centre. An impact hammer instrumented with a piezoelectric force transducer was used to hit the tool and workpiece in order to excite vibrations of the structures; accelerometers were mounted on the tool and workpiece to measure the vibrations, as shown in Figure 7.20. The accelerometers and the force transducer were connected to the Data Physics signal processor, a SignalCalc Mobilyzer, which is actually a Fourier analyzer, to calculate the transfer functions of the tool and workpiece. The transfer functions were then processed by the modal analysis software, STAR, to identify the modal parameters of the tool and workpiece.

7.5.1.1 The Modal Parameters of the Tool
Tables 7.7 and 7.8 list the modal parameters of the tool, which were generated from the modal analysis software and scaled to the units required by the machining dynamics model.

Table 7.7. Identified modal parameters of the tool in X direction

Mode	ω_{dcx} (Hz)	ω_{ncx} (rad/s)	ζ_{cx} (%)	residue $\sigma_{cx} + j\upsilon_{cx}$ (m/N)
1	120.5	757.7963	4.21	8.031267e-006+j1.830470e-006
2	247.05	1553.234	3.54	3.442107e-006+j5.209837e-007
3	341.65	2146.744	0.93693	9.108225e-006+j-1.908909e-006
4	645.48	4055.685	0.26558	8.633025e-007+j-7.830841e-007
5	715.39	4495.076	0.81218	-1.057337e-005+j-7.500871e-006
6	731.75	4597.741	0.29715	1.022476e-006+j-1.673413e-006
7	748.76	4704.604	0.16347	1.016291e-006+j-2.033546e-007
8	761.57	4785.126	0.41044	2.639827e-006+j2.095946e-007
9	811.46	5098.814	1.01	4.304769e-006+j4.423204e-007
10	987.15	6206.627	3.67	3.820509e-005+j3.567028e-006
11	1030	6471.883	0.789	9.331169e-007+j2.193275e-006
12	1160	7288.820	0.94237	-2.194058e-006+j-1.615678e-006
13	1200	7539.899	0.044894	-7.935642e-007+j2.950998e-007

Table 7.8. Identified modal parameters of the tool in Y direction

Mode	ω_{dcy} (Hz)	ω_{ncy} (rad/s)	ζ_{cy} (%)	residue $\sigma_{cy} + j\upsilon_{cy}$ (m/N)
1	49.21	309.1962	0.20605	-1.733888e-008+j1.712200e-008
2	94.72	595.1473	0.36509	2.465764e-008+j-6.332934e-008
3	160.38	1007.997	2.44	-4.392552e-007+j1.364358e-007
4	187.55	1178.641	1.97	2.269027e-006+j-1.301338e-006
5	238.66	1500.043	2.58	1.201795e-006+j-7.724067e-007
6	344.69	2166.074	1.72	3.348905e-006+j-5.251082e-006
7	439.92	2764.451	1.6	-7.110383e-007+j2.433423e-008
8	488.25	3067.772	0.218	-2.326689e-008+j8.504934e-008
9	586.86	3687.390	0.46732	1.274008e-007+j1.548505e-007
10	684.19	4299.626	1.85	2.273516e-005+j-9.893721e-006
11	751.8	4724.287	1.58	1.434414e-005+j-3.420460e-005
12	894.23	5620.215	2.39	-3.932861e-006+j-1.978372e-006

By substituting $\alpha = -\zeta\omega_n\sigma + \omega_d\upsilon$ and $\beta = -\sigma$ (see appendix), the transfer functions of the tool are obtained:

$$G_{xx}^c = \sum_{k=1}^{n} \frac{\alpha_{cx,k} + \beta_{cx,k}s}{s^2 + 2\zeta_{cx,k}\omega_{ncx,k}s + \omega_{ncx,k}^2} \quad (7.86)$$

$$G_{yy}^c = \sum_{k=1}^{n} \frac{\alpha_{cy,k} + \beta_{cy,k}s}{s^2 + 2\zeta_{cy,k}\omega_{ncy,k}s + \omega_{ncy,k}^2} \quad (7.87)$$

By replacing $s = j\omega$ and sweeping all frequencies of interest, the curve-fitted transfer functions of the tool can be reconstructed, as shown in Figure 7.21.

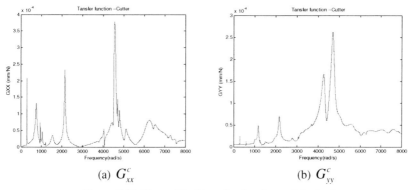

Figure 7.21. Curve-fitted transfer functions of the tool

The results show that the frequency characteristics of the transfer functions of the tool in both directions are similar, but with different amplitudes. The stiffness in X direction is weaker than that in Y direction, because the spindle is strongly supported by the machine's column in Y direction.

7.5.1.2 The Modal Parameters of the Workpiece

Similar to the modal parameters of the tool, Tables 7.9 and 7.10 list the modal parameters of the workpiece.

Table 7.9. Identified modal parameters of the workpiece in X direction

Mode	ω_{dwx} (Hz)	ω_{nwx} (rad/s)	ζ_{wx} (%)	residue $\sigma_{wx} + j\upsilon_{wx}$ (m/N)
1	43.44	273.0853	3.25	-1.258373e-006+j-5.699113e-008
2	79.39	506.8606	17.73	-3.764000e-006+j3.820564e-007
3	148.75	935.5075	4.34	-1.309762e-006+j-1.194027e-007
4	212.22	1333.515	1.21	-4.555126e-008+j1.650306e-008
5	330	2077.282	6.07	-3.609850e-006+j-1.209621e-006
6	383.6	2410.231	0.10927	-1.287093e-009+j9.626157e-009
7	398.76	2505.493	0.28086	-3.883548e-009+j5.595112e-008
8	419.8	2637.722	0.558	-2.171050e-008+j-6.944830e-009
9	454.45	2855.452	0.63879	-1.301600e-008+j7.290193e-009
10	484.83	3046.894	2.01	-4.375980e-008+j3.178899e-008
11	537.7	3378.544	0.6671	2.516786e-008+j1.832542e-008
12	575.21	3614.285	0.86247	6.845461e-008+j-2.399351e-008
13	637	4002.405	0.27934	4.345913e-009+j-1.189243e-008
14	654.58	4117.619	4.81	1.140337e-006+j1.759127e-007
15	703.16	4418.09	0.16402	2.009257e-011+j4.120233e-010
16	698.58	4401.554	7.45	5.674624e-007+j-6.033365e-007
17	839.54	5275.109	0.68471	-1.492584e-009+j-1.441823e-010

Table 7.10. Identified modal parameters of the tool in Y direction

Mode	ω_{dcy} (Hz)	ω_{ncy} (rad/s)	ζ_{cy} (%)	residue $\sigma_{cy} + j\upsilon_{cy}$ (m/N)
1	46.86	294.4607	1.44	3.730523e-006+j-9.531062e-007
2	164.19	1032.036	2.78	2.529632e-006+j-1.402137e-007
3	393.51	2472.770	1.49	-5.422878e-007+j2.234437e-008
4	494.99	3110.123	-0.23898	-2.573799e-007+j-7.380437e-008
5	555.13	3487.985	-0.06252	4.213739e-008+j-8.862332e-009
6	671.28	4218.528	1.89	2.500610e-006+j9.253670e-007
7	781.81	4912.594	1.17	-1.036336e-006+j-1.012902e-006
8	859.35	5400.222	1.68	4.027783e-008+j-1.915551e-007
9	1010	6346.330	0.99336	-2.449498e-007+j2.901953e-007
10	1090	6850.590	2.36	-8.814406e-007+j5.726684e-009

Similarly, we can get the transfer functions of the workpiece:

$$G_{xx}^w = \sum_{k=1}^{n} \frac{\alpha_{wx,k} + \beta_{wx,k} s}{s^2 + 2\zeta_{wx,k}\omega_{nwx,k} s + \omega_{nwx,k}^2} \quad (7.88)$$

$$G_{yy}^w = \sum_{k=1}^{n} \frac{\alpha_{wy,k} + \beta_{wy,k} s}{s^2 + 2\zeta_{wy,k}\omega_{nwy,k} s + \omega_{nwy,k}^2} \quad (7.89)$$

and the curve-fitted transfer functions of the workpiece, as shown in Figure 7.22.

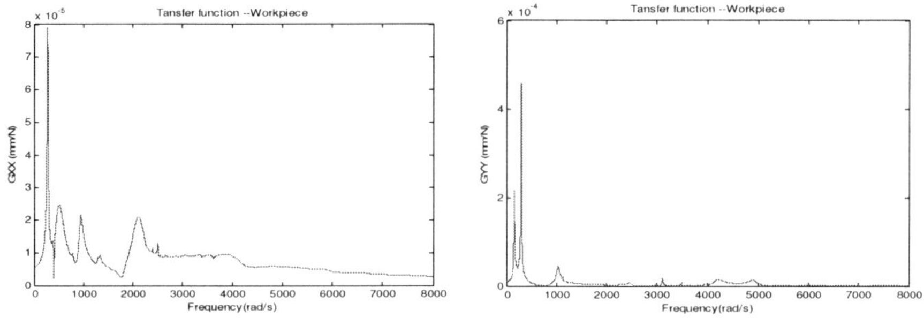

(a) G_{xx}^{w} (b) G_{yy}^{w}

Figure 7.22. Curve-fitted transfer functions of the workpiece

Compared to the tool, the peak values of the workpiece's transfer functions locate at lower frequencies. The stiffness of the workpiece in X direction is much stronger than that in Y direction, due to the structural characteristics of the fixture, as shown in Figure 7.20 b, where the accelerometer is mounted in Y direction.

7.6 The Application of the Machining Dynamics Model

The following example demonstrates the application of the machining dynamics model.

7.6.1 The Machining Setup

The dynamic cutting forces and the vibrations of the tool and workpiece should be recorded during the machining process. However, it is practically impossible to very precisely monitor the tool's vibration during the machining process. An indirect way is to examine the vibration of the spindle head. The dynamic cutting forces are recorded by the Kistler table dynamometer, the tool's vibrations are monitored using piezoelectric accelerometers mounted on the spindle head, and the workpiece vibrations are examined by the accelerometers mounted on the workpiece itself. Figure 7.23 illustrates the experimental configuration of the cutting trials, and Table 7.11 lists the cutting conditions and parameters of these trials.

Table 7.11. Cutting conditions and parameters (peripheral milling)

Tool: solid carbide end-mill, $R = 10$ mm, $\alpha_r = 5°$								
Work material: carbon steel EN8; cutting condition: with fluid								
No.	m	β	δ_e (mm)	n (rpm)	SF (Hz)	TPF (Hz)	f_t (mm)	b_a (mm)	d (mm)
13	2	30°	0.002	1592	26.53	53.07	0.05	5.03	11
14	3	30°	0.005	1751	29.18	87.55	0.05	4.92	11

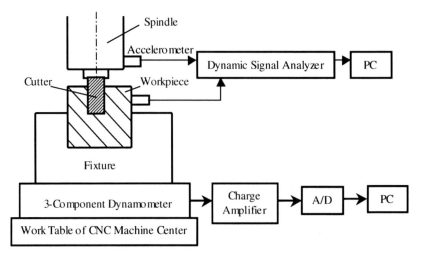

Figure 7.23. Experimental configuration of cutting trials

7.6.2 Case 1: Cut 13

Figures 7.24-7.26 compare the measured and predicted dynamic cutting forces, and the vibrations of the tool and workpiece. (Note: the vibrations of the spindle head substitute the measured vibrations of the tool.)

Figure 7.24 shows that both the amplitude and frequency characteristics of the cutting forces have a good agreement between the measured and predicted values. Both of the measurements and simulations demonstrate FFT peaks at the TPF, 2TPF and 3TPF for Fx, and at the TPF and 2TPF for Fy. The amplitude of the measured Fy shows the influence of the small amount of cutter run-out (2 μm) on the cutting forces; however, it is indistinguishable from the measured Fx. We should ignore the difference of the absolute value of the FFT peaks between the measurement and simulation, because their sample numbers are different. Attention should be given to the unit of frequency; for the measurement it is Hz, and for the simulation it is rad/s.

Both of the measurements and simulations show that there are dominant forced vibrations activated on the tool by the cyclic cutting forces at frequencies TPF, 2TPF and 3TPF. In X direction, the measurement indicates that the spindle head has potential chatter around frequency 7TPF (371 Hz or 2331 rad/s), and the simulation demonstrates an unstable frequency for the tool at about 2150 rad/s, which is close to one of the nature frequencies ω_{ncx} of the tool (see Table 7.7 and Figure 7.21 a). In both X and Y directions, the measurement shows potential chatter of the spindle head at frequencies 660~800 Hz (4147~5026 rad/s), and the simulation indicates potential chatter of the tool at frequencies 4000~5000 rad/s, which can be explained from the transfer functions as shown in Figure 7.21.

Machining Dynamics in Milling Processes 215

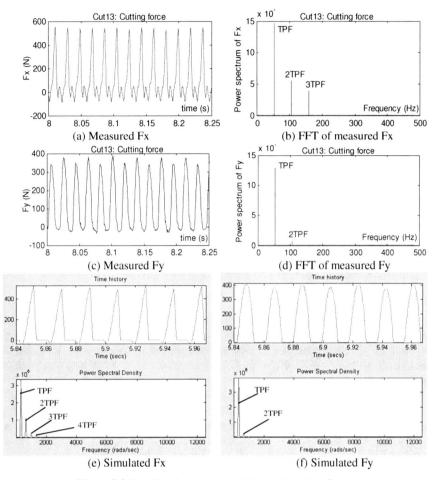

Figure 7.24 Predicted and measured dynamic cutting forces

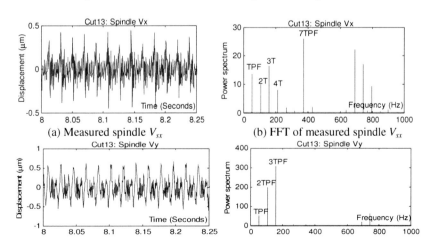

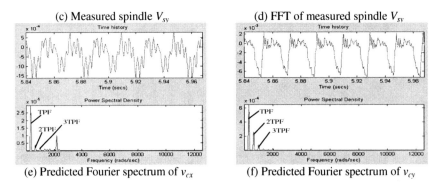

Figure 7.25 Predicted and measured vibrations of the tool/spindle

The difference between the measurement and simulation is that the dominant FFT peaks of the measurement (spindle head) are at the 3TPF, but at the TPF for the simulation (tool) which correlates directly to the FFT peaks of the dynamic cutting forces. Attention should be given to the fact that the vibrations of the spindle head and the tool are not the same.

Figure 7.26 shows that the frequency characteristics of the vibrations of the workpiece have a good agreement between the measurement and the simulation, and both of them reveal that there are only forced vibrations activated on the workpiece by the cyclic cutting forces at frequencies TPF, 2TPF and 3TPF. In X direction, both the measurement and simulation show that the dominant peak is at 3TPF (159.21 Hz or 1000 rad/s), and there are also small peaks at the TPF and 2TPF. This result indicates that the workpiece is more sensitive at 3TPF in X direction and complies with the transfer functions as shown in Figure 7.22 a, even though the FFT peaks of Fx are not dominant at this frequency. In Y direction, both the measurement and simulation demonstrate obvious forced vibrations at frequencies TPF, 3TPT and then 2TPF. Obviously, it complies well again with the transfer functions as shown in Figure 7.22 b. Attention should be paid to the amplitude of the measured vibration, which is released by integration twice from the accelerator and low frequency filtration once that makes its value much smaller than the predicted one.

Now let us look at the surface characters. Figure 7.27 illustrates the surface topography and section profile measured with the Form Talysurf PGI after the cutting process. Figure 7.27 a shows the measured surface topography, Figure 7.27 b illustrates a section profile of the surface along the feedrate direction, Figure 7.27 c presents the FFT results of the section profile in terms of time history, and Figure 7.27 d rebuilds the section profile in terms of time history after filtering the low frequency components.

Figure 7.28 illustrates the simulated surface profiles and their FFT results in terms of time history. The first section profile represents that of the end of the tool ($\varphi = 0$, shown in Figure 7.28 a), the second one stands for the middle section of the surface ($\varphi = \psi/2$, as shown in Figure 7.28 c), and the third is the upper section of the surface ($\varphi = \psi$, as shown in Figure 7.28 e).

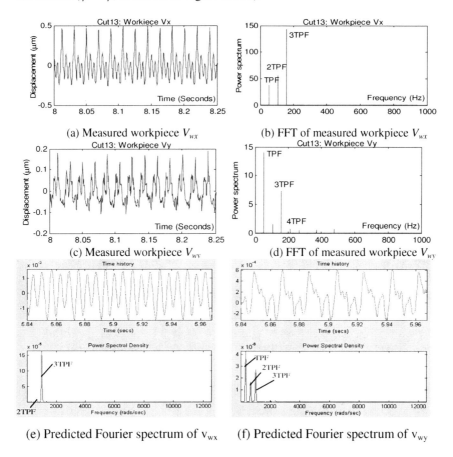

(a) Measured workpiece V_{wx}
(b) FFT of measured workpiece V_{wx}
(c) Measured workpiece V_{wy}
(d) FFT of measured workpiece V_{wy}
(e) Predicted Fourier spectrum of v_{wx}
(f) Predicted Fourier spectrum of v_{wy}

Figure 7.26. Predicted and measured vibrations of the workpiece

The machined surface profile after high-pass filtration shown in Figure 7.27 d illustrates the roughness of the surface profile along the feedrate direction, the filtered most lowest frequencies' components represent the machining errors, and the other low frequencies' components describe the surface waviness/roughness. Figures 7.28 a, 7.28 c and 7.28 e illustrate the roughness of the simulated surface profiles and the machining errors. The results demonstrate that there is a reasonable agreement between the measured and predicted surface roughness Rz along the feedrate direction. Both of the measured and predicted Rz are amount to about 0.15μm. Their frequency characters are obviously different. The dominant FFT peaks of the measured surface profile are found to be at the low frequencies, such as 1-5Hz, 11.5Hz, 14.5Hz and 20Hz, in addition to the small peaks at the SF. The average static dynamic cutting forces may introduce the peaks at the most lowest frequencies. The other low frequencies' waviness/roughness may be introduced by the low frequencies' structural modes of the cutting system/machine base, which is hard to identify by the experimental modal analysis. There are also non-linear factors of the cutting process, such as built-up edge (BUE) and non-homogeneous distribution of the micro-hardness of the workpiece material, which may cause these low frequencies' roughness/waviness. Further investigation is needed to understand the nature of the low frequencies' waviness/roughness.

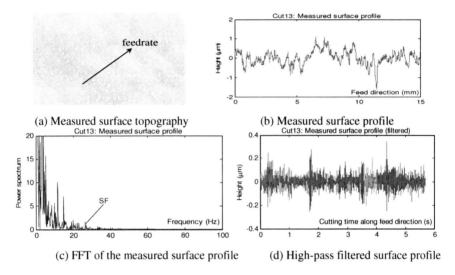

Figure 7.27. Measured surface topography and section profile (Cut 13)

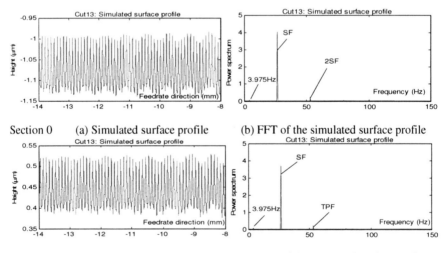

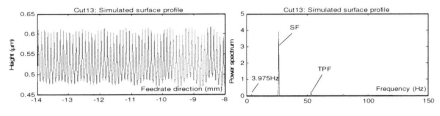

Section 9   (e) Simulated surface profile   (f) FFT of the simulated surface profile

Figure 7.28. Simulated surface profiles (Cut 13)

The predicted surface profiles show dominant peaks at the SF, whereas there are small peaks in the measured surface profile. The result reveals that the cutter run-out (corresponding to the SF) is one of the main sources dominating the roughness of the surface; however, the non-linear factors of the cutting process or the low frequencies' structural modes of the cutting system may introduce more significant low frequencies' waviness/roughness. It is almost impossible to distinguish the roughness at the TPF in both the measurement and simulation.

It is worth noting that both the measurement and the simulation do not indicate forced vibration at the SF in the cutting process for both the tool and workpiece, even though the small cutter run-out (2μm) has an influence on the cutting forces (see Figure 7.24).

The simulated section profiles imply the machining errors of the profiles, which amounts to - 1.06 μm, 0.46 μm and 0.54 μm, respectively for the three profiles, where '-' denotes an over-cut machining error and '+' stands for an undercut error. The machining errors are supposed to result from the deflections of the tool and workpiece, but the result reveals that the cutter run-out and the helix angle may contribute together to this complex machining error distribution along the tool axis (axial depth of cut) direction. The simulation shows peaks at the TPF; however, these peaks cannot be seen from the measured surface profile due to the dominant peaks at the low frequencies.

In view of the vibrations of the tool and workpiece, as shown in Figures 7.25 and 7.26, both of the measured and predicted surface profiles demonstrate that the amplitude of the vibrations of the tool and workpiece are far bigger than the surface roughness Rz. This means that the cutting process has an ease effect on the surface profile and removes the influence of the high frequency vibrations of the tool and workpiece, and the high frequency structural vibration does not contribute directly to the surface waviness/roughness.

7.6.3 Case 2: Cut 14

This case uses another tool, and the difference includes the teeth number and cutter run-out value (see Table 7.11). Because the tool is slightly different and the workpiece is almost exactly the same as the previous one, it is assumed that they have the same transfer function. Figures 7.29–7.31 shows the measured and predicted dynamic cutting forces, and the vibrations of the tool and the workpiece.

Once again Figure 7.29 shows that both the amplitude and frequency characteristics of the cutting forces have a reasonable agreement between the measured and predicted values. Obviously the cutter run-out contributes to the cyclically changed amplitude of the cutting forces. Attention should be paid to the difference of the amplitude of the cutting forces between the measurement and the simulation. It may imply that low frequency modes of the cutting system (particularly the workpiece, as shown in Figure 7.22) control the vibrations of the system, which in turn influence (amplify the difference of) the cutting forces at the low frequencies, such as SF (dominantly) and 2SF.

The measurement and simulation show that there are dominant forced vibrations activated on the tool by cyclic cutting forces at frequencies TPF, 2TPF and 4TPF.

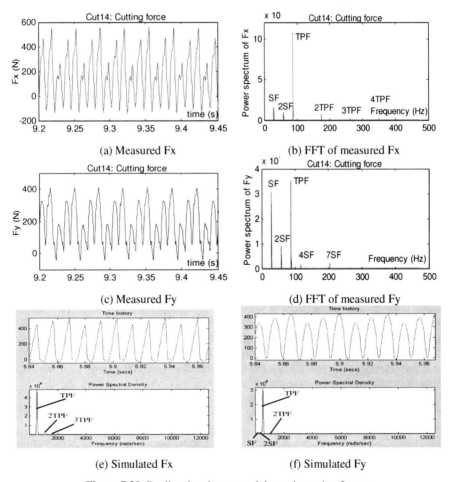

Figure 7.29. Predicted and measured dynamic cutting forces

The difference between the measurement and the simulation is that the measured V_{sx} dominates at 2TPF, but the simulated v_{cx} at TPF. This may be caused by a specific structural mode of the spindle head. In Y direction, the simulated v_{cy} shows the forced vibrations at the TPF and 4TPF, and the measured V_{sy} has more peaks at the 4SF, and then 2TPF in addition to the peaks of the simulated v_{cy}. This result reveals that the spindle head is very sensitive around 4SF. Again the vibrations of the spindle head and the tool are not the same and the predicted v_{cx} and v_{cy} are based on the transfer functions of the tool.

The results reveal that there are only forced vibrations activated on the workpiece by the cyclic cutting forces. In X direction, the measurement shows the dominant

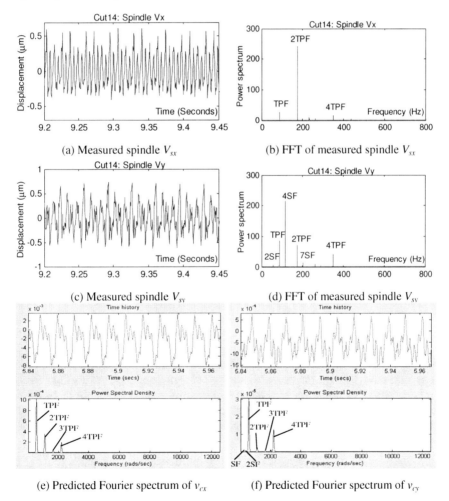

Figure 7.30. Predicted and measured vibrations of the tool/spindle

peak is at 2TPF, there is a small peak at the TPF, and the simulation gives dominant peaks at the TPF and 2TPF. This result indicates that the workpiece is more sensitive at 2TPF (175 Hz or 1100 rad/s) in X direction (see Figure 7.22 a). In fact, the cutting forces (both measurement and simulation) dominate at the TPF rather than 2TPF (see Figure 7.21). In Y direction, both the measurement and simulation demonstrate an obvious forced vibration at the TPF, not 2TPF, which complies with the transfer functions as shown in Figure 7.22 b. Again the measured vibration, which is released by integration twice from the accelerator and low frequency filtration once, makes its value smaller than the predicted one.

Now let us analyses the amplitude of the vibrations. Both the measured V_{wx} and predicted v_{wx} undulate cyclically at the SF. Again the analysis will be based on the simulation. As shown in Figure 7.31 e, the amplitude difference is about 3.2 μm. Bearing in mind that in this case the cutter run-out is 5μm. Interestingly it is hard to distinguish the amplitude difference at the SF in the previous case due to small cutter run-out (2 μm). In this case the amplitude is 7.5 μm and the previous one is 3.5 μm. Therefore it is fair to say that the amplitude difference at the SF is generated directly by the forced vibration from the cyclic cutting forces (Figure 7.29) at the SF due to the cutter run-out.

Now let us look at the surface characters. Figure 7.32 shows the surface topography and section profile along the feedrate direction, and Figure 7.33 illustrates the corresponding simulated surface profile. Similar to the previous case, the results demonstrate again that there is a reasonable agreement between the measured and predicted surface roughness Rz along the feedrate direction. Both of the measured and predicted Rz at the SF are equal to about 0.3 μm.

Once again both of the measured and predicted surface profiles demonstrate that the amplitude of the vibrations of the tool and workpiece are far bigger than the surface roughness Rz. Also the dominant FFT peaks of the measured surface profile are found at low frequencies, such as 9 Hz, 10 Hz, 16 Hz and 20 Hz, in addition to the peak at the SF. The predicted surface profile shows a dominant peak at the SF. It is interesting to discern that the FFT peak of the measured surface profile at the SF (see Figure 7.32 c) seems more standing out comparing to the previous case (see Figure 7.27 c). In this case, the surface roughness Rz (0.3μm) is doubled compared to the previous case (0.15 μm; see Figures 7.27 and 7.28). From the analysis of the vibrations, obviously the cutter run-out contributes not only to forced vibrations at the SF but also to the surface roughness at the SF in X direction. Therefore the surface roughness and cutter run-out value are correlated to each other. This can be used to explain why the FFT peak of the measured surface profile at the SF in this case is more predominant compared to the previous case.

Unlike the previous case, there are no FFT peaks at a low frequency in the predicted surface profile. Therefore it is unlikely that the FFT peak at a low frequency in the simulated surface profile shown in the previous case (see Figure

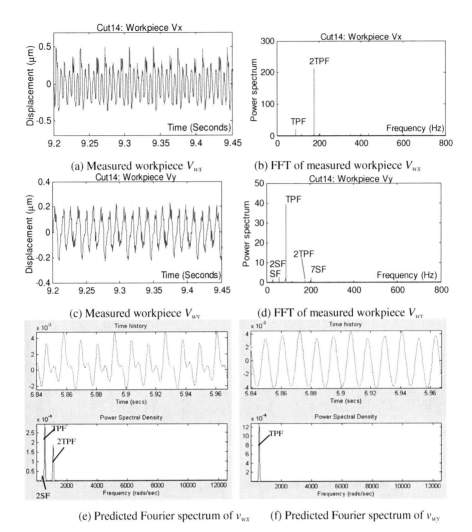

(a) Measured workpiece V_{wx} (b) FFT of measured workpiece V_{wx}

(c) Measured workpiece V_{wy} (d) FFT of measured workpiece V_{wy}

(e) Predicted Fourier spectrum of v_{wx} (f) Predicted Fourier spectrum of v_{wy}

Figure 7.31. Predicted and measured vibrations of the workpiece

7.28) can explain the dominant peaks of the measured surface profile at low frequencies. That means roughness/waviness at the low frequencies must come from some other sources.

It is worth noting that even though a potential chatter is just going to be activated in the previous case, particularly to the spindle head and tool (see Figure 7.25), it seems to not contribute to the surface roughness. However, in this case, even though obviously there is not any sign of potential chatter, the surface roughness Rz is doubled compared to the previous case. Therefore, it is fair to say that high frequency vibrations do not influence the surface roughness, if there is not obvious chatter activated in the cutting process.

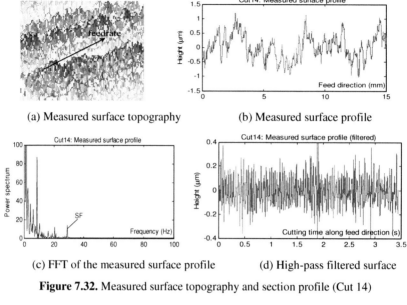

(a) Measured surface topography (b) Measured surface profile

(c) FFT of the measured surface profile (d) High-pass filtered surface

Figure 7.32. Measured surface topography and section profile (Cut 14)

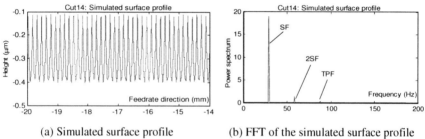

(a) Simulated surface profile (b) FFT of the simulated surface profile

Figure 7.33. Simulated surface profile (– Section 4) (Cut 14)

7.7 The System Identification of Machining Processes

Section 7.5 introduces experimental modal analysis for the identification of modal parameters of a machining system. The process is based on a "static" excitation using an impact hammer instrumented with a piezo-electric force transducer rather than the "dynamic" cutting process. The identified system transfer functions and modal parameters could be different from the real transfer functions and modal parameters due to the following reasons:

- During the cutting process, the spindle's dynamic stiffness could be different from its static stiffness;
- When the tool is engaged with the workpiece in the cutting process, the dynamic damping condition differs from the static damping condition;

- During the cutting process, the machining system structure changes continuously, particularly for the machining of a large workpiece;
- During the cutting process, the damping condition could change continuously;
- The use of coolant has a significant influence on the machining dynamics.

This raises the issue that the prediction of machining dynamic behaviours based on the modal parameters identified by experimental modal analysis could be very different from the real case; particularly, when the resonance modes are close to each other, the prediction could be totally wrong. So, what measures can be used for improvement, particularly when an optimal cutting process is essential for the achievement of high productivity and high quality for high speed machining? Let's consider the system identification in a broad way.

7.7.1 The System Identification

System identification (SI) is a process for the modelling of a dynamic system or process based on measured data. It is a mechanism that processes finite, partial, and inexact information of a physical system to yield an abstract or mathematical description to represent the system. You adjust the parameters of a given model until its output coincides as well as possible with the measured output. Therefore SI generally consists of the following four steps:

- Data acquisition
- Characterisation
- Identification/estimation
- Verification

The first and most important step is to acquire the input/output data of the system/process to be identified. This involves careful planning of the inputs to be applied so that sufficient information about the system dynamics is obtained. If the inputs are not well designed, then they could lead to insufficient or even useless data.

The second step defines the system's structure, for example, the type and the order of the differential equation relating the input to the output. Any prior knowledge of the system to be identified, such as its physical structure, its characteristics and properties, is helpful to select a suitable system's model structure, *e.g.*, auto-regressive with exogenous input (ARX), auto-regressive moving average with exogenous input (ARMAX), *etc.*

The third step is identification/estimation, which involves adjusting the parameters of the system's model structure until its output coincides as well as possible with the measured output, *i.e.*, to minimize the error between the system's output and its model's output. Common estimation methods are least squares, instrumental-variable, maximum-likelihood, *etc.*

The final step, verification, consists of relating the system to the identified model responses in the time or the frequency domain. Residual (correlation) analysis, Bode plots and cross-validation tests are generally employed for model validation. A good test is to compare the output of the model to measured data that was not used for the fit (estimation).

Since the available information is inherently incomplete and contains measurement error, identification is inevitably an approximate procedure, and a measure of accuracy is then desired for quantifying the mismatch between the resultant description and the actual system. Intuitively, the description should be tasked oriented, since identification is intended to yield a model for subsequent analysis and design, and as such a certain specific form representation may be more advantageous than others.

The above-mentioned features of system identification can be illustrated in Figure 7.34, where u is the input, y is the output or response, e is the disturbance, y' is the output of the model to the same input u, and ε is the error (residuals) between the model output and the system output. The objective of system identification is to minimize the sum-squared residuals ε. More details about system identification can be found in [10].

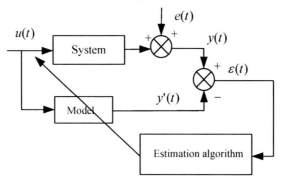

Figure 7.34. System identification

7.7.2 The Machining System and the Machining Process

For the analysis and design of a machining system, such as a milling machine, experimental modal analysis, one of the fundamental system identification methods, is a feasible and practical approach. It can be used to identify the machine structures' dynamic characteristics, particularly the stiffness and natural modes of the column and the machine bed. However, it targets for machining system design and analysis rather than machining process optimisation, such as maximal productivity through selection of machining parameters and conditions. For this purpose, operational or machining process based observation is essential.

For a machining process, such as the milling process, given the machine and machining setup (tool assembly and work assembly), the direct inputs to the process are cutting parameters (spindle speed, axial depth of cut and radial depth of cut) and cutting conditions (down or up milling and coolant), and the expected ultimate outputs are high quality work and high productivity. To achieve these targets, it is vital to keep the vibrations of the machining system under control, which include the vibrations of the machine structures, such as the column and the machine bed, and the vibrations between the tool and workpiece. The vibrations of the machine structures can be modelled and analysed through experimental modal analysis, and can be controlled by carefully choosing the spindle speed. In this section, the focus will be on the control of the vibrations between the tool and the workpiece.

In one way, tremendous efforts have been made to identify the machining process model based on the cutting parameters/conditions and the vibrations between the tool and workpiece. However, as a mechanical system, the cutting parameters/conditions are not the direct factors inducing the vibrations between the tool and workpiece, and such vibrations are excited by the dynamic cutting forces acting on the two components. Therefore it is logical to use the dynamic cutting forces as the input data and the vibrations of the tool and workpiece as the output data under the operation of the machining process for the system identification for the machining process, as shown in Figure 7.35.

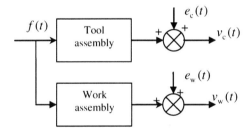

Figure 7.35. Data for system identification of the machining process

Obviously it is a task-oriented description to model the machining dynamics of the tool assembly and work assembly through system identification. Together with the dynamic cutting force model (see Section 7.2), the machining dynamics can be related to the cutting parameters and conditions.

7.7.3 A Case Study

Here we will use the same cases described in Section 7.6, because both the input (dynamic cutting forces) and output (vibrations of the workpiece[3]) for the system identification of the work assembly were recorded during the machining process. Consider case 2 - Cut14. Figure 7.29 a and c record the dynamic cutting forces

---

[3] Because it is difficult to measure the vibrations of the tool, here we only deal with the work assembly.

$f_x(t)$ and $f_y(t)$, Figure 7.31 a and c record the vibrations of the workpiece $v_x(t)$ and $v_y(t)$, let's use these data for the system identification of the work assembly under the operational condition of the machining process.

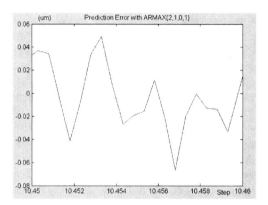

Figure 7.36. Sum-squared errors ε with ARMAX(2, 1, 0, 1)

The second step is to select a suitable system's model structure. As discussed in Section 7.2, the machining system, *i.e.*, the tool assembly and work assembly, can be roughly modelled as a spring-damper vibratory system, as shown in Figure 7.3, which is a typical second order linear system. Therefore we can use ARMAX($2n$, n, 0, 1) model as the work assembly's structural model, *i.e.*, ARMA($2n$, n) model, where $n = 1, 2, 3, \ldots$

The third step is identification/estimation using the SI toolbox in Matlab [11]. For this case, the sampling frequency is 2KHz. Table 7.12 lists the first order modal parameters and transfer functions identified with ARMAX(2, 1, 0, 1), and Table 7.13 lists the second order modal parameters and transfer functions identified with ARMAX(4, 2, 0, 1). Figures 7.36-7.38 show the sum-squared errors ε from ARMAX(2, 1, 0, 1), ARMAX(4, 2, 0, 1) and ARMAX(6, 3, 0, 1), respectively.

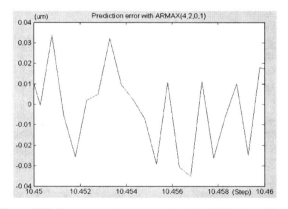

Figure 7.37. Sum-squared errors ε with ARMAX(4, 2, 0, 1)

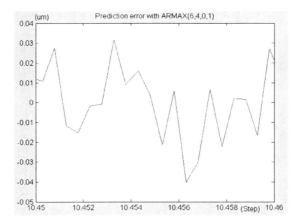

Figure 7.38. Sum-squared errors ε with ARMAX(6, 3, 0, 1)

Table 7.12. Modal parameter and the transfer function identified with ARMAX(2,1,0,1)

Direction	ω_r (rad/s)	ζ_r	Transfer function H(s)
X	980.1	0.2	$H(s) = \dfrac{0.6304e-003s}{s^2 + 398.5s + 960513}$
Y	717.9	0.67	$H(s) = \dfrac{0.5717e-003s}{s^2 + 962s + 515450}$

Table 7.13. Modal parameter and the transfer function identified with ARMAX(4,2,0,1)

Direction	ω_r (rad/s)	ζ_r	Transfer function H(s)
X	$\omega_{r1} = 2607$	$\zeta_{r1} = 0.1155$	$H(s) = \dfrac{1.0e-003s}{s^2 + 602s + 6.796e006}$
	$\omega_{r2} = 840.1$	$\zeta_{r2} = 0.1031$	$+ \dfrac{-2.566e-006s}{s^2 + 173.2s + 7.05739e005}$
Y	$\omega_{r1} = 2723$	$\zeta_{r1} = 0.0986$	$H(s) = \dfrac{-6.9127e-002s}{s^2 + 1678.8s + 1.2614e007}$
	$\omega_{r2} = 511.4$	$\zeta_{r2} = 0.3735$	$+ \dfrac{4.7554e-004s}{s^2 + 642.8s + 7.7964e005}$

Figures 7.36-7.38 indicate that the sum-squared errors of ARMAX(4, 2, 0, 1) are smaller than that from ARMAX(2, 1, 0, 1); however, ARMAX(6, 3, 0, 1) does not improve the accuracy any more, but takes much longer to calculate. Therefore, it is fair to use the modal parameters identified from the ARMAX(4, 2, 0, 1) model to represent the work assembly under the operational condition of the machining process. Figure 7.39 shows the simulated vibrations of the workpiece based on the identified modal parameters using the ARMAX(4, 2, 0, 1) model.

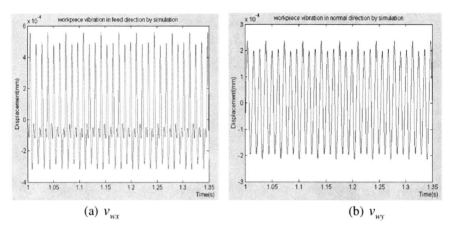

Figure 7.39. Simulated workpiece vibrations with modal parameters based on ARMAX(4, 2, 0, 1)

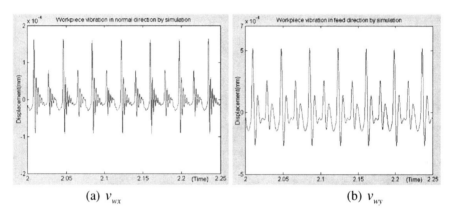

Figure 7.40. Simulated workpiece vibrations for case 1 - Cut13 using the same modal parameters as for case 2

The final step is model validation, *i.e.*, to compare the output of the model to measured data that was not used for the system identification. For this purpose, let's use the workpiece vibrations of case 1 - Cut13 discussed in Section 7.6. Figure 7.40 shows the simulated vibrations of the workpiece for case 1 - Cut13 based on the same modal parameters identified from case 2 - Cut14 using ARMAX(4, 2, 0, 1) model.

Comparing the simulated vibrations to the measurement shown in Figure 7.26 a and c, we can see that there is a reasonable agreement between the simulation and the measurement.

Comparing the simulated vibrations to the measurement shown in Figure 7.31, it is perceived that there is a good agreement between the simulation and the measurement.

7.7.4 Summary

The case study result reveals that system identification based on the measurement from the operational machining process is a feasible alternative to estimate the modal parameters and transfer functions of machining systems. This approach overcomes the problems encountered by the traditional experimental modal analysis, which was analyzed in the beginning of this section. However a through investigation is underway to systemize the methodology "system identification of machining processes", which is to include the tool assembly and to look at the whole machining system as a whole, and further its application in the research of machining dynamics. It is worth mentioning that neural networks can be another alternative to identify the machining system under the operational condition of machining processes, even though there is not yet a mature case study to support this statement.

References

[1] J. Tlusty (1985) *Machine Dynamics*. In Robert I. King edited *Handbook of High Speed Machining Technology*. New York: Chapman and Hall.
[2] D. A. Stephenson, and J. S. Agapiou (1997) *Machining Dynamics*, Chapter 12 in *Metal Cutting Theory and Practice*. New York: Marcel Dekker.
[3] Y. Altintas (2000) *Manufacturing Automation: Metal Cutting Mechanics, Machine Tool Vibrations, and CNC Design*. Cambridge: Cambridge University Press.
[4] M. C. Shaw (2005) *Metal Cutting Principles*, Second Edition. Oxford: Oxford University Press.
[5] X. Liu, K. Cheng, D. Webb and X. Luo (2002) Improved dynamic cutting force model in peripheral milling – Part 1: Theoretical model and simulation. *International Journal of Advanced Manufacturing Technology*, 20:631–638.
[6] X. Liu, K. Cheng K, D. Webb, A. P. Longstaff, M. H. Widiyarto and D. Ford (2004) Improved dynamic cutting force model in peripheral milling – Part 2: experimental verification and prediction. *International Journal of Advanced Manufacturing Technology*, 24:794–805.
[7] X. Liu, K. Cheng, A. P. Longstaff, M. H. Widiyarto and D. Ford (2005) Improved dynamic cutting force model in ball end milling – Part 1: Theoretical modelling and experimental calibration. *International Journal of Advanced Manufacturing Technology*, 26:457–465.
[8] Liu X, K. Cheng, D. Webb, A. P. Longstaff, M. H. Widiyarto and D. Ford (2008) Improved dynamic cutting force model in ball end milling – Part 2: Verification and refinement. *International Journal of Advanced Manufacturing Technology* (in press).
[9] X. Liu, K. Cheng (2005) Modelling the machining dynamics of peripheral milling. *International Journal of Machine Tools & Manufacture*, 45:1301–1320.
[10] L. Ljung (1999) *System Identification, Theory for the User*, 2nd edition. New York: Prentice Hall PTR.
[11] http://www.mathworks.com/products/sysid/: *The System Identification Toolbox 6.1.2*

8

Machining Dynamics in Grinding Processes

Xun Chen

School of Mechanical, Materials and Manufacturing Engineering,
The University of Nottingham
Nottingham NG7 2RD, UK

8.1 Introduction

Grinding has traditionally been a finishing operation, as it is today for many precision parts. With the advance of technology, grinding has extended its application to a roughing process. Today the grinding process has been successfully applied to almost all types of material removing processes with an extremely high material removal rate of more than 3,000 mm^3/(mm s) and ultra-precision accuracy to the nanometre lever surface finish. Due to the high hardness of abrasives used as cutting media, grinding is often the first choice for removing materials of a high hardness. Grinding is becoming one of the most popular choices of machining processes, which comprises 25% of total machining processes [1]. A recent survey [2] shows that the application of grinding is still increasing and is becoming the most popular material removal process in industry. The wide use of the grinding process is becoming an important feature of manufacturing process in modern industry.

As a materials removal process, the grinding process is similar to other cutting processes in many respects; for example, materials are removed in the form of chips. However, grinding is distinct from other machining processes.

Firstly, abrasive grits on the wheel surface are of an irregular shape and are randomly positioned. This makes the cutting action of each grit involved in grinding very different. Therefore statistical analysis is commonly used in the investigation of grinding.

Secondly, the undeformed cutting depth of each grit or cutting edge in grinding is very thin, down to the sub-micrometer level. Such a small cutting depth enables grinding to generate a very high accuracy and very low surface roughness. However it normally requires a much higher specific energy for materials removal.

Thirdly, the grinding cutting speed is much higher than with other machining process. Normally it is 10 times higher. A high grinding speed provides an increase of materials removal rate on the one hand but on the other hand it also introduces problems with high grinding temperatures and coolant delivery.

Finally, a special feature in grinding which has to be mentioned is the self-sharpening action of the grinding wheel. When a cutting tool is worn or damaged during the machining process, it must be replaced with a new one or re-sharpened by means of grinding. The wear of an individual grit may not have a large effect on the grinding wheel performance. Such wear may increase the force on the worn grit so as to fracture the grit itself creating new, sharp cutting edges or to force the grit drop off the wheel surface, allowing new grit in a lower layer to engage in the grinding process. Because of the self-sharpening action, grinding is able to grind materials of a similar hardness to the abrasive grits.

Considering such distinct features of grinding, a different consideration is required in practice. Systematic consideration of a grinding process requires knowledge of inputs and outputs of the grinding process. A grinding process is a material removal process utilizing a grinding wheel, which is made of a large number of randomly positioned grits. The grits subsequently penetrate the workpiece to remove the material in front of their paths. Understanding a grinding process starts from the material removal by the individual grits.

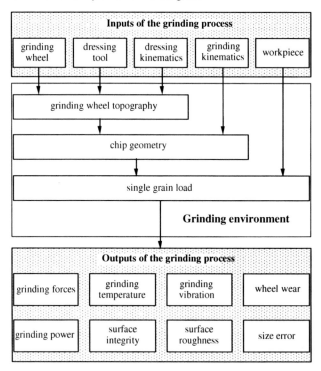

Figure 8.1. A systematic view of the inputs and the outputs of a grinding process

Focusing on grit and workpiece interactions, the relationship between influential parameters and grinding performance can be summarized in Figure 8.1. The factors in the top layer are the primary inputs to a grinding process and the outputs are presented at the bottom. The environment includes the atmosphere, the coolant delivery and the machine structure which holds the wheel and workpiece at the correct position.

The grinding system behaviour is primarily considered from the viewpoints of geometry, kinematics, mechanics, energy consumption and material properties. The performance of a grinding wheel is usually assessed using force, vibration, temperature, surface roundness and surface integrity. A good grinding with a well-dressed wheel presents low force and temperature, consistent accuracy, satisfactory workpiece integrity and surface roughness. Large variations in grinding behaviour are not acceptable.

The actual grinding performance can be considered as the summation of the performance of individual grits. Based on the relationships illustrated in Figure 8.1, the improvement and the stabilization of the grinding performance may be achieved by considering grinding wheel surface topography, chip formation and materials properties as well as the grinding environment including the machine structure, coolant delivery, and so on.

Grinding involves a large number of interacting variables. Before the grinding operation, it is necessary to decide which variables should be considered in the condition selection process. Consideration is therefore given to the relationships and interactions between these variables. The basic relationships between grinding variables in a grinding control system are presented in Figure 8.2.

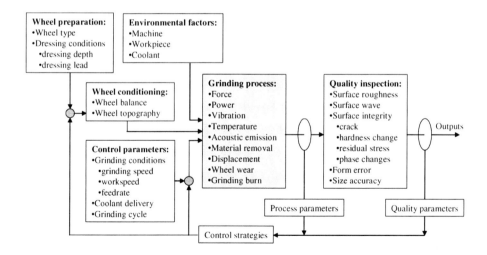

Figure 8.2. Inputs and outputs of a grinding system [3]

The output variables of the system comprise the workpiece quality, the productivity and the cost, which should meet the design and manufacturing requirements. The output variables are therefore the main variables to be controlled.

Process variables include power, force, temperature and vibration. The process variables are determined by the grinding conditions and affect the output variables. When a grinding wheel engages with a workpiece, the grinding force depends on the kinematic relation between the individual grit on the wheel surface and on the workpiece. The force affects other process variables and output variables. The higher the forces applied, the faster is the material removal rate. The force also affects the surface roughness, the deflection of the system, the grinding temperature and the onset of thermal damage. However, process variables are intermediate variables that provide evidences of the relationships between input and output variables.

The input conditions may be divided into the grinding conditions that are selectable and other conditions that are uncontrolled. Uncontrolled variables, *e.g.*, material properties, cannot be changed by the operator but have a significant effect on the grinding process and output variables. The grinding conditions consist of the grinding wheel, the coolant, the dressing conditions and the grinding kinematic conditions, of which the kinematic conditions are the most commonly used controllable parameters in the grinding operation. The grinding conditions should be determined by the operator before starting a grinding process.

8.2 The Kinematics and the Mechanics of Grinding

8.2.1 The Geometry of Undeformed Grinding Chips

The removal of materials during grinding depends on the cutting action of each grit involved in the grinding operation. A ground surface is created as results of a series of cutting actions of grits. During grinding, randomly spaced grits on the wheel surface engage with the workpiece in sequence. Figure 8.3 shows the performance

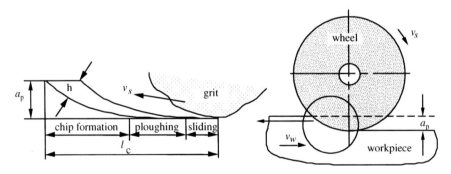

Figure 8.3. Three stages of the generation of a grinding chip

of a grit on the wheel surface during grinding. Depending on the cutting depth, the grit may experience rubbing, ploughing and cutting in three stages engaging with workpiece materials. The proportion of rubbing, ploughing or cutting in a grit pass depends on the amount of grit engagement that is determined by the grit position on the wheel surface and grinding kinematics. The engagement of a grit in grinding can be illustrated by undeformed grinding chip shape, which is commonly described by the average chip thickness and chip length.

Considering a surface grinding process, the contact length between the wheel and workpiece can be calculated by:

$$l_c = (1 \pm \frac{v_w}{v_s})\sqrt{d_s a_p} \tag{8.1}$$

where l_c is the contact length, v_w is the workspeed, v_s is the wheel speed, d_s is the diameter of the grinding wheel and a_p is the grinding depth of cut. The + sign is for up-grinding and – sign is for down-grinding. In grinding, the following relationship for steady removal rate should be satisfied by:

$$\lambda b_c v_s V_c = a_p v_w b_c \tag{8.2}$$

where λ is the density of the cutting edge distribution on the wheel surface, b_c is the average width of chips, and V_c is the average volume of a chip. Therefore:

$$V_c = \frac{a_p v_w}{\lambda v_s} \tag{8.3}$$

If the undeformed chip length is equal to the contact length l_c, the mean chip cross-sectional area A_m is:

$$A_m = \frac{V_c}{l_c} = \frac{a_p v_w}{v_s \lambda l_c} = \frac{h_{eq}}{\lambda l_c} \tag{8.4}$$

where h_{eq} is the equivalent grinding depth that equals to av_w/v_s. If the average chip thickness is h, the average width of the chips is b_c, and the average chip volume is:

$$V_c = hb_c l_c \tag{8.5}$$

let the chip shape ratio be:

$$r = b_c/h \tag{8.6}$$

The average chip thickness can then be calculated from Equations 8.3, 8.5 and 8.6:

$$h = \sqrt{\frac{a_p v_w}{r\lambda v_s l_c}} \tag{8.7}$$

The chip shape ratio is determined by the shape of the grits that engaged with the workpiece. By defining the grinding speed ratio q as v_w/v_s and combining Equations 8.1 and 8.7, the average chip thickness can therefore be expressed as:

$$h = \sqrt{\frac{q}{\lambda r(1\pm q)}} \sqrt{\frac{a_p}{d_s}} \tag{8.8}$$

This equation shows that if the wheel diameter is reduced by half, the undeformed chip thickness will increase by 19%. This will significantly affect grinding behaviours. For the cylindrical grinding process the diameter of the workpiece should also be taken into account. Therefore the wheel diameter is replaced by the equivalent wheel diameter d_e.

$$\frac{1}{d_e} = \frac{1}{d_s} + \frac{1}{d_w} \tag{8.9}$$

Usually, the grinding speed is much higher than the workspeed. Therefore the speed ratio q is much smaller than 1 in grinding, so that:

$$h = \sqrt{\frac{v_w}{\lambda r v_s}} \sqrt{\frac{a_p}{d_e}} = \sqrt{\frac{h_{eq}}{\lambda r l_c}} \tag{8.10}$$

This equation shows that the formation of grinding chips is determined by the grinding wheel speed, the workspeed, the depth of cut, the wheel diameter, the workpiece diameter and the wheel surface characteristics, which represented by the cutting edge density and the chip shape ratio. These parameters are the fundamental factors for the analysis of grinding mechanics and grinding surface formation.

A grit engaged with the workpiece can also be considered as a cutting tool. It may be assumed that the average cutting force on a grit is proportional to the mean cross sectional area A_m of the idealized chip:

$$f_t = k_c A_m \tag{8.11}$$

where f_t is the tangential force on a grit. From a survey of experiments on the grinding force, Ono discovered that the grinding force coefficient k_c can be empirically expressed as a power function [4]. That is:

$$k_c = k_0 A_m^{-\eta} \tag{8.12}$$

where k_0 and η are constant and η ranges from 0.25 to 0.5. The tangential grinding force F_t is the sum of the forces on each individual grit. Therefore:

$$F_t = bl_c \lambda f_t = bl_c \lambda k_c A_m = k_0 bl_c \lambda A_m^{1-\eta} \tag{8.13}$$

where b is the grinding width. The mean chip cross sectional area A_m can be obtained by combining Equations 8.1, 8.4 and 8.9.

$$A_m = \frac{v_w}{v_s \lambda (1 \pm q)} \sqrt{\frac{a}{d_e}} \approx \frac{v_w}{v_s \lambda} \sqrt{\frac{a}{d_e}} \tag{8.14}$$

Normally q is much smaller than 1. From Equations 8.1, 8.13 and 8.14:

$$F_t = k_0 b \lambda^\eta a^{\frac{1-\eta}{2}} (\frac{v_w}{v_s})^{1-\eta} d_e^{\frac{\eta}{2}} \tag{8.15}$$

According to Equation 8.15, the grinding force is influenced by both the kinematic parameters of grinding and the characteristics of the wheel. Equation 8.15 can be rearranged into:

$$F_t = b[\lambda k_0 \sqrt{ad_e}]^\eta [k_0(\frac{v_w a}{v_s})]^{1-\eta} \tag{8.16}$$

At one extreme, $\eta = 1$:

$$F_t = \lambda b k_0 \sqrt{ad_e} = k_0 \lambda bl_c \tag{8.17}$$

where l_c is the geometrical contact length. According to Equation 8.17, the grinding force is directly related to the contact area bl_c and the cutting edge density λ. This implies that at this extreme the grinding force is generated by friction, which is related to the contact situation. At the other extreme, $\eta = 0$:

$$F_t = bk_0(\frac{v_w a}{v_s}) = k_0 b h_{eq} \tag{8.18}$$

Equations 8.16 and 8.18 indicate that when $\eta = 0$, the grinding force is directly related to the equivalent chip thickness and the specific energy is constant with the increasing depth of cut. This is close to the situation where the grinding force is generated mainly by chip formation. In practice, $0.1 < \eta < 0.7$ [5]. In a practical situation, Equation 8.16 applies which is consistent with the assertion that the grinding force is a combination of a cutting force and a friction force. A large value of η indicates a high proportion of friction in the grinding process.

Idealised chip size is also determined by the wheel topography. Under the same grinding conditions, a large number of cutting edges on the wheel surface will reduce the size of the chips. Therefore, a higher density of cutting edges on the wheel surface increases the grinding force and the grinding power [3].

One commonly used measure of grinding effectiveness is the specific energy of grinding which is defined as:

$$e = \frac{P}{Q_w} = \frac{F_t v_s}{abv_w} \tag{8.19}$$

From Equations 8.16 and 8.19, the grinding specific energy can be:

$$e = k_0 (\lambda \sqrt{ad_e})^\eta (\frac{v_w a}{v_s})^{-\eta} = k_0 (\lambda l_c)^\eta h_{eq}^{-\eta} \tag{8.20}$$

It can be seen that the specific energy increases with the decrease of the equivalent grinding depth. This phenomenon is called size effect. Combining with Equations 8.7, 8.10 and 8.20, the grinding specific energy is:

$$e = k_0 (rh^2)^{-\eta} = k_0 (hb_c)^{-\eta} \tag{8.21}$$

This equation illustrates the core issue of the size effect in a grinding is the chip shape and the chip size. When comparing Equations 8.21 and 8.12, it can be seen that the specific energy has the same value as the grinding force coefficient.

Grinding force can be estimated by considering the interactions of the grains that pass through the grinding zone. Shaw suggested that the theoretical analysis of grinding force on a grit can be simplified as an indentation force [6]:

$$R = \frac{1}{4} \pi b_{ci}^2 H \frac{C'}{3} \tag{8.22}$$

where R is the indentation force, b_{ci} is the grit cutting width that is similar to the chip width, H is the hardness of the workpiece materials and C' is a constraint factor, which is defined as the ratio of the average pressure in the contact area to the uniaxial flow stress. In most case, C' is about 3 [6].

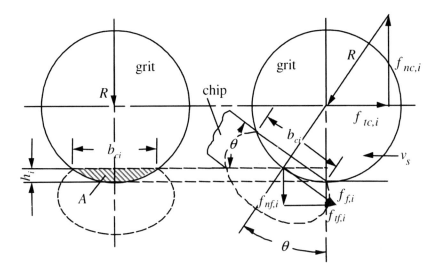

Figure 8.4. Grinding force on a grit

Due to the irregular shape of a grit, the grit may be modelled as a sphere as illustrated in Figure 8.4. In the most case, the grit size is much larger than the grit cutting depth; therefore, the area A in Figure 8.4 can be approximated to

$$A = \frac{2}{3} b_{ci} h_i \qquad (8.23)$$

where h_i is the grit cutting depth. Assuming the coefficient of friction at the contact surface is μ, the force on each grit can be expressed as:

$$f_{t,i} = \frac{3\pi}{8} \frac{b_{ci}}{h_i} H(\frac{C'}{3}) A(\sin\theta + \mu\cos\theta) \qquad (8.24)$$

$$f_{n,i} = \frac{3\pi}{8} \frac{b_{ci}}{h_i} H(\frac{C'}{3}) A(\cos\theta + \mu\sin\theta) \qquad (8.25)$$

where f_{ti} is the tangential force on the grain i, f_{ni} is the normal force on the grain i, H is the hardness of the workpiece, C' is a constraint factor, A is the cross-sectional area of the undeformed chip, μ is the friction coefficient and θ is the angle of the indentation force from the normal.

$$\theta = \arcsin\frac{b_{ci}}{d_{gi}} = \arcsin\frac{2\sqrt{h_i(d_{gi} - h_i)}}{d_{gi}} \qquad (8.26)$$

where d_{gi} is the diameter of the grit. The spherical grain model takes account of grain sharpness with grain depth of cut. A larger depth of cut h_i results in larger indentation angle θ, which leads to a smaller ratio of f_{ni}/f_{ti}. This makes the grit appear to be sharper. This effect can also contribute to the grinding size effect.

The total grinding force is the sum of the individual forces on each of the grits that pass through the grinding zone to remove the materials with the given depth of cut. So there are:

$$F_t = \sum_{}^{n} f_{t,i} \qquad (8.27)$$

$$F_n = \sum_{}^{n} f_{n,i} \qquad (8.28)$$

8.3 The Generation of the Workpiece Surface in Grinding

A ground workpiece surface is created when the abrasive grits on the wheel surface pass through the workpiece removing materials. During grinding, each grit that cuts the workpiece removes a portion of the workpiece material and leaves a trace on the workpiece surface as shown in Figure 8.5. The final workpiece surface is determined by accumulating the interactions of grains that pass the workpiece. When the workpiece passes through the grinding contact zone of length l_c, the distance travelled by the surface of the wheel is l_s.

$$l_s = \frac{v_s}{v_w} l_c \qquad (8.29)$$

This equation indicates that all the active grits within the length l_s will engage with the cross section of the workpiece ABB'A'. When a cutting edge passes through a cross section of the grinding zone, the material engaged with the grit is removed. For example, the grit $G_{i,j,k}$ passes through the grinding zone at abb'a', the shaded area will be removed. After the grinding wheel passes through the cross section ABB'A', the remaining surface contour is the ground surface of the workpiece.

The local elastic deflection of a grit reduces the real cutting depth of a grain. Saini, Wager and Brown revealed that the local deflections of the wheel when a grain is in contact with steel are of the same order of magnitude as the undeformed chip thickness [7]. The variation of the deflection of the grain centre δ_c has a trend and scale similar to the total deflection. Therefore the deflection of the grain centre is considered as the local deflection in the analyses of both the dressing and the grinding process. Nakayama, Brecker and Shaw described the deflection of the grain centre as following the form of a Hertz distribution [8]:

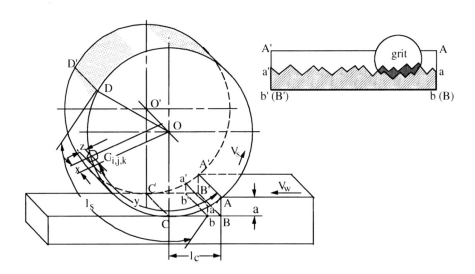

Figure 8.5. Generation of the workpiece surface

$$\delta_c = C f_n^{2/3} \tag{8.30}$$

where C is a constant and f_n is the normal force acting on the grit.

During grinding, ploughing displaces material to the side of the grain. The amount of displaced material depends on the workpiece hardness [9]. The proportions of rubbing, ploughing and cutting also depend on the grinding depth. Lortz observed a dead zone in the area ahead of the grain [10]. If the cutting depth of a grain is smaller than a critical value, no material is removed as the grain passes through the grinding zone. The critical depth of the dead zone depends on the grain shape and the friction between the grain and the workpiece. Due to the dead zone, the proportions of cutting, ploughing and rubbing vary with the depth of cut. A large grinding depth increases the proportion of cutting. The proportion of ploughing and rubbing increases with the decrease of cutting depth. If the grains are dull, the process will be dominated by inefficient sliding and ploughing phases of a grain interaction rather than the efficient cutting phase. Ploughing squeezes the engaged material to the sides of the grain to form ridges. The ridges are removed by subsequent grains.

In grinding, only a proportion of the material in the interference zone is removed by a grain. The main influences are elastic deflection and plastic deformation. The elastic deflection is represented by Equation 8.30. The remaining plastically deformed material piles up at the sides of the grain. It can be rationally assumed that the area of the material removed within a groove formed by a grain is proportional to the undeformed cutting area. The ratio of the volume of the material removed to the volume of the undeformed chip is defined as the cutting efficiency ratio β. The material piled up alongside the groove will be the proportion $(1 - \beta)$ of the volume of the undeformed chip. Therefore the total ridge area A_p of the material remaining on the surface is:

$$A_p = (1-\beta) \cdot A \qquad (8.31)$$

where A is the cross-sectional area of the undeformed chip. The shape of the displaced material can be approximated by a parabola superimposed at each side of the groove as shown in Figure 8.6.

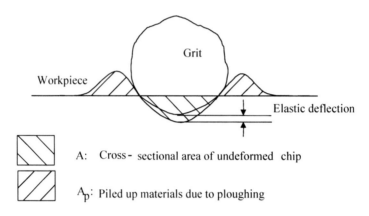

Figure 8.6. Materials removal by a grit during grinding

During grinding, randomly spaced grits on the grinding wheel surface are engaged with workpiece in sequence. The mechanism of material removal consists of elastic deformation (rubbing) and plastic deformation (ploughing and cutting). Each grit transfers its shape onto the workpiece surface, although only a proportion of the material in front of the grit is removed. Due to elastic deformation and plastic ploughing, the final ground workpiece surface presents a very complicated pattern as the results of kinematic cutting paths, grit shape, grit distribution, elastic deflection and stretched ridges induced by ploughing. Because of the rubbing and ploughing actions, grinding consumes more energy than other machining processes.

The smaller grinding chip size requires more cutting paths to move the same amount of materials. For a grinding chip, whose length is much larger than its thickness as shown in Figure 8.3, the surface of the chip is approximately given by S_c:

$$S_c = 2b_c l_c + 2hl_c = 2(r+1)hl_c \qquad (8.32)$$

Combining Equations 8.5, 8.7, 8.10 and 8.32, the ratio of the chip surface to the chip volume is:

$$\frac{S_c}{V_c} = \frac{2(r+1)}{rh} = \frac{2(r+1)}{\sqrt{\frac{v_w r}{v_s \lambda}}\sqrt{\frac{a_p}{d_e}}} \tag{8.33}$$

The above equation implies that if the grinding depth a_p is reduced to half, the surface of the chip surface area will increase by about 19% for the same material's volume. Considering the rubbing and ploughing actions involved in grinding surface creation, more energy will be consumed when grinding depth decreases. The size effect of grinding reflects the influence of rubbing and ploughing actions in grinding.

By considering each grain involved in the grinding process, the generation of a grinding surface can be simulated by a computer simulation program [11]. The basic procedure of the computer program to simulate the process of surface generation is illustrated in Figure 8.7. Three functions are represented in the grinding simulation. The first is the dressing process, which provides the location of the grains and the shape of the cutting edges. The second is the grinding process by the grains according to the grinding kinematics. The third function is to represent the action of each individual grain as it contributes to the total grinding forces and the surface profile, where the elastic and plastic deformations in the grinding are considered. The simulation demonstrates the grinding performance by accumulating the influences of the grain shape, and the elastic and plastic deformation. Both fracture wear and attritious wear of the grinding wheel is considered when the wheel surface is created. The effect of wheel wear is simulated by taking account of the change of grit shape and grit density during grinding. A simulated ground surface is illustrated in Figure 8.8 and is compared with an experimental result. It can be seen that the simulated surface contains features which bear a resemblance to the experimental surface.

In the analysis of grinding kinematics and mechanics, it has been demonstrated that the grinding chip thickness was influenced by grit shape and grit density on the wheel surface as indicated in Equation 8.10. The grit shape and grit density depend not only on the wheel intrinsic structure but also on the dressing operation. For a dressing operation using a single point diamond, a helix groove will be created on the wheel surface. A larger dressing depth and lead will generate a coarser wheel surface with fewer cutting edges on it. Koziarski and Golabczak suggested that the cutting edge density λ_0 of grinding wheel surface after dressing might be expressed as [12]:

$$\lambda_0 = c_0 a_d^x f_d^y \tag{8.34}$$

where a_d is the dressing depth, f_d is the dressing lead, c_0, and x and y are the constants. By replacing the λ_0 to λ in Equation 8.10, the effects of dressing on grinding chip thickness can be presented. Therefore the effects of dressing on grinding behaviours can be analysed. On the other hand, the effects of dressing on grinding behaviours can also be analysed by the simulation of dressing actions based on the dressing kinematics [13, 14]. Figure 8.9 illustrates the effects of dressing on grinding behaviours. It can be seen that larger dressing depth and lead gives larger surface roughness and lower grinding power.

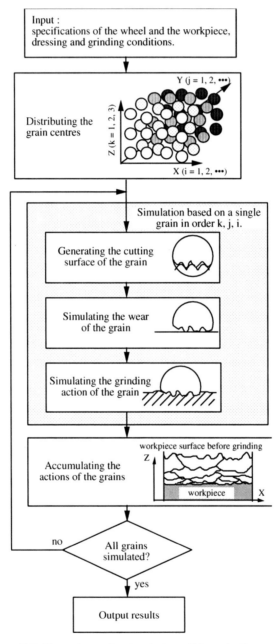

Figure 8.7. Flow chart of a simulation of the grinding process

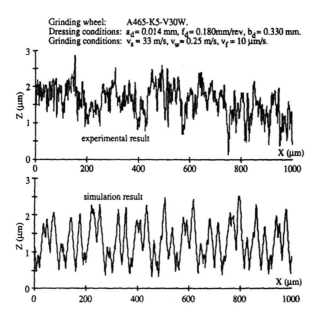

Figure 8.8. Comparison of the ground surface from the experiment and the simulation

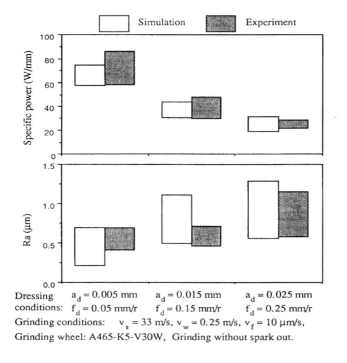

Figure 8.9. Variation of grinding performance under different dressing conditions

8.4 The Kinematics of a Grinding Cycle

Grinding behaviour in a grinding cycle is strongly affected by the deflections between the grinding wheel and the workpiece, resulting from the grinding force and grinding system stiffness. Typically the deflection is of the same order of magnitude as the depth of cut in precision grinding. Sometimes the deflections measured are considerably larger than the depth of cut. The selection of optimum grinding cycle parameters depends on the knowledge of the deflection performance of a grinding cycle. Figure 8.10 illustrates a simple plunge grinding cycle that consists of infeed and dwell stages. During the infeed stage, the wheel is fed into the workpiece at a constant rate. Due to the deflection of the grinding system, the reduction of workpiece size lags behind the indicated infeed position. This difference depends on the compliance of the grinding system and on the grinding force. In the dwell stage, which is also called spark-out, the residual stock continues to be removed until the wheel is retracted. The roundness and roughness also improve during the dwell period.

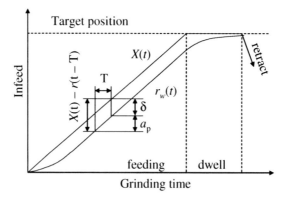

Figure 8.10. A simple plunge grinding cycle

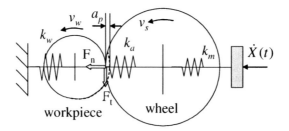

Figure 8.11. Idealised model of cylindrical grinding

A grinding machine can be simplified as Figure 8.11, where the machine structure supports the wheel with a linear spring of stiffness k_m and the workpiece with a linear spring of stiffness k_w. The stiffness of the grinding wheel at the contact with the workpiece is k_a. An expression for the overall effective stiffness k_e is:

$$\frac{1}{k_e} = \frac{1}{k_m} + \frac{1}{k_w} + \frac{1}{k_a} \tag{8.35}$$

when the infeed rate $v_f = \dot{X}$ is applied to the cross slide of the machine, and the wheel advances towards the workpiece. The wheel grinds the workpiece, generating the grinding interface force and the deflection in the grinding system. The deflection δ is given by:

$$\delta = F_n / k_e \tag{8.36}$$

where F_n is the normal component of the grinding force. Assuming the normal force is approximately proportional to the real depth of cut [15]:

$$F_n = k_{cn} a_p = \frac{k_{cn} \cdot \dot{r}_w}{n_w} \tag{8.37}$$

where k_{cn} is the grinding force coefficient, a_p is the real depth of cut, $\dot{r}$ is the rate of reduction of the workpiece radius and n_w is the workpiece rotational speed. If the grinding wheel wear is neglected, the difference between the command infeed velocity $\dot{X}$ and the actual infeed velocity $\dot{r}_w$ can be attributed to the changing radial elastic deflection $\dot{\delta}$ in the grinding system.

$$\dot{X} - \dot{r}_w = \dot{\delta} \tag{8.38}$$

The parameter r_w is the reduction in the radius of the workpiece and thus expresses the radial shape. Combining Equations 8.36, 8.37 and 8.38 leads to the controlling equation of the grinding system.

$$\dot{X} - \dot{r}_w = \frac{k_{cn}}{k_e n_w} \ddot{r}_w \tag{8.39}$$

So that:

$$\tau \frac{\partial^2 r_w}{\partial t^2} + \frac{\partial r_w}{\partial t} = v_f \tag{8.40}$$

where τ is the removal rate time constant of the grinding system. The removal rate time constant is a measure of the effects of the system stiffness k_e, the grinding force coefficient k_{cn} and the workpiece rotational speed n_w.

$$\tau = \frac{k_{cn}}{k_e n_w} \tag{8.41}$$

This first order model of material removal can be used to control grinding power, workpiece size and roundness. The effect of the grinding system compliance and the wheel sharpness are jointly represented by a single parameter, the removal rate time constant. By using the Equation 8.40, the reduction of the workpiece radius during the plunge stage can be expressed as:

$$r_w(t) = v_f(t - \tau + \tau \cdot e^{-t/\tau}) \tag{8.42}$$

When the grinding continues more than three time constant, it achieves a steady state, where the real depth of grinding cut is approximately equal to the nominal value. Therefore the grinding depth of cut is approximately as:

$$a_p = v_f/n_w \tag{8.43}$$

Grinding force and grinding power will be approximately constant. When grinding dwells from the moment t_1, the solution of Equation 8.40 is:

$$r_w(t) = v_f(t_1 - \tau \cdot e^{-(t-t_1)/\tau}) \tag{8.44}$$

Material removal during the dwell period depends on the deflection that built up during the plunge stage. According to Equations 8.36 and 8.37, the deflection of the grinding system in the steady-state before the commencement of dwell period is:

$$\delta = \frac{F_n}{k_e} = \frac{k_{cn} a_p}{k_e} \tag{8.45}$$

During the dwell stage, the depth of cut for the first spark-out pass is:

$$a_1 = \delta_0 - \delta_1 \tag{8.46}$$

where δ_0 is the initial deflection and δ_1 is the deflection in the first spark-out pass. For the subsequent spark-out passes, the real depths of cut are:

$$a_1 = \frac{a_p}{1 + \frac{k_e}{k_{cn}}}$$

$$a_2 = \frac{a_1}{1 + \frac{k_e}{k_{cn}}}$$

Generally, the real depth of cur for pass i will be:

$$a_i = \frac{a_{i-1}}{1 + \frac{k_e}{k_{cn}}} = \frac{a_p}{\left(1 + \frac{k_e}{k_{cn}}\right)^i} \tag{8.47}$$

The value of k_e/k_{cn} can be obtained by time constant measurement.

$$\frac{k_e}{k_{cn}} = \frac{1}{\tau n_w} \tag{8.48}$$

A commonly used time constant measuring method is based on the grinding force or power measurement [16]. Combining Equations 8.36 and 8.43, the grinding force during grinding dwell can be expressed as:

$$F_n(t) = F_n(t_1) \cdot e^{-\frac{t-t_1}{\tau}} \tag{8.49}$$

Therefore:

$$\tau = \frac{\log F_n(t) - \log F_n(t_1)}{t - t_1} \tag{8.50}$$

Considering the grinding performance in a grinding cycle, the grinding power is normally proportional to the grinding force. Therefore the grinding power can also be used to determine the system removal rate time constant. By integrating the force signal, Equation 8.49 becomes:

$$\int_{t_1} F_n(t) = \tau F_n(t_1)(1 - e^{-\frac{t-t_1}{\tau}}) \tag{8.51}$$

This refinement automatically filters noise, so it is a convenience when used with data logging on a computer. When t is large enough, the time constant can be approximated as:

$$\tau = \frac{\int_{t_1}^{t} F_n(t)}{F_n(t_1)} \tag{8.52}$$

Allanson used the weighted-least-mean-square method to estimate the time constant from power signals, which can automatically cope with a wide range of applications with a high noise environment [17].

$$\tau = \frac{\sum w_i (t_i - t_1)^2}{\sum w_i (t_i - t_1) \log_e \left(\frac{P(t_i)}{P(t_1)} \right)} \tag{8.53}$$

where w_i is the given weight depending on the variation of the signals, and t_i is the time when the power $P(t_i)$ is measured.

By using the simulation program shown in Figure 8.7, the improvements of the surface finish in the grinding spark-out stage are illustrated in Figures 8.12. The simulated results matches well with the experimental results.

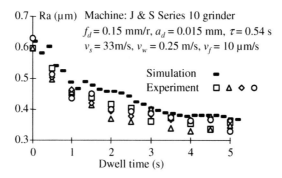

Figure 8.12. Roughness improvement during spark-out

Similar to the performance of the grinding force and the grinding power, the improvement of surface roughness during spark-out may also be described approximately by a time constant. A commonly used model is:

$$R(t) = (R_0 - R_\infty) e^{-t/\tau_R} + R_\infty \tag{8.54}$$

where $R(t)$ is the roughness at the time t from the commencement of spark-out, R_0 is the roughness at the commencement of the spark-out, R_∞ is the asymptotic value of roughness during the spark-out and τ_R is the time constant of the roughness decay curve. The roughness time constant is not usually equal to the removal rate time constant of the grinding system. As shown in Figure 8.12, the system removal rate time constant is 0.54 seconds, but the roughness improvement time constant is 1.21 seconds. The difference between the two time constants for the removal rate and for the surface roughness demonstrates that surface roughness is not related proportionally to the system deflection. This fact implies that the surface roughness follows a non-linear relationship with the grinding depth. The rate at which the residual stock is removed is slowed down due to plastic pile-up and therefore slows down the improvement in surface roughness. The simulation results also show that the time constant for the surface roughness is larger than the removal rate time constant. Salje, Mushardt and Scherf [18] also noticed the roughness improvement in spark-out did not necessarily follow an exponential decay. Rowe [19] demonstrated that the roughness improvement curve depended on the stiffness of the grinding system. Analysed by the simulation program in Figure 8.7, Table 8.1 shows the effect of the removal rate time constant on the improvement of the surface roughness. A small removal rate time constant provides less improvement of surface roughness during spark-out. As indicated in Equation 8.41, the removal rate time constant represents the effects of the wheel sharpness and the machine system stiffness. For the same wheel sharpness, a small time constant represents a stiff machine, and less residual stock is left for spark-out to remove. This demonstrates that a small time constant leaves less margin for surface roughness improvement.

Table 8.1. Simulation results of the improvement of surface roughness

System time constant	Roughness time constant	$R\infty$	$R0 - R\infty$
0.10 second	0.43 second	0.46 μm	0.15 μm
0.55 second	1.28 second	0.37 μm	0.32 μm
1.00 second	2.27 second	0.33 μm	0.35 μm

8.5 Applications of Grinding Kinematics and Mechanics

The quality of the grinding operation depends on a well-designed grinding cycle and the correct selection of grinding parameters. With the models of grinding kinematics and mechanics, grinding control strategies can be developed for different applications. For example, in order to improve the grinding productivity, the grinding infeed rate should be as high as possible. However the higher infeed rate requires a higher power consumption and may cause a grinding burn problem. A simulation of the grinding performance using available grinding models will provide a suitable strategy to solve such problems.

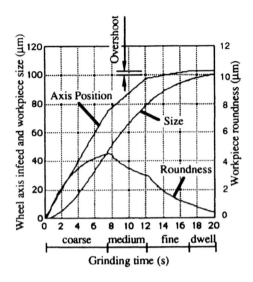

Figure 8.13. Multi-infeed and dwell grinding cycle simulation

Figure 8.13 illustrates the simulation results of a standard grinding cycle consisting of coarse infeed, medium infeed, fine infeed and dwell stages. The response of the system in bringing the workpiece to size was simulated and the overall cycle time, the final size errors and the final roundness were predicted. The grinding cycle was optimised to control errors within the required tolerances for size and roundness by adjusting the infeed rate, the infeed period, the overshoot and the dwell period. All changes caused by adjusting parameters were clearly evident in graphical representations of the grinding cycle.

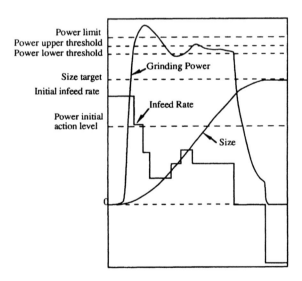

Figure 8.14. Grinding behaviours under a non-optimal power control strategy

In principle, the best strategy is to control the power level rather than the infeed rate in order to achieve a minimum cycle time consistent with specified quality requirements. This strategy requires that the power is increased to the set power limit as quickly as possible and the controller adjusts the infeed rate adaptively to maintain the power setting [15, 20, 21]. However the implementation of in-process power control presents a difficulty, because of deflections in the system. It is difficult to know when and how the infeed rate should be reduced to prevent the grinding power overshooting the power setting. This problem is illustrated in Figure 8.14, which indicates that the grinding power takes a considerable time to react to the decreases in infeed rate and a power "overshoot" occurs. This situation could be unacceptable if the power limit setting has been set to avoid poor size accuracy and possible burn. To avoid the occurrence of this phenomenon, the infeed rate needs to be reduced at appropriate times before the power limit is reached. This was achieved by setting an initial action level that was lower than the lower power threshold. The grinding process begins with the fastest infeed rate. When the grinding power is higher than the initial action level, the infeed rate decreases with set change steps. When the grinding power is between the required power control limits, the infeed rate is varied to constrain the power within this power control band. This is achieved by decreasing the infeed rate when the power is higher than the upper power threshold and increasing the infeed rate when the power is lower than the lower power threshold. In this way, the shape of the grinding power curve was adjusted to prevent a power overshoot.

The grinding cycle was optimised by adjusting the upper power threshold, the lower power threshold, and the initial action level together with infeed rate change steps. Figure 8.15 illustrates a simulation of a grinding cycle developed using the above procedure. The grinding power was controlled to remain under the set limit of 1.5 kW. It was found that the power reached its limit in about 1 second with no overshoot. Providing the data base values are appropriate, the simulation indicated that burn will not occur and the cycle time will be close to the shortest possible. In principle, the dwell time can also be minimized, if the system time constant is known. The effectiveness of the control strategy developed by the simulation technique was evaluated by using the parameters determined by these means on a real machine fitted with an appropriate adaptive control system.

Figure 8.16 illustrates the grinding process control achieved in a real grinding experiment using the values developed by simulation. Power overshoot was avoided and maximum power was reached in the shortest time possible for the particular system. In this trial the wheel speed was 33 m/s and the workpiece speed was 0.25 m/s. Because the infeed rate was frequently varied during the power control cycle, the control of the size accuracy and the roundness was worse than that of the cycle using a constant infeed rate. This means a longer dwell period was required to achieve the same accuracy. This problem could be solved by using an overshoot and retract strategy [22, 23]. Correct selection of dwelling time becomes an important issue.

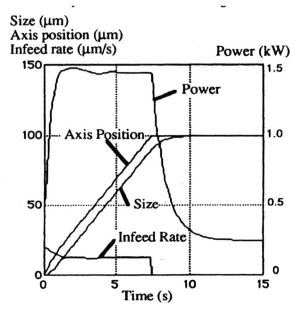

Figure 8.15. The simulated grinding cycle after optimisation

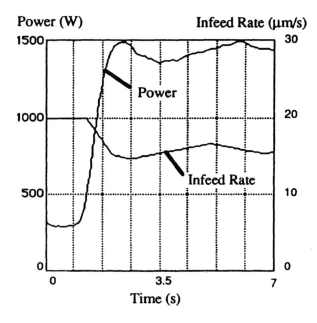

Figure 8.16. Experimental results of the grinding power control cycle determined by simulation

An adaptive control strategy was developed to determine a suitable grinding dwell time for grinding a long slender component of a 25:1 aspect ratio [17]. The workpieces were originally rough and finish ground using a traverse grinding cycle. As the parts had a large initial runout a workpiece steady could not be applied during rough grinding. This resulted in excessive time spending in air grinding. Initial tests using a standard multiplunge grinding cycle with a fixed dwell resulted in either an excessive cycle time or the generation of a stepped workpiece due to incomplete spark-out. As large diameter steps prevented finish traverse grinding, the workpieces were an ideal application for using a multi-plunge grinding cycle. Each plunge cycle relied on the adaptive dwell strategy to select the correct dwell time. The grinding parameters used for the tests ware listed in Table 8.2.

Table 8.2. Conditions of a multi-plunge grinding case

Wheel speed	33 m/s
Workpiece speed	15 m/s
Infeed rate	10 μm/s
Dwell time	3 time constants
Stock removal	250-300 μm off-diameter
Grinding wheel	White aluminium oxide 60 grit, dressed before each workpiece
Wheel diameter	440 mm
Wheel width	50.0 mm
Workpiece	Stainless steel, diameter approximately 25 mm, length approximately 600 mm

It was found that the dwell control strategy automatically adjusted the dwell period to compensate for the change in the workpiece stiffness at the different plunge positions. The typical variation in the estimated time constant is illustrated in Figure 8.17. It can be seen that the time constants changed from approximately 3 seconds at the plunge position closest to the tailstock to 62 seconds at mid-span (corresponding to the 20:1 range). Close to the workhead the time constant was 7.3 seconds. The multiplunge adaptive dwelling control cycle reduced the barrelling error from 150-180 μm to 25 μm in diameter. After the multiplunge cycle, the workpieces were finish ground using a traverse grinding cycle and a diameter gauge to the required tolerances. The use of the adaptive dwell cycle allowed the workpiece to be rough ground close to the finished size and greatly reduced the time previously caused by the need to eliminate barrelling during the finish grinding operation. This resulted in a more than 50% reduction in cycle time and a reduction in the need for operator intervention.

Variation in grinding performance comes not only from the variation in grinding kinematic parameters but also from the change of cutting edges on the wheel surface. As analyzed previously, the grinding chip thickness depends on the shape and distribution of cutting edges, which is influenced by the dressing operation and wheel wear situation. Generally, grinding behaviour immediately after dressing is strongly affected by dressing than subsequently [24]. Although the initial stage is relatively short, it can substantially affect the redress life, the wheel and the stability of grinding. An inadequate dressing operation may cause more than 100% variation of grinding performance in a wheel redress life cycle. Figure 8.18 shows the grinding behaviours in a wheel redress life cycle. With the same grinding kinematical condition, the variation of grinding behaviour can only contribute to the variation of grit shape and grit distribution. Umino and Shinozaki suggested that the initial cutting edge density of active grits after dressing will converge towards a critical density in the grinding process [25]. The variation of the cutting edge density $\lambda(t)$ was expressed as

$$\lambda(t) = \lambda_e + (\lambda_0 - \lambda_e)e^{-mn_s t} \qquad (8.55)$$

where λ_0 is the initial cutting edge density after dressing, λ_e is the critical cutting edge density at the end of wheel redress life, m is the probability of grain fracture in one revolution of the wheel, n_s is the wheel rotational speed, and t is the wheel grinding time after dressing.

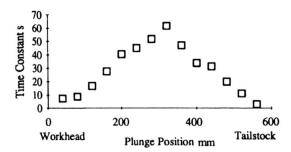

Figure 8.17. Time constant estimated during multiplunge grinding of a long slender part

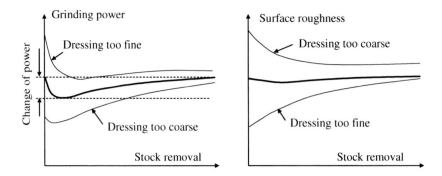

Figure 8.18. Variation of grinding power and surface roughness in a wheel redress life cycle

Based on the observation of grinding performance in a wheel redress life cycle, Chen [26] introduced a grinding power model for the selection of dressing and grinding conditions. The model is expressed as Equation 8.56, which takes account of grinding kinematics, dressing operation and wheel wear:

$$F_t(t) = \{k_2 + (k_1 a_d^\alpha f_d^\beta - k_2)e^{-mn_s t}\} b\left(\frac{v_s}{v_w}\right)^{\frac{\eta}{2}} d_e^{\frac{\eta}{2}} h_{eq}^{1-\frac{\eta}{2}} + k_3 b n_s t \quad (8.56)$$

where k_1, k_2, k_3, α and β are constants; other parameters have been defined previously. It should be noted that the constants in this model are subject to other nonmentioned parameters. For example the wear of diamond dresser will have a strong influence on the k_1 and wheel wear performance will govern the k_2 and k_3. So far, the mechanism of wheel wear and diamond wear is not well understood. Therefore this model has to be used adaptively according to the grinding situations.

In order to minimize the variation of grinding in a wheel redress life cycle, a suitable strategy is to make the grinding behaviour immediately after dressing the same as or similar to that at the final steady stage of redress life. The effort can be focused on the grinding behaviours at the beginning and the end of the redress life cycle. The grinding behaviours at the steady stage of the grinding redress cycle are dominated by the grinding kinematical conditions, which may be presented by the equivalent chip thickness. The grinding performance at the beginning of wheel life cycle can be controlled by adjusting the dressing parameters. It is noticed that the dressing depth has a stronger influence on grinding power but less influence on surface roughness. Conversely, dressing lead gives strong effects on surface roughness than on grinding power. An adaptive control strategy was developed to minimize the variation of grinding behaviour in the wheel redress life cycle [26, 27]. It was suggested that dressing depth was used to control grinding power immediately after dressing; dressing lead was selected to control surface roughness and equivalent chip thickness was determined to balance the grinding behaviours at the steady stage.

8.6 Summary

Kinematics and the mechanics of grinding presents relationships of grinding control variables and process output parameters. It has demonstrated that a grinding process performance can be analysed by considering the grinding as a system. All grinding behaviours can be related to the core of grinding process - the action of individual grit. Focusing on the individual grit load, the relationship between grinding inputs and outputs can be easily defined. The behaviour of the grinding process can be presented as the average actions of the grits on the grinding wheel surface or illustrated by accumulating the individual action of the grits that are involved in a grinding process.

A special grinding behaviour, the so-called size effect, can be explained by considering the rubbing and ploughing actions in grinding. When the grinding chip size reduces, the chip surface area per unit of material removal increases. This leads to a higher proportion of rubbing and ploughing in grinding. More energy is required to remove the same amount of materials.

The stiffness of the grinding system has strong influences on the grinding behaviour in a grinding cycle. The correct selection of a grinding cycle is important to ensure the quality and efficiency. Investigation shows that the spark-out is the most important stage for the reduction of surface roughness in a plunge grinding cycle. The reduction of surface roughness does not follow the changes of the system deflection. To achieve the best surface roughness requires a longer spark-out than that for size and roundness control. The ratio of roughness time constant to the removal rate time constant varies. Therefore for the control of the surface roughness, the dwell time should be determined according to the roughness time constant, which depends on the effects of plastic deformation and the non-linear performance of the grinding system.

Most aspects of grinding can be simulated using grinding kinematics and mechanics and the following list is a summary of the areas in which the technique has been successfully used to study grinding cycle performance:

- observing the effects of variability in the system time constant;
- predicting the overall cycle duration;
- predicting final size errors;
- predicting roundness errors for a given dwell period;
- observing the effects of changing infeed rates after various time durations;
- designing the grinding cycle for optimal results under particular constraints;
- demonstrating effects of the dressing condition in a grinding cycle;
- demonstrating surface roughness improvement during the spark-out stage;
- communicating and illustrating ideas for the purposes of training;
- presenting graphical representations of improvement in the grinding cycle.

The kinematics and mechanics models provide a good guidance for the grinding process control. As indicated in this chapter, the core of grinding is the grinding chip shape, which is determined by grinding kinematical parameters and the cutting edge shape and distribution. The shape and distribution of cutting edges on the wheel surface are governed by the dressing operation and wheel wear performance. Therefore the main control parameters should include grinding kinematical parameters and dressing parameters. As a system the grinding environmental conditions should also be considered. This includes the coolant delivery and machine tool structure. To minimise the variation of grinding performance in a wheel redress life cycle, the selection of dressing conditions is very important. Optimal grinding performance may be achieved by balancing grinding behaviour immediately after dressing and at the end of the grinding steady stage. With the help of system time constant identification, the variance of grinding system stiffness can be compensated by using suitable adaptive control strategies. With a better understanding of grinding kinematics and mechanics, grinding is no longer a black art.

References

[1] Tonshoff H K, Karpuschewski B, Mandrysch T. Grinding process achievements and consequences on machine tools challenges and opportunities, *Annals of the CIRP*, 1998. Vol. 47, No. 2, 651–668.
[2] Blunt L. Development and dissemination of SOFTGAUGES for surface topography, National Measurement System Science & Technology Programme in Length Metrology, NMS Project Ref No: Length Metrology (2002–2005) 4.4.
[3] Chen X. Systematic consideration of grinding process monitoring, Proceedings of the 8th Chinese Automation & Computer Science Conference in the UK, 2002. 175–178.
[4] Ono K. Analysis on the grinding force, *Bulletin of the Japan Society of Grinding Engineers*, 1961. No. 1, 19–22.
[5] Werner G. Influence of work material on grinding force, *Annals of the CIRP*, 1978. Vol. 27, No. 1, 243–248.
[6] Shaw M C. A new theory of grinding, Proceedings of International Conference on Science in Australia, Monash University, Australia, 1971. 1–16.
[7] Saini D P, Wager J G, Brown R H. Practical significance of contact deflections in grinding, *Annals of the CIRP*, 1982. Vol. 31, No. 1, 215–219.
[8] Nakayama K, Brecker J, Shaw M C. Grinding wheel elasticity, *Transactions of the ASME: Journal of Engineering for Industry*, 1971. Vol. 93, No. 5, 609–614.
[9] Hamed M S. Grinding Mechanics – Single Grit Approach, *PhD Thesis*, 1977. Leicester Polytechnic, UK.
[10] Lortz W. A model of the cutting mechanism in grinding, *Wear*, 1979. Vol. 53, 115–128.
[11] Chen X. Strategy for the Selection of Grinding Wheel Dressing Conditions, *PhD Thesis*, 1995. Liverpool John Moores University, UK.
[12] Koziarski A, Golabczak A. The assessment of the grinding wheel cutting surface condition after dressing with the single point diamond dresser, *International Journal of Machine Tools and Manufacture*, 1985. Vol. 25, No. 4, 313–325.
[13] Chen X, Rowe W B. Analysis and simulation of the grinding process – part I, generation of the grinding wheel surface, *International Journal of Machine Tools and Manufacture*, 1996. Vol. 36, No. 8, 871–882.
[14] Chen X, Rowe W B, Mills B, Allanson D R. Analysis and simulation of the grinding process – part IV, effects of wheel wear, *International Journal of Machine Tools and Manufacture*, 1998. Vol. 38, No. 1–2, 41–49.
[15] Hahn R S. Controlled-force grinding – a new technique for precision internal grinding, Transactions of the ASME: Journal of Engineering for Industry, 1964. Vol. 86, No. 8, 287–293.
[16] Chen X, Allanson D, Thomas A, Moruzzi J L, Rowe W B. Simulation of feed cycles for grinding between centres, *International Journal of Machine Tools and Manufacture*, 1994. Vol. 34, No. 5, 603–616.
[17] Allanson D R, Rowe W B, Chen X, Boyle A. Automatic dwell control in computer numerical control plunge grinding, Proceedings of the Institution of Mechanical Engineers – Part B: Journal of Engineering Manufacture, 1997. Vol. 211, 565–575.
[18] Salje E, Mushardt H, Scherf E. Optimization of short grinding cycles, *Annals of the CIRP*, 1980. Vol. 29, No. 2, 477–495.

[19] Rowe W B. An experimental investigation of grinding machine compliances and improvements in productivity, *Proceedings of the 14th International MTDR Conference*, 1973. Manchester, UK, 479–486.
[20] Malkin S. Thermal aspects of grinding, Transactions of the ASME: Journal of Engineering for Industry, 1974. Vol. 96, 1184–1191.
[21] Amitay G, Malkin S, Koren Y. Adaptive control optimization of grinding, Transactions of the ASME: Journal of Engineering for Industry, 1981. Vol. 103, 103–108.
[22] Malkin S and Koren Y. Optimal infeed control for accelerated spark–out in plunge grinding, Transactions of the ASME: Journal of Engineering for Industry, 1984. Vol. 106, 70–74.
[23] Cai G Q, Allanson D, Rowe W B, Morgan M. Accuracy in plunge grinding cycles with overshoot and back–off, The CIRP International Conference on Precision Engineering, Tianjin University, P. R. China, 1991.
[24] Rowe W B, Chen X, Morgan M N. The identification of dressing strategies for optimal grinding wheel performance, *Proceedings of the 30th International MATADOR Conference*, 1993. Manchester, UK, 195–202.
[25] Umino K, Shinozaki N. One aspect of variation of grinding wheel surface based on grinding force analysis – studies on wear and redress life of grinding wheel (1st report), Semishu Kikai, 1976. Vol. 42, No. 4, 299–305.
[26] Chen X, Rowe W B, Allanson D R, Mills B. A grinding power model for selection of dressing and grinding conditions, Transactions of the ASME: Journal of Manufacturing Science and Engineering, 1999. Vol.121, 632–637.
[27] Rowe W B, Chen X, Mills B. Towards an adaptive strategy for dressing in grinding operations, *Proceedings of the 31st International MATADOR Conference*, 1995. Manchester, UK, 415–420.

Materials-induced Vibration in Single Point Diamond Turning

W.B. Lee, C.F. Cheung and S. To

Ultra-Precision Machining Centre
Department of Industrial and Systems Engineering
The Hong Kong Polytechnic University
Hung Hom, Kowloon, Hong Kong

9.1 Introduction

With the advances in the micro electronic and computer technologies, the design and fabrication of ultra-precision machines have been improving dramatically over the past several decades. However, these have not been parallel with the development and the understanding of the physical mechanism of the micro and nano-cutting process and how to obtain a quality surface with the surface finish down to nanometre level. Such knowledge is indispensable in the design and control of advanced ultra-precision machine tools.

Ultra-precision machines with an extreme high stiffness have been constructed [1-4]. A mirror-like machined surface can be obtained without the need for any post-machining polishing. However, there is a limit to which the machine vibration can be reduced without leaving a signature on the machined work surface. Vibrations of precision machining system comes from various sources. These include machine design, tool wear and chatter, and ambient factors. The one which is intrinsic and could not be removed is the vibration caused by the changing microstructure of the workpiece being cut, unless the work material is perfectly homogeneous. Such variation in the material properties will induce a change in the micro-cutting force and hence a change in the dynamics of the cutting system.

In single point diamond turning (SPDT), the depth of cut is normally of the order of a few micrometers, which is normally smaller than the grain size. In other words, cutting is performed within a single grain. As a result, the crystalline structure has a great influence on the quality of the machined surface. In conventional cutting, there is no need to consider the fluctuation of the cutting force induced by the changes in grain orientation. Although most of the materials used in diamond turning are polycrystalline in nature, the use of single crystal materials is not uncommon. Examples include the use of a KDP single crystal in laser systems and single crystal silicon for high precision infrared optical systems.

Single crystal materials are known to be highly anisotropic in their physical and mechanical properties. The local variation in machinability due to the variation of the crystallographic orientation induces a local variation in the surface qualities [5]. There is also strong experimental evidence of significant variation in local surface roughness in diamond turned aluminium alloy. Such local a variation in surface quality has been attributed to the anisotropy of each grain which makes up the surface to be machined. Although there is research work studying the effect of crystallographic orientation on surface roughness in SPDT, most of this work focused on qualitative analysis [5]. Only a limited number of studies [6-8] have been found on the development of a quantitative model for explaining the effect of crystallographic orientation on the cutting force variation and surface roughness [9]. The vibration induced by the changing crystallography of the machined workpiece on the cutting forces imposes a limit on the performance that can be obtained from an ultra-precision machine even when the machine can be made vibration free.

The variation of micro-cutting forces caused by the changing crystallographic orientation of the material being cut will induce vibration on the machine-tool-work system. It is a kind of self-excited vibration which is inherent in any cutting system for crystalline materials. This kind of vibration results in a local variation of surface roughness of the diamond turned surface. In this chapter, a model-based simulation system is adopted to predict the materials induced vibration and its effect on the surface generation in ultra-precision machining. The model takes into account the effect of machining parameters, the tool geometry, the relative tool-work motion as well as the crystallographic orientation of the materials being cut.

9.2 A Model-based Simulation of the Nano-surface Generation

As shown in Figure 9.1, the modular-based simulation system is basically composed of several model elements which include a microplasticity model [8], a dynamic model and an enhanced surface topography model. The microplasticity model is used for the prediction of the periodic fluctuation of micro-cutting forces due to the changing crystallographic orientation of the materials being cut. Apart from the fine tool-work vibration due to the machining environment, such a periodic variation of micro-cutting forces also induces additional tool-work vibration in the cutting system which is referred to in the present study as materials induced vibration. In the model-based simulation system, a dynamic model is deployed to determine the materials induced vibration. The influence of this vibration on the surface roughness of the workpiece is predicted by an enhanced surface topography model which takes into account process parameters, tool geometry and the resultant vibration between the tool and the workpiece.

9.2.1 A Prediction of the Periodic Fluctuation of Micro-cutting Forces

In the machining process, the shear angle affects the cutting forces [10] and hence the surface roughness of the workpiece. There are various classical shear angle theories to correlate the relationship between the magnitude of the shear angle, the cutting tool geometry, and the friction angles, the notable ones being the Merchant Theory and the slip-line field theory. However, the magnitude of the shear angle cannot be determined directly and has to be deduced from the measurement of the chip thickness obtained from the cutting experiment.

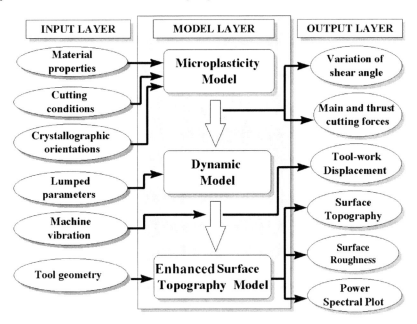

Figure 9.1. The architecture of the model-based simulation system

Sato *et al.* [6] has attempted to use the continuum yield theory to analyse the shear stress and shear angle with material anisotropy. However, his attempt has been deemed to be unsuccessful since the value of the shear angle was in the reverse phase to shear stress. From the extensive experiments done by the research team [11-18], there clearly exists a strong relationship between the crystallographic texture or orientation and the surface roughness obtained in diamond turning (Figure 9.2). Based on the crystal plasticity, the shear angle can be inferred from the value of the Taylor factor for a given crystallographic orientation. The shear angle is predicted to occur where the minimum value of the effective Taylor factor determined from the load instability criterion is a minimum and the number of dislocation slip systems is also the smallest [8, 19, 20]. Figure 9.3 shows the shear angle obtained from cutting two crystals of different orientations with a tool with a zero rake angle. There is a good agreement between the experimental results and the ones predicted from the theory (Table 9.1).

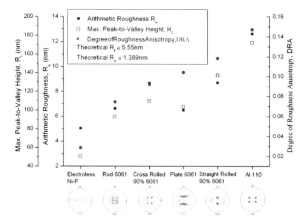

Figure 9.2. The effect of crystallographic orientation on surface roughness

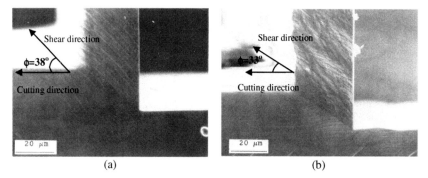

Figure 9.3. SEM micrograph of shear zone formation of a copper single crystal when cutting was done along **a** [001] direction and **b** [$\bar{3}$ 24] on (230) plane

The shear angle thus calculated can be incorporated into various cutting mechanics models to predict the corresponding changes in the cutting force [21-27]. During diamond turning, the crystallographic orientation of the workpiece material change with the cutting directions and this brings about the periodic fluctuations in the cutting forces. These fluctuations depend on the symmetry of the crystallography of the substrate materials and the spindle speed. The effect of the cutting force variation can be measured by a force transducer and analyzed by a power spectrum analysis.

Table 9.2 shows the predicted shear angle variation, and the corresponding variation of the cutting force in diamond turning aluminium single crystals with (001) and (111) planes. It is found that the shear angle varies with the crystallographic orientation of the material being cut. There seems to exist a fundamental cyclic frequency of variation of the cutting force for each workpiece revolution which is determined to be four for (001) and three or six for (111) and (111) crystals. These relate closely to the crystallographic orientation of the crystals being investigated. The predicted and the measured spectral plots for the variation of cutting forces in diamond turning of (001) and (111) crystals are also shown. The predicted spectra are found to agree well with the measured spectra.

Table 9.1. The measured and predicted shear angle for different cutting direction on a (230) plane

Cutting plane	Cutting direction	Measured shear angle	Predicted shear angle
(230)	[001]	38	40
(230)	[$\bar{3}$24]	33	37

Table 9.2. Comparison of the variation of shear angle, cutting forces, and predicted and measured models of the cutting force in a diamond turning aluminium single crystal with different crystallographic orientations

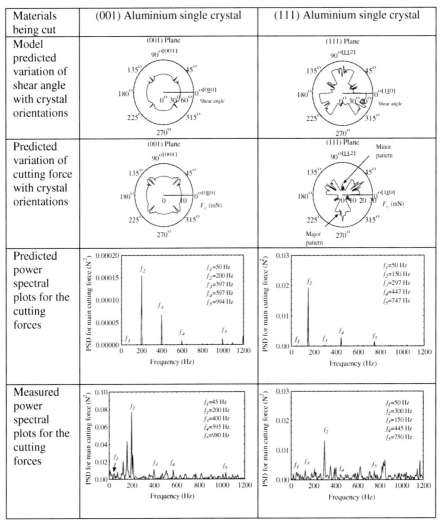

Table 9.3. Comparison between the predicted and measured dominant frequency components of the variation of the cutting forces

Specimen no.	Dominant frequency component f_2 (Hz)			
Aluminium	Cutting force		Thrust force	
Single crystal	Predicted	Measured	Predicted	Measured
(001) plane	200	200	200	200
(111) plane	150	300	150	300
(111) plane*	297	150	297	140

Note : * Frequency component at f_3

Table 9.4. Comparison between the predicted and measured fundamental cyclic frequency of cutting forces variation

Specimen no.	Fundamental cyclic frequency of cutting forces variation (cycles per workpiece revolution)			
Aluminium	Cutting force		Thrust force	
Single Crystal	Predicted	Measured	Predicted	Measured
(001) plane	4.0	4.0	4.0	4.0
(111) plane	3.0	6.0	3.0	6.0
(111) plane*	5.94	3.0	5.94	2.8

Note: * Frequency component at f_3

When comparing (001) and (111) single crystals, distinctive patterns of frequency distribution are observed for different crystals. Both the predicted and the measured spectra exhibit a dominant frequency component (f_2) which can be correlated to the crystallographic orientation of the work material being cut. As shown in Tables 9.4 and 9.5, the captioned component appears at almost the exact fundamental cyclic frequency for (001) crystals. For the (111) single crystal, the measured dominant frequency component (300 Hz) appears almost double to that of the predicted dominant frequency component (150 Hz) and is almost identical to its first harmonic (297 Hz). The cyclic cutting forces are shown to be made up of two patterns hereby referred to as the major pattern, with a higher amplitude, and a minor pattern, with a lower amplitude. The presence of six peaks is well predicted except that there is a discrepancy in the amplitude of the peak. This suggests that the (111) single crystal might possess two fundamental cyclic frequencies which are three for the major pattern of the cutting force variation and six for both of the major and the minor patterns of the cutting force variation. The overall results support the argument that the variation of the cutting forces is related closely to the crystallographic orientation of the crystals being investigated. The microplasticity model is proven to be helpful in explaining the variation of micro-cutting forces in diamond turning crystalline materials. The main features of the cutting forces patterns are well predicted and confirmed by the cutting tests. There is a good agreement between the experimental findings and the predicted results.

9.2.2 Characterization of the Dynamic Cutting System

The influence of vibration on the surface roughness machined with a single point tool has been previously studied by a number of researchers [28-31]. Mitsui and Sato [32] concluded that the surface roughness affected by the vibration was nearly equal to the whole amplitude of the vibration. A surface finish better than the whole amplitude of vibration, however, is often obtained in ultra-precision machining. Since some of the chatter marks were removed by the following cuts, Tai et al. [33] emphasized that the height of chatter marks in the cutting direction became smaller than the whole amplitude of chatter vibration. However, the surface roughness of the machined surface in the transverse direction is more dominant than it is in the cutting direction. This is due to the very low frequency vibration present in ultra-precision machining. In other words, the surface roughness in the infeed cutting direction is more dominant than it is in the cutting direction, as the relative displacement of the tool in the cutting direction does not engage in the surface generation [34]. As a result, only the relative displacement between the workpiece and the tool in the infeed direction is considered and the cutting system is modelled with one degree of freedom in consideration of the cutting tool and the workpiece.

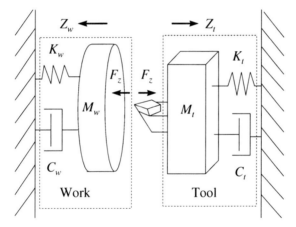

Figure 9.4. A lumped-parameter representation of the dynamic cutting system

As shown in Figure 9.4, the complexity of the machine structure is represented in terms of a lumped-parameter for the purposes of the analytical study. The lumped parameters in such a representation are the equivalent mass, the rigidity and the damping for one of the modes of oscillation of the machine structure. The tool and work systems are represented by springs and dashpots connected in parallel. The tool and the workpiece are acted upon by the periodic fluctuation of thrust force F_z (t) as determined by the microplasticity model [21]. The parameters of the dynamic cutting system can be estimated by the covariance equivalent ARMA (2, 1) model [24, 35, 36]. Figure 9.5 shows the results of the characterization of the dynamic characteristics of the tool and work systems. It is found that the predicted displacement values conform well to the measured values.

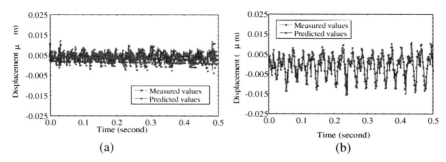

Figure 9.5. A comparison between the measured and the predicted displacement values for **a** the tool and **b** the work systems, respectively

Figures 9.6 and 9.7 show the predicted materials induced vibration and the corresponding power spectra in the diamond turning of (110) and (111) aluminium signal crystals, respectively. It is observed that the vibration induced by the change of the crystallographic orientation of work materials is not a simple harmonic. Each crystal has a distinctive vibration pattern. As shown in the spectral plots of the vibration in Figures 9.6 b and 9.7 b, the spectra are composed mainly of the harmonics of the rotational frequency of the spindle, *i.e.*, 33.3 Hz. However, the dominant frequency components vary with the crystallographic orientation of the work materials [36]. They are identified to be 66.7 Hz and 100.0 Hz for the (110) and (111) crystals, respectively. These components are equivalent to the first harmonic and the second harmonic of the rotational frequency of the spindle. It is interesting to note that the locations of these components closely match the fundamental cyclic frequency of the variation of the cutting forces. The effect of the materials induced vibration on the tool loci is simulated in Figure 9.8.

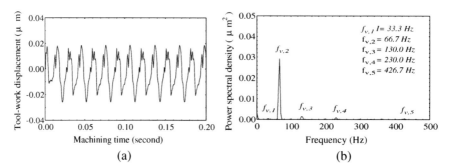

Figure 9.6. a Predicted materials induced vibration and **b** its power spectrum for machining aluminium single crystal with (110) as the cutting plane

There is a difference between the frequency of surface modulation induced by the materials induced vibration and that induced by the relative vibration between the tool and the workpiece. The former depends on the crystallographic orientation of the work material and is thus independent of the phase relationships. The machine tool vibration, on the other hand, induces surface modulation with a frequency which depends on the ratio of frequency of vibration to the spindle rotational speed. In other words, the former is material dependent whereas the latter is machine dependent.

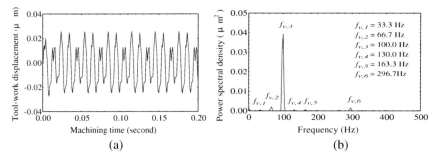

Figure 9.7. a Predicted materials induced vibration and **b** its power spectrum for machining aluminium single crystal with (111) as cutting plane

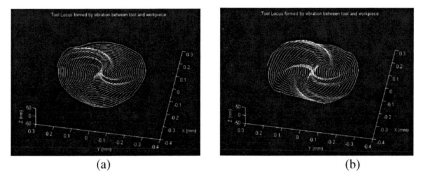

Figure 9.8. Simulated tool loci for face cutting of aluminium single crystals on: **a** (110) plane and **b** (111) plane

9.2.3 A Surface Topography Model for the Prediction of Nano-surface Generation

SPDT is a complex metal removal process. The achievement of a super mirror finish still depends much on the experience and skills of the machine operators through an expensive trial and error approach when new materials or new machine tools are used. Since the surface quality plays a major role in the functional performance of a workpiece, the process of surface generation in machining operations has attracted a lot of research interest [33, 37]. Pandit and Revach [38] made use of the data dependent systems (DDS) approach to decompose the surface roughness profiles into wavelength components. Sata *et al.* [31] used the Fast Fourier Transform (FFT) spectrum to analyse the factors affecting surface generation. Attempts have been made in developing models for the simulation of surface generation in different machining processes. Bael *et al.* [34], Montgomery and Altintas [39], and Ismall *et al.* [40] put forward a model for the milling process. Recently, Lin and Chang [41] have developed a two-dimensional surface topography model for lathe turning.

Although some attempts [3, 42] reported in the three-dimensional (3D) analysis of surface generation, their emphases were placed on the synthesis of surface topography from measurement data captured by the interferometric type of instruments. They failed to predict the surface generation based on the cutting mechanics, a factor affecting the surface generation and the machine dynamics. Relatively little research work [44-46] has reported in analysing the generation of 3D surfaces based on machine kinematics and cutting theories.

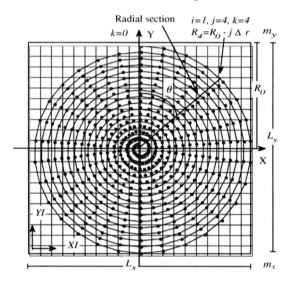

Figure 9.9. Tool locus and linear mapping of surface data on a cross lattice

In the modelling of surface topography, the real trick to the model is its new approach to the deterministic modelling and simulation of the 3D surface topography of a diamond-turned surface [18, 24, 36, 47-49]. It uses the surface roughness profiles predicted at a finite number of equally spaced radial sections of the workpiece to construct the surface topography of a diamond turned surface. The roughness data are determined from the surface roughness profiles predicted at a finite number of equally spaced radial sections on the workpiece as shown in Figure 9.9. The surface roughness profiles are predicted based on a 2D surface topography model [50].

In the 2D surface topography model, the surface roughness profile is formed by successive movements of the tool profiles at intervals of the tool feed along the tool locus as shown in Figure 9.10. Surface roughness depends on the location of the succeeding cutting edges which start removing chips from the surface profile formed by the preceding cutting edges. Since the minimum cutting edge profile below the intersecting points of each tool profile constitutes the surface roughness, a surface roughness profile can be constructed by trimming the lines above the intersecting points. The plot illustrates the movement of the tool tip at each feed position across the workpiece. Such a movement can be visualised as successive movements of tool profiles across the workpiece as shown in Figure 9.11. As shown in Figure 9.12, a good agreement between the simulated and the measured profiles is observed. The difference between the predicted and measured arithmetic roughness values is less than 10%.

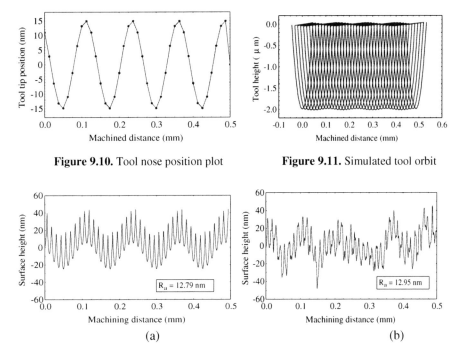

Figure 9.10. Tool nose position plot   **Figure 9.11.** Simulated tool orbit

(a)   (b)

Figure 9.12. A comparison between **a** the simulated and **b** measured surface roughness profiles

These surface roughness data from different radial sections of the surface are mapped on the surface elements of a cross lattice (see Figure 9.9). The surface elements are used to build the mesh and the parametric surfaces [47] which are best fitted to the surface topography data. The contour levels of the parametric surfaces are proportional to the surface height. This is a departure from conventional surface topography approaches that are based on the measurement data captured by interferometry or scanning electron microscopy. As the effect of material anisotropy caused by the changing crystallographic orientation of the work material is taken into account, an additional displacement between the tool and the workpiece is introduced into the cutting system. Since the force function $F_z(t)$ varies with the crystallographic orientation of the workpiece, the patterns of the variation of the cutting force is not a simple harmonic function. Instead, it can be treated as an arbitrary function which varies with the crystallographic orientation of the work materials. The resultant displacement between the tool and the workpiece is an arbitrary function which varies with the crystallographic orientation of the work materials being cut. The new approach of 3D surface modelling is particularly useful for the prediction and analysis of local variations of surface roughness in SPDT of highly anisotropy materials.

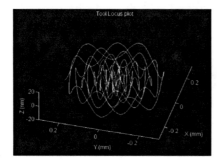

Figure 9.13. 3D plot of the tool locus

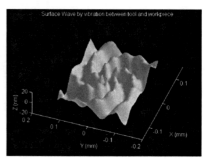

Figure 9.14. Virtual surface waviness

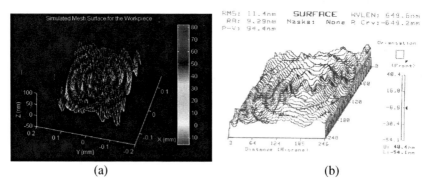

Figure 9.15. a Simulated and **b** measured 3-D surface topographies

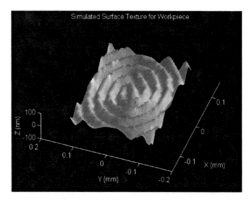

Figure 9.16. Virtual surface topography of the turned workpiece

Figure 9.13 shows the locus of the tool generated under the relative displacement between the tool and the workpiece. The tool moves with a spiral locus on the X-Y plane towards the centre of the workpiece. Simultaneously, the tool vibrates with a simple harmonic motion in the infeed cutting direction, *i.e.*, the Z-axis direction. The kinematics of the tool motion generates modulation/waviness on the machined surface as shown in Figure 9.14. A simulated 3D surface topography of the diamond turned surface is produced by adding the tool geometry component to the model as shown in Figure 9.15. The model is found to give an estimate of the surface roughness value R_t of 92.6 nm which is closest to the measured value of 94.4 nm. The experimental results agree well with the predicted results. Figure 9.16 shows a virtual surface topography of a diamond turned workpiece. The use of virtual surface representation allows us to visualise and examine in detail the various features of the surface topography.

9.2.4 Prediction of the Effect of Tool Interference

In SPDT, high spindle rotational speed together with fine feed rate are usually adopted for improving the surface roughness quality. Under these conditions, a phenomenon, the so-called interference of the tool [30, 36, 47, 48, 51] occurs at which the preceding movement of the tool has already removed the chip which should be cut away by some of the succeeding tool movements. This results in the non-existence of intersecting points at two or more successive cutting tool profiles. In this case, there is/are non-existing point(s) of intersection.

Takasu *et al.* [30] has established criteria for the occurrence of the tool interference. However, these criteria did not determine the exact locations at which this phenomenon actually takes place. In the simulation model, the exact locations of tool interference are determined by continual checking for the existence of intersecting points at each tool feed movement. Should a tool profile not intersect with its previous and succeeding tool profiles, it is skipped and the next closest intersection of tool pattern is used instead, in the estimation of the surface roughness profile. This allows the effect of tool interference to be reflected in the simulated surface roughness profile and the predicted surface roughness value [48].

As shown in Figure 9.17, the cutting tool cuts sequentially the active cutting edge formed by the preceding cutting edge in each tool feed under a non-interference condition [51]. The formation of the surface roughness profile can be treated as the trimming of the lines above the intersecting points of the minimum edge profile. Therefore, a clear tool mark and sharp peaks with an exact periodicity of the feed components can be found in the surface roughness profile and its spectrum, respectively. As the interference of the tool occurs, the cutting edges of the preceding cuts remove the matcrials that should be cut by the succeeding cuts. Taking Cut #1 and Cut #2 in Figure 9.18 as an example, the cutting edges in these cuts have already removed the material that should be cut by Cut #3. Therefore, only two tool marks were formed in these three cuts. Figure 9.19 depicts the effect of the tool interference on the simulated 3D surface topographies of a diamond turned surface.

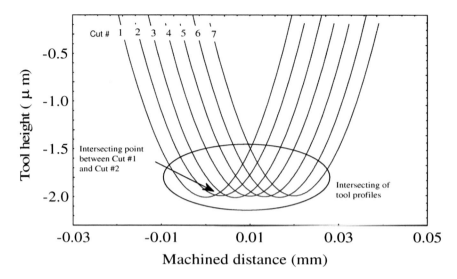

Figure 9.17. Graphical illustration of non-interference of tool under the following conditions: a feed rate of 10 mm/min; a spindle speed of 3,000 rpm; a depth of cut of 2 μm; a tool nose radius of 0.1 mm; and an amplitude and frequency of tool-work vibration of 0.01 μm and 30 Hz, respectively

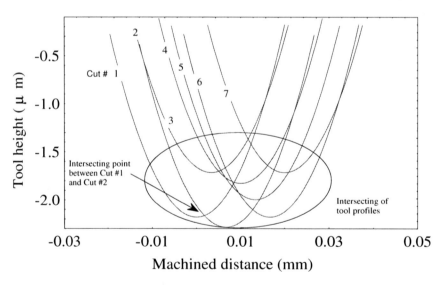

Figure 9.18. Graphical illustration of interference of tool under the following conditions: a feed rate of 10 mm/min; a spindle speed of 3,000 rpm; a depth of cut of 2 μm; a tool nose radius of 0.1 mm; and an amplitude and frequency of tool-work vibration of 0.3 μm and 30 Hz, respectively

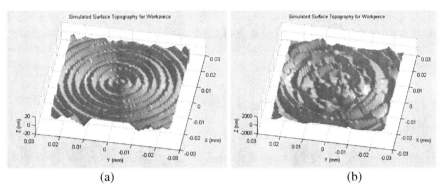

Figure 9.19. A comparison between the simulated 3D surface topographies generated under **a** non-interference and **b** interference of tool conditions as in Figures 9.17 and 9.18, respectively.

9.2.5 Prediction of the Effect of Material Anisotropy

As the effect of material anisotropy caused by the changing crystallographic orientation of the work material is taken into account, an additional displacement between the tool and workpiece is introduced into the cutting system. Since the force function $F_z(t)$ varies with the crystallographic orientation of the workpiece, the patterns of the variation of the cutting force is not a simple harmonic function but can be treated as an arbitrary function which varies with the crystallographic orientation of the work materials. The resultant displacement between the tool and the workpiece varies with the crystallographic orientation of the work material being cut. The surface topography model is modified to accomplish the additional displacement due to the changes induced by the crystallographic orientation of the material being cut.

Table 9.5. Comparison between the predicted and the measured arithmetic roughness

Specimen no.	Mean arithmetic roughness $\overline{R}_a$ (nm)	
	Predicted	Measured
(110) Aluminium single crystal	13.38	17.18
(111) Aluminium single crystal	15.28	19.99

Figures 9.20 and 9.21 show the simulated and the measured surface topographies for the diamond turned (110) plane and (111) plane, respectively. The simulated surface topographies are found to be similar to those obtained from the WYKO interferometric microscope for the work materials being investigated. The topographies exhibit combined features of surface modulation caused by the materials induced vibration and the relative displacement between the tool and the workpiece. Most of the surface features found in the measured patterns are reflected in the simulated patterns. Table 9.5 tabulates the predicted and the measured arithmetic roughness R_a values of the machined surfaces. The predicted results were found to agree well with the measured results. As a whole, the model-

based simulation system is proven to be helpful in explaining the additional roughness due to the variation of the crystallographic orientation of the workpiece. There is a good agreement between the experimental findings and the simulation results.

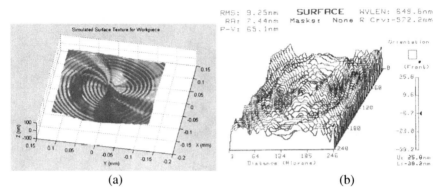

Figure 9.20. a Simulated and **b** measured surface topographies for the face cutting of an aluminium single crystal with (110) plane (combined effects)

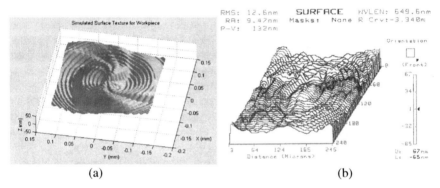

Figure 9.21. a Simulated and **b** measured surface topographies for the face cutting of an aluminium single crystal with (111) plane (combined effects)

9.3 Conclusions

The present study is the first to come up with a mesoplasticity model which unifies the different length scales of study from grain orientation, and cutting geometry to machine characteristics and comes up with a 3D simulation model that can predict the surface topology of the surface with details down to the nano-level. Moreover, a model-based simulation system is presented for the quantitative determination of the variation of local surface roughness due to materials induced vibration in SPDT. The system is based on several model elements which include a microplasticity model, a dynamic model and an enhanced surface topography model. The microplasticity model is used for the prediction of the variation of micro-cutting forces caused by the changing crystallographic orientation of the workpiece materials being cut. The surface roughness and the topography of the machined surface are predicted by an enhanced surface topography model. A software package was developed to implement the simulation system, and the performance of the system has been evaluated through a series of cutting experiments. Experimental results indicate that the variation of the cutting forces and the surface roughness are related closely to the crystallographic orientation of the crystals being cut. As the depth of cut increases, the influences of the crystallographic orientation of the single crystal materials on micro-cutting forces could be pronounced. Overall, the simulation results are found to agree well with the experimental ones.

This research is a new attempt in which the microplasticity theory, the theory of system dynamics and the machining theory are integrated to address the materials problems encountered in ultra-precision machining. Indeed, this is a new attempt to link up the microplasticity theory to macro-mechanisms in metal cutting. The successful development of the model-based simulation system allows the prediction of the magnitude and the effect of periodic fluctuation of the micro-cutting force and its effect on the nano-surface generation in SPDT of crystalline materials. It also helps to explain quantitatively the additional roughness caused by the variation of the crystallographic properties of the workpiece, and leads us to a better understanding of the further improvement of the performance of ultra-precision machines.

Acknowledgements

The authors would like to express their sincere thanks to the Research Grants Council of the Hong Kong Special Administrative Region of the People's Republic of China for financial support of the research work under the project no. PolyU 5189/02E.

References

[1] McKeown, P.A., Structural design for high precision measurement and control, SPIE, 1978; 153: 101.
[2] Carlisle, K. and Shore, P., Experiences in the development of ultra stiff CNC aspheric generating machine tools for ductile regime grinding of brittle materials, Proceedings of the 6th International Seminar on Precision Engineering, Braunschweig, Germany, 1991:85.
[3] McKeown, P.A., Carlisle, K., Shore, P., and Read, R.F.J., Ultraprecision, high stiffness cnc grinding machines for ductile mode grinding of brittle materials, SPIE, Infrared Technology and Application, 1990; 1320: 301–313.
[4] Mizumoto, H., Matsubara, T., Yamamoto, H., Okuno, K. and Yabuya, M., An infinite–stiffness aerostatic bearing with an exhaust–control restrictor, Proceedings of the 6th International Seminar on Precision Engineering, Braunschweig, Germany, 1991:315.

[5] Nakasuji, T., Kodera, S., Hara, S., and Matsunaga, H., Diamond turning of brittle materials for optical components, Annals of the CIRP, 1990; 39 (1): 89.
[6] Sato, M., Kato, K., and Tuchiya,K., Effect of material and anisotropy upon the cutting mechanism, Transactions of JIM., 1978; 9: 530.
[7] Yuan, Z.J., Lee, W.B., Yao, Y.X., and Zhou, M., Effect of crystallographic orientation on cutting forces and surface quality in diamond cutting of single crystal, Annals of the CIRP, 1994; 43 (1): 39–42.
[8] Lee, W.B. and Zhou, M., A theoretical analysis of the effect of crystallographic orientation on chip formation in micro–machining, International Journal of Machine Tools and Manufacture, 1993; 33 (3): 439–477.
[9] Lee, W.B., Cheung, C.F., and To, S., Materials induced vibration in ultra–precision machining, Journal of Materials Processing Technology, 1999; 89–90: 318–325.
[10] Arcona, C. and Dow, Th. A., An empirical tool force model for precision machining, journal of Manufacturing Science and Engineering, 1998; 120: 44, 700–707.
[11] Cheung, C.F., Influence of cutting friction on the anisotropy of surface properties in ultra–precision machining of brittle single crystals, Scripta Materialia, 2003; 48(8): 1213–1218.
[12] Cheung, C.F. and Lee, W.B., Study of factors affecting the surface quality in ultra–precision diamond turning, Materials and Manufacturing Processes, 2000; 15 (4): 481–502.
[13] Cheung, C.F., To, S. and Lee, W.B., Anisotropy of surface roughness in diamond turning of brittle single crystals, Materials and Manufacturing Processes, 2002; 17 (2): 251–267.
[14] To, S., Lee W.B. and Chan, C.Y., A study on ultraprecision diamond turning of aluminium single crystals, Journal of Materials Processing Technology, 1997, 63 (1–3): 157–162.
[15] To, S., Lee, W.B. and Chan, C.Y., Effect of machining velocity on the crystallographic texture in a diamond turned aluminum single crystal, Textures and Microstructures, 1999; 13 (4): 249–261.
[16] To, S., Lee, W.B. and Cheung, C.F., Orientation changes of substrate materials of aluminium single crystal in ultra–precision diamond turning, Journal of Materials Processing Technology, 2003; 140 (1–3): 346–351.
[17] Lee, W.B., To, S. and Chan, C.Y., Deformation band formation in metal cutting, Scripta Materialia, 1999; 40 (4): 439–443.
[18] Lee, W.B., Cheung, C.F. and Li, J.G., Prediction of 3–D surface topography in ultra–precision machining, China Mechanical Engineering, Transactions of CMES, 2000; 11 (8): 845–848.
[19] Yang, W. and Lee, W.B., Mesoplasticity and its Applications, Springer–Verlag, London: 1993.
[20] Lee, W.B. and Chan, K.C., A criterion for the prediction of shear band angles in F.C.C. metals, Acta Metallurgica et Materialia, 1991; 39 (3): 411–417.
[21] Ernst, H., and Merchant, M.E., Chip formation, friction, and high quality machined surfaces, Surface Treatment of Metals, ASM, 1941; 29: 299.
[22] Shaw, M.C., Metal Cutting Principles, Clarendon Press, Oxford: 1984.
[23] Shaw, M.C., Shear strain in cutting, Crookall, J.R. and Shaw, M.C., Metal Cutting Principles, Clarendon Press, Oxford: 1984; 168.
[24] Lee, W.B. and Cheung, C.F., A dynamic surface topography model for the prediction of nano–surface generation in ultra–precision machining, International Journal of Mechanical Sciences, 2001; 43 (4): 961–991.
[25] Lee, W.B., Cheung, C.F. and To, S., A microplasticity analysis of micro–cutting force variation in ultra–precision diamond turning, Transactions of the ASME: Journal of Manufacturing Science and Engineering, 2002; 124 (2): 170–177.

[26] Lee, W.B., Cheung, C.F. and To, S., Friction induced fluctuation of cutting forces in diamond turning of aluminium single crystals, Proceedings of The Institute of Mechanical Engineers, Part B: Journal of Engineering Manufacture, 2003; 217 (B5): 615–631.
[27] Lee, W.B., To, S. Sze, Y.K. and Cheung, C.F., Effect of materials anisotropy on shear angle prediction in metal cutting – a mesoplasticity approach, International Journal of Mechanical Sciences, 2004; 45 (10): 1739–1749.
[28] Cheung, C.F. and Lee, W.B., A multi–spectrum analysis of surface roughness formation in ultra–precision machining, Precision Engineering, 2000; 24 (1): 77–87.
[29] Cheung, C.F. and Lee, W.B. An investigation of cutting dynamics in single point diamond turning, JSME International Journal–Series C: Mechanical Systems, Machine Elements and Manufacturing, 2000; 43 (1): 116–126.
[30] Takasu, S, Masuda, M and Nishiguchi, T., Influence of study vibration with small amplitude upon surface roughness in diamond machining, Annals of the CIRP, 1985; 34 (1): 463–467.
[31] Sata, T., Li, M., Takata, S., Hiraoka, H., Li, C.Q., Xing, X.Z. and Xiao. X.G., Analysis of surface roughness generation in turning operation and its applications, Annals of the CIRP, 1985; 34 (1): 473–476.
[32] Mitsui, K. and Sato, H., Frequency characteristic of cutting process identified by a in–process measurement of surface roughness, Annals of the CIRP, 1978; 27 (1): 67–71.
[33] Tai, T.P., Yang, Y.C., Hwong, Y.C. and Ku, C.H., A new concept of cutting marks formation in metal cutting vibration, Proceedings of the 20th MTDR, London: The Macmillan Press Ltd, 1979; 449–456.
[34] Bael, D.K., Ko, T.J. and Kim, H.S., A dynamic surface roughness model for face milling, Precision Engineering, 1997; 20 (2): 171–178.
[35] Pandit, S.M. and Wu, S.M., Time Series and System Analysis with Applications, New York, Wiley, 1983.
[36] Cheung, C.F. and Lee, W.B., Surface Generation in Ultra–precision Diamond Turning: Modelling and Practices, Engineering Research Series, Professional Engineering Publishing Limited of Northgate Avenue, Bury St Edmunds, Suffolk, IP32 6BW, UK, 2003.
[37] DeVries, W.R., Autoregressive time series models for surface profile characterization, Annuals of the CIRP, 1979; 28 (1): 437–440.
[38] Pandit, S.M. and Revach, S., A data dependent systems approach to dynamics of surface generation in turning, Transactions of the ASME: Journal of Engineering for Industry, 1981; 103B: 437–445.
[39] Montgomery, M. and Altintas, Y., Mechanism of cutting force and surface generation in dynamic milling, Transactions of the ASME: Journal of Engineering for Industry, 1991, 113: 160–168.
[40] Ismall, F., Elbestawi, M.A., Du, R. and Urbasik, K., Generation of milled surfaces including tool dynamic and wear, Transactions of the ASME: Journal of Engineering for Industry, 1993; 115: 245–252.
[41] Lin, S.C. and Chang, M.F., A study on the effects of vibrations on the surface finish using a surface topography simulation model for turning, International Journal of Machine Tools and Manufacture, 1998; 38: 763–782.
[42] Sato, H. and O–hori, M. Characteristics of two dimensional surface roughness – taking self–excited chatter marks as objective, Annals of the CIRP, 1981; 30 (1): 481–486.
[43] Bispink, T. Performance analysis of feed–drive systems in diamond turning by machining specified test samples, Annals of the CIRP, 1992; 41 (1): 601–604.
[44] Byne, G. New approach to the theoretical analysis of surface generation mechanisms in machining, Annals of the CIRP, 1992; 41 (1): 67–70.

[45] Tsai, M.D., Takata, S., Inui, M., Kimura, F. and Sata, T. Prediction of chatter vibration by means of a model–based cutting simulation system, Annals of the CIRP, 1990; 39 (1): 447–450.
[46] Weck, M., Hartel, R. and Modemann, K. Performance assessment in ultra–precision micromachining, Annals of the CIRP, 1988; 37 (1): 499–502.
[47] Cheung, C.F. and Lee, W.B., Modelling and simulation of surface topography in ultra–precision diamond turning, Proceedings of the Institution of Mechanical Engineers, Part B: Journal of Engineering Manufacture, 2000; 214 (B6): 463–480.
[48] Cheung, C.F. and Lee, W.B., Prediction of the effect of tool interference on surface generation in single–point diamond turning, International Journal of Advanced Manufacturing Technology, 2002; 19 (4): 245–252.
[49] Yin Z.Q. and Li, S.Y., Simulation of surface topography in diamond turning under the effect of vibration, Proceedings of the Second International Conference on Precision Engineering and Nano Technology, Changsha, Hunan, China, 2002; 106–112.
[50] Cheung, C.F. and Lee, W.B. A theoretical and experimental investigation of surface roughness formation in ultra–precision diamond turning, International Journal of Machine Tools and Manufacture, 2000; 40 (7): 979–1002.
[51] Cheung, C.F. and Lee, W.B., Characterization of nano–surface generation in single–point diamond turning, International Journal of Machine Tools and Manufacture, 2001; 41 (6): 851–875.

10

Design of Precision Machines

Dehong Huo[a], Kai Cheng[a] and Frank Wardle[b]

[a]Advanced Manufacturing and Enterprise Engineering (AMEE) Department
School of Engineering and Design
Brunel University
Uxbridge, Middlesex UB8 3PH, UK

[b]UPM Ltd, Mill Lane, Stanton Fitzwarren, Swindon, SN6 7SA, UK

10.1 Introduction

Precision machines are essential in modern industry and directly affect machining accuracy, repeatability, productivity and efficiency. Generally, the design of a precision machine mainly includes the design of its key elements such as mechanical structure, spindles and drive system, control and inspection systems, etc. There is a lot of literature available on the design of machine elements [1-5]; while it is difficult to cover in details on design of precision machine in one chapter. In this chapter, therefore, emphasis is placed on the mechanical and structural design of precision machines, relevant design methodology and tools driven by dynamics. Furthermore, the chapter focuses on the integrated approach for modelling and simulation of the machine and machining dynamics, and thus achieving an optimal design of the machine and its performance in the dynamic machining process.

There are two basic approaches to studying the machining dynamics. One is cutting force modelling that does not contain structural parameters of machines and has been discussed in the previous chapters; the other is the dynamic modelling of the machine and structure or the machining processes. The methodology presented in this chapter falls into the latter approach, but towards an integrated approach to covering the dynamics of both the machine and the machining process. Section 10.2 discusses design principles of precision machine tools, including the machine configuration, tool-workpiece loops, and stiffness, mass and damping issues. The material presented is only a refined formulation with an emphasis on machining dynamics. Section 10.3 formulates the machine design methodology, covering a dynamics-driven design process, modelling and simulation enhancement, and design guidelines. Section 10.4 presents the implementation perspectives on the design of precision machines, and a practical design process based on finite element analysis (FEA). Section 10.5 provides three case studies on the design of a fast tool servo system, a 5-axis bench-type milling machine tool and a precision grinding machine.

10.2 Principles

Precision machine tools are highly dynamic systems in order to sustain the required accuracy, productivity and repeatability. The precision of a machine is affected by the positioning accuracy of the cutting tool with respect to the workpiece surfaces and their relative structural and dynamics loop precisions, which are fundamental and essential for the machine design.

10.2.1 Machine Tool Constitutions

A typical precision machine consists of five major sub-systems. They are mechanical structure, the spindle and drive system, the tooling and fixture system, the control and sensor system, and the measurement and inspection system. These sub-systems are essential and directly contribute to machine tool performance. Figure 10.1 highlights the machine tool constitution and key evaluation criteria for the machine tool's performance. Because of varied machining purposes and different machine configurations, Figure 10.1 cannot be very comprehensive, but rather provides a thorough summary for understanding the machine tool constitution and its performance evaluation related to machining dynamics in particular.

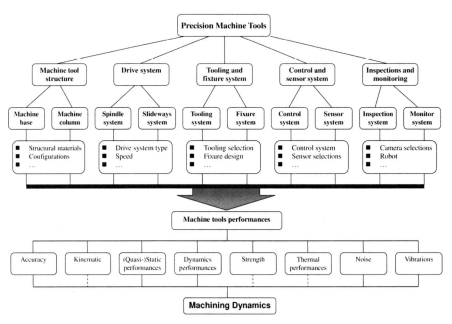

Figure 10.1. Machine tool constitutions and performance

10.2.1.1 Mechanical Structure
Mechanical structure is normally comprised of stationary and moving mechanical bodies. The stationary bodies include machine base, the column and spindle housing, etc. They usually carry moving bodies, such as worktables, slides and carriages. The structural design is critical since the mechanical structure not only provides the support and accommodation for all the machine's components but also contributes to dynamics performance possessed in a machine tool. To achieve a high stiffness, and a damping and thermal stability, two major design issues are involved in mechanical structure design, *i.e.*, the material selection and configuration.

The material selection for a machine tool structure is one of the essential factors in determining final machine performance, with many criteria being considered, such as temporal stability, specific stiffness, homogeneity, easiness of manufacturing and cost, *etc.* [6].

Although there are a number of structural materials available, up to now only a few materials have been chosen to build machine tool structures. Cast iron has widely been used for many years due to inexpensiveness and good damping characteristics to minimize the influence of dynamics loads and transients. There are still many cast iron applications in precision machine tools albeit its high initial cost of fabricating patterns and moulds and the poor environment of operating foundries [7]. Granite is another popular material used to build precision machine base and slideways because of its low thermal expansion and damping capacity. The drawback of granite is that it can absorb moisture so it should be used in a dry environment. For this reason many machine tool builders seal the granite with epoxy resin. The need for higher material damping and light weight in ultra-precision machining, combined with long-term dimensional and geometrical stability, leads to the development and usage of polymer concrete for machine tool structural elements in spite of its low strength [8].

Figure 10.2 illustrates the comparison of relative damping capacity between cast iron and polymer concrete material. The vertical scale is the amplitude of vibration, and the horizontal scale is time. This figure shows that polymer concrete has better vibration absorption capacity compared to cast iron.

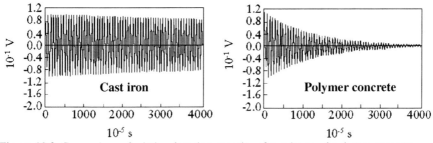

Figure 10.2. Comparison of relative damping capacity of cast iron and polymer concrete

The symmetry and closed loop structural configuration are widely used in precision machine tool design. Among various configurations the "T" configuration is popularly used for most of the precision turning and grinding machines. Recently, a novel tetrahedron structure proposed by the NPL in England has been applied in an internally damped space frame with all the loads carried in a closed loop. The design generates a very high stiffness coupled with exceptional dynamics stiffness [9], albeit its complexity and cost are increased in manufacturing and assembly.

10.2.1.2 The Spindle and the Feed Drive System
Spindle is a key element of the precision machine tool because the spindle motion error will have significant effects on the surface quality and accuracy of machined components. The most often used spindles in precision machine tools are aerostatic spindles and hydrostatic spindles. They both have high motion accuracy and are capable of high rotational speed. An aerostatic spindle has lower stiffness than an oil hydrostatic spindle. Aerostatic spindles are widely used in machine tools with medium and small loading capacity while hydrostatic spindles are often applied in large heavy-load precision machine tools.

Accurate linear motions are generated by the use of slideways. Similarly, aerostatic slideways and hydrostatic slideways have been frequently applied in precision machine design and are replacing contacting surface type slideways.

On the drive side, the electric AC motor and DC brushless motor for high speed spindles are frequently built into the spindle so as to reduce the inertia and friction produced by the motor spindle shaft coupling as well as the dynamic. DC and AC linear motor drives can perform a long stroke direct drive and thus eliminate the need for conversion mechanisms such as lead screws, belt drives, and rack and pinions, with potentially better performance in terms of stiffness, acceleration, speed, smoothness of motion, accuracy and repeatability, *etc.* [10]; however, there are only limited applications in machine tools though linear motors have been available for a long time [7]. Friction drives are very predictable and reproducible due to a prescribed level of preload at the statically determinate wheel contacts, thereby superior in machining optically smooth surfaces [11]. But there are some practical considerations that restrict the application of friction drives in machine tools. One such limitation is referred to as the thermal capacity. Therefore, it is difficult for friction drive to achieve a high speed operation.

10.2.1.3 The Tooling and the Fixture System
Fixture system and tooling are the essential parts of the machining system. They also play significant roles in the machine tool design, because they are at the end of the machine tool-machining loop. The deformation of the tooling and the fixture system both in static and dynamic circumstances will entirely be copied to the workpiece surface and hence influence the workpiece form and dimensional accuracy as well as its surface texture and topography.

In contrast to the machine tool dynamics, the dynamics of tooling and fixture could change significantly, depending on the location of the cutting tool with respect to the workpiece, owing to its localized structure and geometry of the workpiece [12]. In precision machine tool design, it is very important if designers can take this varying dynamics into account in spite of the possible difficulty, because this will be helpful to accurately evaluate the machine dynamics and errors budget. In practice, the dynamics change by the location of the cutting tool with respect to the workpiece can be dealt with by putting a larger safe bandwidth of the machine tool, and the speed of spindle can be limited as designed to decrease this dynamic change.

10.2.1.4 Control and Sensor System
Computer numerical control (CNC) was introduced into the machine tools industry in the early 1970s and since then many companies started to develop their own control systems for machine tools. The control sub-system includes motors, amplifiers, switches and the controlled sequence and time. High speed multi-axis CNC controllers are essential for the efficient control of, not only servo drives in high precision position loop synchronism for contouring, but also thermal and geometrical error compensation, optimized tool setting and direct entry of the equation of shapes [13].

From the dynamics viewpoint, stiffness in the control system indicates the capability to hold a position when dynamic forces try to move it. Therefore, a proper design of the control system and its algorithms can lead to a high servo-stiffness and hence improve machining precision through the machine tools.

10.2.1.5 The Metrology and the Inspection System
The metrology and the inspection systems are the basis for the qualifying assurance of precision machining and enabling the technology to be widely applied in industry. On the other hand, higher level accuracy assurance in metrology and inspection system is also a drive for precision machines towards a higher precision requested for the future engineering industry. Fast and accurate positioning of the cutting tools towards the workpiece surface and monitoring of the tool conditions visually by the operator should be integrated into the inspection system especially for on-line operation purposes.

10.2.1.6 The Machine Tool Performance Evaluation
The overall objective of the design of machine tool sub-systems discussed above is to achieve required machine performances. The performances are evaluated normally in the following aspects:
- Accuracy
- Kinematics
- Static performances
- Dynamics performances
- Strength performances
- Thermal performances

- Noise
- Vibration

These machine performances are collectively reflected on the tool-workpiece loop in terms of stiffness, thermal stability, static and dynamics as shown in Figure 10.1. The following sections will focus on the tool-workpiece loops, in relation to the machine dynamics in particular.

10.2.2 Machine Tool Loops and the Dynamics of Machine Tools

From a machining viewpoint, the main function of a machine tool is to accurately and repeatedly control the point of contact between the cutting tool and the uncut material - the "machining interface". This interface is normally better defined as tool-workpiece loops. Figure 10.3 shows a typical machine tool-workpiece loop. The position loop is the relative position between the workpiece and the cutting tools which directly contributes to the precision of a machine tool and directly leads to the machining errors.

On the other hand, deformations introduced by stiffness/compliance and the thermal loop are two important aspects in tool-workpiece loops. The stiffness loop in a machine tool is a sophisticated system. The stiffness loop of the machine includes the cutting tool, the tool holder, the slideways and stages used to move the tool and/or the workpiece, the spindle holding the workpiece or the tool, the chuck/collet, fixtures, and internal vibration, and other dynamic effects. The physical quantities in the stiffness loop are force and displacement. During machining, the cutting forces at the machining point will be transmitted to the machine tool via the stiffness loop and return to the original point thus closing the loop. Influences outside of the structural loop, which still influence the loop and cause errors, include floor vibration, temperature changes, and cutting fluids.

The thermal dynamic loop, which is similar to the stiffness loop, contains all the joints and structure elements that position the cutting tool and workpiece. The difference lies in that the physical quantity in the thermally dynamic loop is temperature and heat flux which are transmitted in the loop.

Machine tool vibrations play an important role in determining structural deformations and dynamics performance. Furthermore, excessive vibrations accelerate tool wear and chipping, cause poor surfaces, and damage to the machine tool component. These types of vibrations have been discussed in Chapter 2.

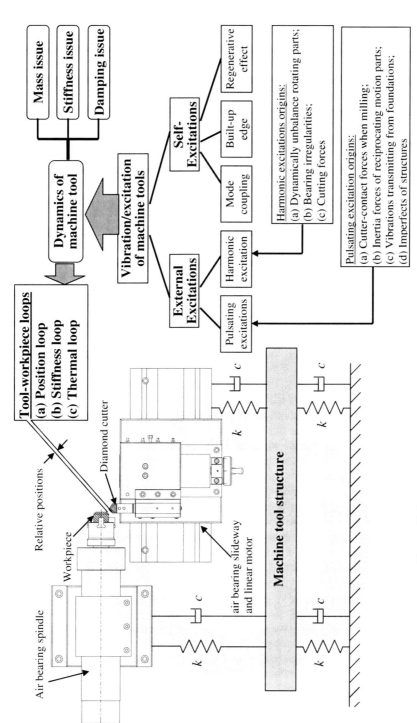

Figure 10.3. Machine tool loops and dynamics of machine tools

10.2.3 Stiffness, Mass and Damping

The dynamic behaviours of machine structures result from the exchange and dissipation of energies, and dynamic loads, which transfer their energy to the structure, which then response via several mechanisms, such as bending or extension [14]. During the process of energies transferring, stiffness, mass and damping are the main measures governing the machine and process dynamics. Furthermore, the dynamics of the machine system behave dynamically differently in the machining process because of the change of the loading and position all the time. Therefore, the dynamics measures of the machine system discussed below are broadly concerned with the machine system in the dynamic machining process context.

10.2.3.1 Stiffness
Stiffness normally can be defined as the capability of the structure to resist deformation or to hold a position under the applied loads. Static stiffness in machine tools refers to the performance of structures under the static or quasi-static loads. Static loads in machine tools normally come from gravity and cutting force, etc. Apart from the static loads, machine tools are subjected to constantly changing dynamic forces and the machine tool structure will deform according to the amplitude and frequency of the dynamic excitation loads, which is termed dynamic stiffness. Dynamic stiffness of the system can be measured using an excitation load with a frequency equal to the damped natural frequency of the structure.

The structural stiffness of a machine tool is one of the main criteria in the design of a precision machine tool. Generally speaking, both the static and dynamic stiffness of the machine tool structure determine the machine's accuracy and productivity. Therefore, high stiffness is required both statically and dynamically, which individually affects different aspects of the machining process.

Static stiffness affects displacement between cutting tools and workpieces during cutting. High static stiffness is required to produce parts to a desired size, shape and form, although finish machining in precision machines often takes place with a depth of cut of a few micron or even smaller and correspondingly light cutting forces. The resulting deflections can still be excessively large if the machine has inadequate static stiffness. The resulting deflection can thus exceed the machining tolerance of workpieces.

The need for high dynamic stiffness in precision machines results mainly from two separate aspects of the machining process. In the first case inadequate dynamic stiffness will result in a poor quality surface finish of the machined parts due to relatively low levels of vibrations occurring during machining processes. In the second case low dynamic stiffness can lead to chatter and even damage the cutting tool and machine structures.

Normally static stiffness is relatively easy to be predicted during the design process, and hence can be optimized so as to obtain the desired machining accuracy. Dynamic stiffness is difficult to predict at the design stage. With some modelling and analysis tools, such as finite element analysis (FEA), however, dynamic stiffness can be analyzed and precisely determined. This is important for the machining process optimization and thus optimally maximizing the machine performance in term of machining accuracy and productivity in particular. Section 5 provides a case study in analyzing static and dynamic stiffness for a 5-axis milling machine tool using FEA.

10.2.3.2 Mass
Mass or inertia is an issue that influences precision machine performance and cost. Generally speaking, reducing mass is an everlasting objective of machine design and highly desirable in precision machines design. A lightweight structure can benefit the precision machines in number of ways:

- less mass can enhance the ability of the machine to respond to high frequency input and action requirements
- less mass can offer a higher natural frequency
- less mass can improve the damping ratio
- less mass can reduce the manufacture cost.

There are two approaches to reduce mass. One is selecting lighter materials with higher performance, the other is structural optimal design in which mass is reduced with the requirement being met at the same time.

There are numerous structural materials available, some with a very high stiffness-to-mass ratio, such as an aluminium-magnesium alloy and a titanium alloy, *etc*. Unlike the aircraft and aerospace industries in which most advanced material has been widely used, precision machine design has to take the cost into account. Up to now only a few practical materials have been chosen to build machine tool structures.

Lightweight design or structural optimization design, on the other hand, became the main approach to reduce mass in precision design. Many structural optimization methods from parameter optimization to shape optimization and even topology optimization have been proposed. Numerical methods, such as the finite element method, have been providing practical solving approaches for these structural optimization methods.

It should also be noted that higher mass helps reduce high frequency noise and reduce vibrations. Therefore, the machine bed in some precision machines is built using high mass materials like solid granite table to reduce vibrations.

10.2.3.3 Damping
Damping defines the ability of a system or structure to dissipate energy. Among stiffness, mass and damping that affect the dynamics of a machine tool, the importance of damping probably is the easiest issue to neglect. However, damping in fact plays a very important role in the dynamic performance of precision machines. Some of the most important issues includes:

- absorbing energy from the process and thus to reduce vibrations
- preventing chatter and damage to the machined surface and machines
- absorbing energy from structural modes excited by the servo and drive system.

Sources of damping can be classified into internal damping and external damping: the internal damping includes material damping and joint damping. Material damping is a property of every material. Compared to other dampers material damping is very low and sometimes can be ignored. Joint damping exists between joint surfaces. The more joint surfaces a machine has, the more damping a machine has. However the stiffness will be reduced because of joint surfaces.

Damping can be from external sources, and it usually is termed dampers. The kinds of dampers for machine tools include tuned mass dampers, constrained layer dampers, velocity feedback in servos and actively controlled damped masses attached to the structure.

Figure 10.4. Squeeze film dampers (Courtesy: UPM Ltd)

Figure 10.4 is the oil lubricated squeeze film dampers proposed by UPM Ltd which are used in an aerostatic slideway to improve its dynamic performance. Dry cutting trials were performed with and without squeeze film dampers fitted to the slide for a workpiece speed of 500 rpm, a tool feed rate of 3 mm/min and a depth of cut of 5 um. Without damping the surface finish obtained was 0.51 μm R_a whereas with damping a finish of 0.1 μm R_a was obtained. The results show that the dampers produce substantial improvements in dynamic stiffness and that sufficient drive stiffness can be obtained to enable the slide to be used in some precision machining applications.

10.3 Methodology

10.3.1 Design Processes of the Precision Machine

As illustrated in Figure 10.5, the design of a precision machine tool requires the basic step of customers' requirements clarification, system functional requirements, conceptual design, analysis and simulation, experimental analysis, detailed design, and design follow up, *etc.*, albeit the full design process is always iterative, parallel, nonlinear, multidisciplinary and open-ended to any innovative and rational ideas and improvements. The functional requirements of a precision machine will address the machine geometry, kinematics, dynamics, the power requirement, materials, the sensor and control, safety, ergonomics, production, assembly, quality assurance, transport, maintenances, cost and services, *etc.* [1]. In this stage assessment of the state-of-the-art technology is much needed to make the design more competitive and the cost economical. The final specifications will be determined after several specification iterations. The resultant conceptual design is important for the innovation of the machine design.

Brainstorming is a method most often thought of for generating conceptual design [1]. In this stage, the selection of key components in precision machines has to be considered. These key components include machine structure and materials, the main spindle and slideway, the feed drive, control units, the inspection unit, the tool and fixtures, etc. The advantages and disadvantages of these components should be compared and evaluated with respect to system functional requirements and other factors such as the cost and design for manufacture, *etc*. Some of these key components in precision machines have been briefly reviewed in the previous section. Several design schemes may be proposed in this stage, which are followed by analysis and simulation processes and experimental analysis processes. From the dynamics point of view, the machine vibrations should be avoided during this stage the first time, by the use of various integrated analysis and testing disciplines, from the component level to the final machine system assembly.

The analysis and simulation includes key component modelling, system modelling, static analysis and dynamics analysis, *etc*. FEA is prioritized as the major analysis and simulation method. The analysis results, together with errors budget and cost estimation, will be used to check the conformance to the machine's specifications. The analysis results also help to identify some of the weakest parts in the machine tool structure and then provide data for structural modification, and hence speed up the decision-making process in a precise manner.

Experimental dynamic analysis on the precision machine involves the selection of testing methods, frequency response function analysis, modal updating and stringer comparisons with simulation results, etc. Through the experimental dynamic analysis, some important dynamic characteristics of key machine components or even the machine system assembly such as modes and shapes, natural frequencies

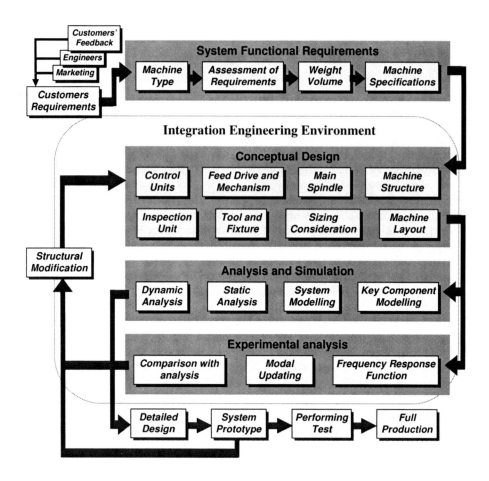

Figure 10.5. The precision machine design procedures

and damping ratio, will be obtained. Experimental analysis results can also be used for the structural modification and the verification of simulation models.

It should be stressed that the processes of structural design, structural dynamic analysis and tests are not necessary linear but interactive to each other. Design of precision machines should involve structural design, analysis and experiments in an integrated engineering environment. First of all, experimental dynamic tests will be in support of dynamic simulation, since many unknowns prevail in a pure analysis and simulation process, especially while dealing with a fully assembled new design configuration. An insufficient understanding of the various simulation procedures, the characterization of new materials, or the use of different construction methods for structures, all generate unknowns and can lead to an inefficient use of simulation, and therefore more iterations. A principal role for structural dynamics tests will provide the necessary feedback data to support the design and analysis process. Data feedback can often be understood in a broad sense. It is usually unnecessary to perform the dynamic testing for the whole assembled precision machine, or it will fall into traditional trial-and-error methods. For instance, data from the dynamic testing of aerostatic slideways can be feedback into finite element analysis to establish an accurate model; data from nanoindentation tests will benefit the development of simulation criteria of nano/micro machining processes modelling. The experimental database should be integrated into the overall analysis and modelling processes to help update or correct the existing analytical models such as FEA models or to build new models based on experimental data.

The need for this integration is driven by the increasing demand of high precision machines. Fortunately advances in hardware and software have been continuously contributing to this integration. The hardware and software available for executing structural testing and analysis have evolved from standalone instruments to computer based systems and usually PC-based systems. A variety of successful commercial CAE software available has been of benefit to designers in dynamic analysis. The data acquisition card and analyzers used for data acquisition and signal processing have become flexible, powerful and customizable to the user requirements.

Once conceptual design, dynamic analysis and tests have been completed a design plan can be formulated for detailed design. At the detailed design stage, all subsystems including the mechanical structure, the spindle and the feed drive systems, the tooling and fixtures, the control and sensor systems, and the metrology and inspection systems will be finalized. If necessary more detailed analysis and simulation may need to be undertaken in light of the detailed design.

After the detailed design is completed, there is still much work that needs to be carried out in order to make the design successful, including the development of test and user support programs, liaisons with manufacturing engineers, the update of design and documentation, *etc.* [1].

10.3.2 Modelling and Simulation

Many researchers and machine tool manufacturers are making efforts to improve the dynamic performance of machine tools; however, due to the complexity of the machine tool structure and the machining process, the experimental measurement of the structural and thermal dynamic performance is difficult with high costs and also time-consuming. Establishing the computational model, therefore, would have a great value in improving the dynamic performance of machine tools. Some of the most important issues are listed below:

- Quantitatively predicting and evaluating structural/thermal deformation distributions of the machine tool structure even in the early design stage for compensation

- Optimizing the machine tool structure for a best dynamic accuracy at the design stage
- Identifying a few structure elements that influence significantly machining accuracy
- Verifying machine performance such as stiffness, chatter, accuracy, and reliability
- Reducing the number of experimental data required and the hardware cost.

From the machine tool design point of view, modelling can be used to simultaneously represent the machining processes and the machine tool structure. The simulation model can therefore establish the relationship between inputs and outputs which enable the static or dynamics performance of the machine tool numerical and graphically illustrated by using the process variables of the input.

To some extent, the simulation model can bridge the gaps among the real machine, its physical model, the mathematical model and its dynamic performance outputs. Figure 10.6 illustrates the general modelling and simulation approach to simulating the dynamic performance of the real machine tool system. The ideal physical model is extracted from the real machine system by rational simplification. Simplification is necessary because of the complexity of machine tools. Some minor factors will be neglected during this stage. The mathematical model is deductively derived from basic physical principles and is established to solve the physical model. The mathematical model can be regarded as an idealization of the ideal physical model; conversely, the ideal physical model can be presented as a realization of the mathematical model. For dynamic analysis, the mathematical model is often an ordinary differential equation in space and time. In practice it is difficult to solve the equation and get the solution directly by the analytic method. Therefore, the discrete model is developed to solve the problem. The discrete model, also termed the simulation model or the numerical method, is the imitation of the discrete value of time of a dynamic process on the basis of a model and generated from the mathematic model; this process is called the discretization method. Although there are various methods to generate discrete models, including the finite element method, the boundary element method, the finite difference method, the finite volume method, and the mesh-free method, *etc.*, because the finite element method is well known among the engineers, nowadays, it still dominates most of the engineering design and analysis. It should be noted that in some practical machine tool design and analysis processes the discrete model (FEA model) may be generated without reference to the mathematical model and directly from the real physical model instead because it may be difficult to establish a mathematic model.

Design of Precision Machines 297

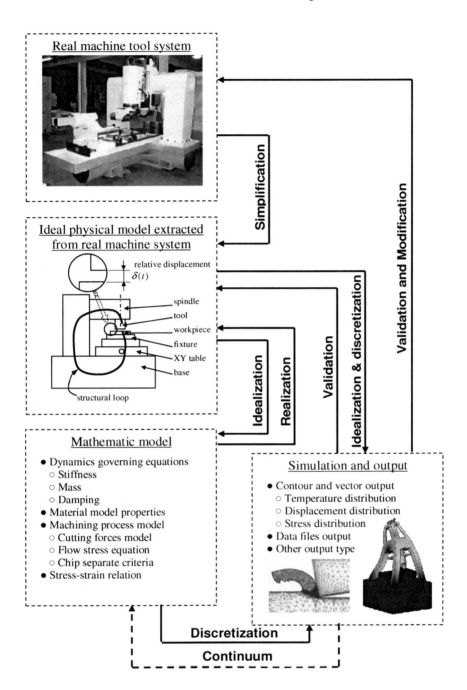

Figure 10.6. A general modelling procedure

10.4 Implementation

The design of precision machine tools has significantly benefited from the development of computer-based techniques to analyze the static and dynamics characteristics of the machine tool. Finite element analysis (FEA) is one of the state-of-the-art tools available in machine tool design. Nowadays FEA has been widely applied to the machine tool design processes and replaced lumped-parameter techniques (LPT) where the structure is modelled as consisting of discrete spring-less mass connected by discrete mass-less springs and mass-less damping for analyzing structures. Prior to that, only actual prototype testing was available for structural analysis.

The FEA process consists of subdividing all systems into individual components or elements whose behaviour is easily understood and then reconstructing the original system from these components. Interactions of elements are determined by the construction of a stiffness matrix.

In this section, different analysis types with FEA are reviewed and Figure 10.7 highlights an overview of machine tool analysis types, followed by a practical structural analysis approach using FEA.

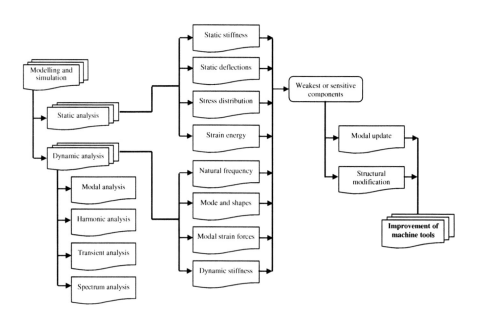

Figure 10.7. Overview of machine tool analysis

10.4.1 Static Analysis

The most common analysis type with the finite element method is static analysis. A static analysis calculates the effects of steady loads or quasi-static loads on a machine structure and inertia and the damping effect is often ignored. But inertia loads can be considered in static analysis, such as gravity and rotational velocity.

In machine tool analysis, static analysis can be used to determine the static stiffness, static deflections, stress distribution, etc. Loads that can be applied to the static analysis of machine tools include:

- Gravity under which static deflections of machine tool structures are determined
- Rotational velocity such as a high speed spindle to determine the effect of centrifugal force effects
- Inertia forces of reciprocating motion parts such as slideways
- Temperature distributions to determine thermally deduced deformations
- A cutting force that can be assumed to vary slowly with respect to time
- Pressure such as hydraulic oil, *etc.*

10.4.2 Dynamic Analysis

Apart from the static loads, machine tools are subjected to constantly varying dynamic loads which must be taken into account. Due to the dynamic excitation loads, the machine tool system is subject to vibration. Previous sections have discussed these dynamic loads. Here some analysis approaches to deal with dynamic loads are presented. There are several dynamic analysis types including modal analysis, harmonic analysis, transient analysis and spectrum, etc.

10.4.2.1 Modal Analysis
Modal analysis or mode-frequency analysis is used to determine the vibration characteristics (natural frequencies and mode shapes) of the machine tool structure while it is being designed. In modal analysis the structure has no time varying forces, displacement, pressure, or temperatures applied which is equivalent to the free vibration condition. Two quantities obtained from modal analysis are natural frequencies and mode shapes which can be the starting point for another, more detailed, dynamic analysis, such as a transient dynamic analysis, a harmonic response analysis, or a spectrum analysis.

10.4.2.2 Harmonic Analysis
Any harmonic type load (which can be represented by sine or cosine functions or their combinations) will produce a sustained harmonic response in a machine tool structural system. Harmonic analysis provides the ability to predict the sustained dynamic behaviour of the structure and to verify whether or not the machine tool design will successfully overcome resonance, fatigue, and other harmful effects of

forced vibrations. All loads in harmonic analysis are assumed to vary sinusoidally at the specific frequencies. Structural harmonic dynamic response at any specific frequency can be determined by harmonic analysis and hence the dynamic stiffness. It is very important to perform a harmonic analysis because the periodic loads in practical machining processes can be represented by its harmonic components by expansion using a Fourier series. Typical harmonic forces in machining processes are milling forces.

10.4.2.3 Transient Analysis
Transient dynamic analysis, also termed time-history analysis, is a technique used to determine the dynamic response of a structure under the action of any general time-dependent loads. In a machine structural analysis, transient dynamic analysis is used to determine the time-varying displacements, strains, stresses, and forces as it responds to any combination of static, transient, and harmonic loads. The response of machine tool structures under some type excitation signals such as step signal and acceleration signal or arbitrary time time-vary signals can be obtained from transient dynamic analysis.

10.4.2.4 Spectrum Analysis
A spectrum analysis is mainly used in time-history analysis to determine the response of structures to random or time-dependent loading conditions such as earthquakes, wind loads, ocean wave loads and environmental vibrations, *etc.* [15]. In spectrum analysis the results of a modal analysis are used with a known spectrum to calculate displacements and stresses in the model. Two typical types of spectrum used in spectrum analyses are the deterministic response spectrum and the nondeterministic random vibration. In machine tool analysis the spectrum analysis can be used to determine to response of vibrations transmitting from foundations.

10.4.3 A General Modelling and Analysis Process Using FEA

Although there are many commercial FEA software packages available, such as ANSYS, ABAQUS, COSMOS *etc.*, and every FEA software has its own features which are different from others, almost all of the general purpose FEA software usually include three modules or procedures, namely, the pre-processor, the solution and the post-processor as shown in Figure 10.8. And in this figure, ANSYS software is taken as an example.

10.4.3.1 Preprocessor
The purpose of the preprocessor is to establish the FEA model for the solution module. A FEA model is one in which elements are specified and are connected by nodes, and loads and boundary are applied. Some of the steps (not strictly necessary) in the preprocessor include the material model definition, establishing the geometry model, the mesh generation, the load and boundary condition definition, and other features that are used to represent the physical system.

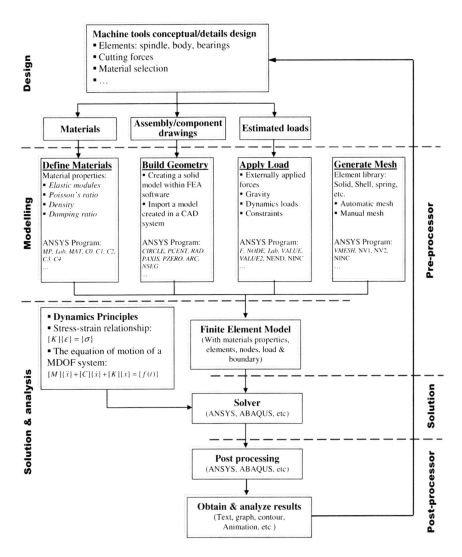

Figure 10.8. A general modelling and simulation process based on FEA

The material model is an approximate of the real material and its accuracy is responsible for the error of FEA results. There are various material models in commercial FEA software which can meet general dynamic analysis and common engineering material. However, with new material advances and the size of the simulation scale decrease, some of the material models provided by commercial software cannot meet the accuracy requirement. Therefore it is necessary to develop user material model for FEA in some particular conditions.

The geometry is meshed with a mapping algorithm or an automatic free-meshing algorithm. The implementation of good algorithms for mesh generation is very important in machine tool analysis. It must be possible to specify desired element sizes within each design element, and the mesh must remain adequate during the domain shape updating process when a shape optimization is performed. It may be advantageous to include adaptive mesh generation in some simulations.

Loads include forces, pressures and heat flux etc. It is preferable to apply boundary conditions to the CAD geometry, with the FEA package transferring them to the underlying model, to allow for a simpler application of adaptive and optimization algorithms. It should be noted that probably the large error in the entire process often results from boundary conditions. Running multiple cases as a sensitivity analysis may be helpful.

10.4.3.2 Solution
While the pre-processing and post-processing modules of the FEA are interactive and time-consuming, the solution is often a batch process, and is demanding of computer resources. The governing equations are assembled into matrix form and are solved numerically. The assembly process depends not only on the type of analysis (*e.g.*, static or dynamic), but also on the model's element types and properties, material properties and boundary conditions.

In the solution phase of the analysis, the computer takes over and solves the simultaneous set of equations that the finite element method generates. The results of the solution are a nodal degree of freedom values, which form the primary solution, and derived values, which form the element solution.

10.4.3.3 Postprocessor
Following establishing the model and finishing the solution step, the postprocessor in FEA is used to review the results of an analysis and checks the validity of the solution. In this step, designers are trying to understand how the applied loads affect the machine tool performance and how good the finite element mesh is, and so on. Normally there are kinds of analysis outputs available for reviewing the results including a text file, a graph, a contour chart, and animation *etc*. All of these outputs will benefit to evaluate the performance of the machine tools.

10.4.3.4 FEA Software Packages
The following general purpose software has been reported to be able to analyze the machine tool structures and machining processes. They include:

- ANSYS[16-20]
- SDRC/I-DEAS[21-23]
- NASTRAN and PATRAN[24]
- ABAQUS/Standard and ABAQUS/Explicit[25-29]
- ALGOR[30]
- Deform 2D/3D[31]

10.5 Applications

10.5.1 Design Case Study 1: A Piezo-actuator Based Fast Tool Servo System

The piezoelectric actuator is a kind of short stroke actuator. It is very promising for applications in the rotary table drive and slideways drive because of its high motion accuracy and wide response bandwidth. Currently, piezoelectrics have been applied in the design of the fine tool-positioner in order to obtain a high precision motion of the cutting tool. The piezoelectric actuator combined with mechanical flexure hinges is used for positioning control of the diamond cutting tool. More recently, the Fast Tool Servo (FTS) system has been introduced for diamond turning components and products with structured and non-rationally symmetric surfaces such as laser mirrors, ophthalmic lenses moulds, *etc.* [32].

10.5.1.1 Design Requirements
The piezo-actuator based FTS system is designed to perform precision positioning of the tool during a short stroke turning operation which has been implemented in a test turning machine tool set up at Brunel University in the UK.

The FTS is designed to perform turning operations and hold diamond tools. Static deformation of the FTS structure caused by cutting forces during rough and finish machining, must be minimized to reduce the form and dimensional error of the workpiece at nanometere scales. Therefore, high stiffness, particularly in the feed direction which affects the machined surface directly, is required. A high first natural frequency is required in the FTS structure to prevent resonance vibrations of FTS structure under cutting forces.

On the other hand, the high stiffness of the FTS structure will reduce the effective stroke of the piezo actuator to some extent. For this reason a compromise is made between the high stiffness and actuator stroke reduction.

The requirements for the FTS are as follows:

- First natural frequency of the flexure hinge: >1000Hz
- Maximum piezo actuator stroke reduction: <10%(PPA10M), 25%(PPA20M) and 50%(PPA40M)
- Maximum recovery force by the flexure hinge must be less than the maximum output of the piezo actuator
- Strength criterion of the flexure hinge must be met with a high safety factor
- The geometry of the flexure hinge has a good machinability by EWM
- The FTS can house different piezo actuators (different length, strokes and stiffness)
- Precision position measurement and a control algorithm are required to compensate the positioning error

- Self-contained and simple interfacing with other components of machine tools

10.5.1.2 Design Details

Figure 10.9 shows the schematic of the FTS that comprises of a piezoelectric actuator (Cedrat PPA10M, PPA20M, or PPA40M), a Flexure hinge, two cover plates, a tool holder, a capacitive sensor, and a piezoelectric adjustment screw.

The piezo actuator is housed under preload within the flexure hinge made from spring steel. Three piezo actuators, 18mm(PPA10M), 28mm(PPA20M) and 48mm(PPA40M) length, respectively; are used. Three different adjustment and preload screws with corresponding lengths are designed to house the three actuators in the same flexure hinge. The actuator displacement in actuated/feed direction is transmitted to the tool holder via four symmetric solid flexure hinges which have a circular hinge profile as shown in Figure 10.9. The tool holder is designed to be integrated with flexure hinges in a single part. The overall envelope of FTS is 90mm × 80mm × 55mm, with the tool holder extending 35mm. the design is compact and self-contained for ease of mounting on the slideway and tool posts in other machine tools.

10.5.1.3 Finite Element Analysis (FEA)

To determine the flexure hinge dimension, both static and modal finite element analysis for the flexure were conducted, in which static stiffness and natural frequencies were obtained. In both analyses, the hinge radius r, the hinge thickness t and the hinge length were chosen as optimization variables, and optimization FE analysis results were obtained based on the requirement of stiffness, natural frequencies and strokes reduction.

Figure 10.10 shows a static FE optimization analysis result. In this analysis, a force of 216 N was applied in the piezo actuator feed direction, and a maximum deformation of 0.0223 mm was obtained, corresponding to a static stiffness of 9.7 N/μm.

The effective stroke of the piezo actuator housed in the flexure hinge can be calculated by:

$$\Delta L_{eff} = \Delta L_{piezo} \frac{K_{piezo}}{K_{piezo} + K_{hinge}}$$

where ΔL_{eff} is the effective stroke, ΔL_{piezo} is the nominal stroke (8, 20 and 40 μm), K_{piezo} is the nominal stiffness of the piezo actuator, and K_{hinge} is the stiffness of the flexure hinge (an optimal value of 9.7 N/μm from FEA).

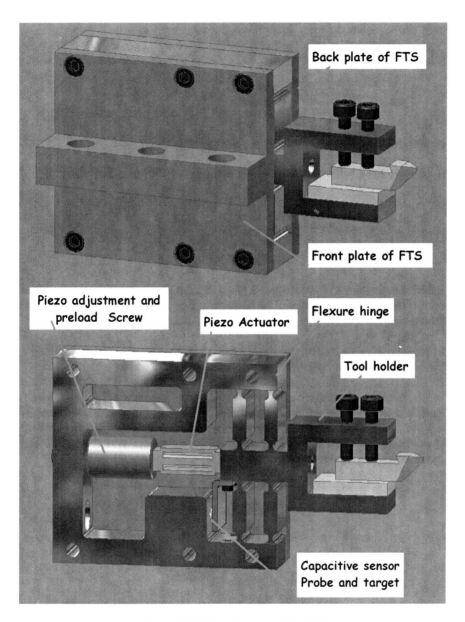

Figure 10.9. The schematic of the FTS

Piezo-actuator 1 (PPA10M): nominal stiffness is 100 N/μm and stroke is 8 μm.

$$\Delta L_{eff} = 8 \times \frac{100}{100+9.7} = 7.29 \mu m$$

and the stroke reduction = (8 - 7.29)/8=8.875%.

The piezo-actuator 2 (PPA20M): nominal stiffness is 40 N/μm and stroke is 20 μm.

$$\Delta L_{eff} = 20 \times \frac{40}{40+9.7} = 16.09 \mu m$$

and stroke reduction = (20 - 16.09)/20=19.55%.

The piezo-actuator 3 (PPA40M): nominal stiffness is 20 N/μm and stroke is 40 μm.

$$\Delta L_{eff} = 40 \times \frac{20}{20+9.7} = 26.94 \mu m$$

and stroke reduction = (40 - 26.94)/40=32.65%.

Stroke reductions in three cases are below the design requirement and it should be noted that stroke reduction will be higher if taking piezo-actuator preload effect into account.

The maximum Von Mises stresses predicted by FEA under the maximum load are 45MPa, which are well below the strength value of the spring steel.

The FE modal analysis was used to predict natural modes of the flexure structure. Figure 10.11 shows the first three natural modes. The first three natural frequencies are 1,262, 2,086, and 2,791 Hz. The lowest frequency mode is the translational motion along the actuated/feed direction with a natural frequency of 1262 Hz, which is above the design requirement of 1,000 Hz.

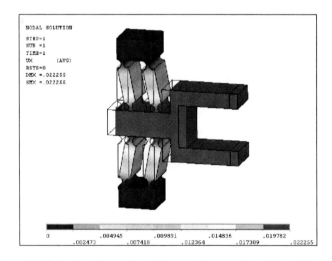

Figure 10.10. Static deformation of flexure hinge predicted from FE analysis

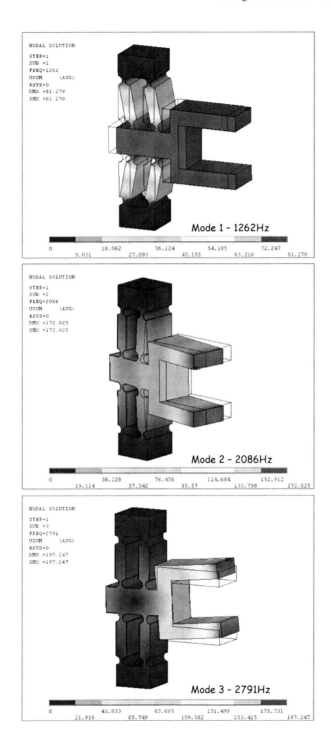

Figure 10.11. Flexure hinge natural modes predicted from FE analysis

10.5.1.4 Experiments
The FTS was set up as shown in Figure 10.12.

Figure 10.12. The FTS assembly

Both static and dynamic performance tests have been performed to obtain the static gain, dynamic gain and system bandwidth.

The experimental system for vibration tests includes:
- LabVIEW software which provides an environment for generating testing signals and collecting response signals, and also data display and analysis
- A DAQ card (data acquisition card, PCI 6024E) was used for D/A and A/D conversion purposes
- A capacitive position sensor (S 601-0.05, micro-epsilon) with 50μm strokes and 10nm resolution which has been used to measure the deformation of FTS. The bandwidth of the sensor is 6 kHz (-3db)
- A capacitive sensor system (capaNCDT 600, micro-epsilon), including a power supply and amplifier, *etc.* which provides signal amplifying and conditioning functions
- A piezo actuator (PPA10M, PPA20M and PPA40M, CEDRAT)
- A piezo-actuator drive system (LC75C and LA75C, CEDRAT) which consists of a power supply and amplifier
- A FTS system

In static tests, the input voltage was applied from -1V to 7.5 V (which corresponds to -20V to 150V in piezoactuator's input), stepped by 0.5V. The interval time between two input voltage is 10s to perform a quasi-static measurement. The deformation of FTS was measured via the capacitive sensor. A random white noise signal was generated in LabVIEW and output to the piezo-actuator in order to excite the FTS. The bias voltage was set to 3V.

Static performance testing aims at measuring strokes, the resolution and the linearity of FTS. The static performance results can also be used in accurate compensation.

Figure 10.13 shows the displacement of FTS vs. the input voltage curve. It can be seen from this figure that the loading curve and the unloading curve are not coincident. Hysteresis phenomenon can be observed clearly. There is a better linearity in the input voltage range of 40V-100V for PPA 40, 20 and 10. The repeatability of return-to-zero in the test of PPA 40M is less than 0.05µm.

Table 10.1 gives the strokes (from 0V to 150V) in three tests, and results are in reasonable agreement with the FEA results.

Table 10.1. The strokes of FTS obtained from experiments and FEA

Piezo stack code	Experimental results (µm)	FEA results (µm)	Discrepancy
PPA 40M	24.55	26.09	1.54
PPA 20M	12.45	16.94	4.49
PPA 10M	3.80	7.29	3.49

There are some discrepancies between experiments and FEA results. The discrepancies come from the following:

- Manufacturing errors of the flexure hinges
- A FTS preload in FEA which has not been taken into account, which will reduce strokes
- A nominal stiffness of the piezo-actuator from the manufacturer's catalogue are used in FEA.

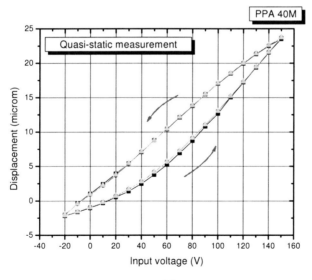

(a)

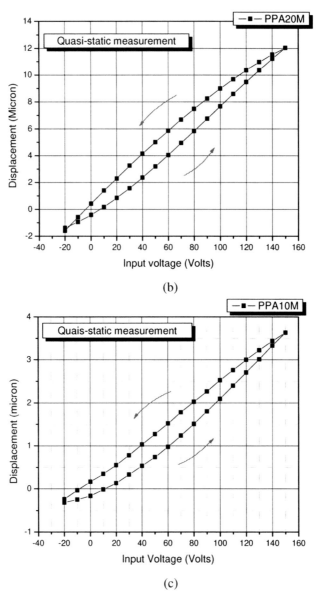

Figure 10.13. Displacement of FTS vs. input voltage curve

Figure 10.14 presents the frequency-domain response of FTS under the random excitation signal. From the figure, the natural frequency of FTS in the motion direction is about 2.3k Hz. FTS with the other piezo-actuators (PPA20 and PPA10) were also tested using the same method, and similar results were obtained.

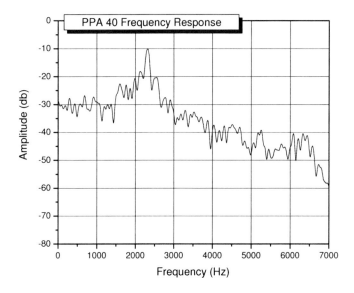

Figure 10.14. The FTS response under the random excitation signal (PPA 40M)

After finishing vibration tests of the FTS, some preliminary diamond cutting trails were conducted using the FTS. In the cutting trails, FTS was mounted on a test machine tool with an aerostatic spindle and aerostatic slideway and is shown in Figure 10.15.

Figure 10.15. The diamond turning machine tool equipped with the FTS

The cutting parameters and conditions (finish cutting) used in cutting trails include:

- Cutting speed: 10,000 rpm
- Feed rate: 5 mm/min
- Depth of cut (finish cutting): 5 micron
- Workpiece material: aluminium alloy
- Lubricant: alcohol
- Cooling air pressure: 2 Bar

Mirror-like finish surfaces were obtained as shown in Figure 10.16 a. The surface roughness was measured by a Zygo microscope and a Ra value of 10 nm was obtained (Figure 10.16 b). The cutting results show that the FTS has high static and dynamic stiffness and further improvement will be made by introducing more effective control algorithms for FTS.

(a) Finished surfaces

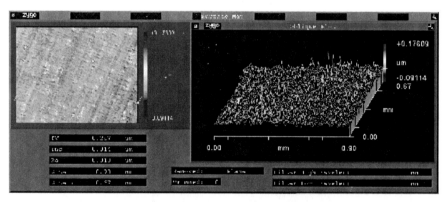

(b) The surface roughness profile from Zygo microscope

Figure 10.16. Finished surfaces of the Al workpiece and the surface profiles from a Zygo 3D surface profiler

10.5.2 Design Case Study 2: A 5-axis Micro-milling/grinding Machine Tool

This case study describes the conceptual design process of a bench-type 5 axes micro-milling machine, which is currently being developed by the authors with the support of the EU MASMICRO project [33].

10.5.2.1 Definition of Machine Tool Specifications
The machine aims at manufacturing the miniature and micro components in various engineering materials; potential applications are listed as follows:

- MEMS (silicon) including complex microstructures and micro-grooves, *etc.*
- Optical components (glass, polymer, aluminium) including complex shape or freeform optics (lenses, mirrors, prisms) and lens shutters, display, light guiding microstructures, *etc.*
- Medical components (polymer, glass) including ophthalmics and dentistry, *etc.*
- Mechanical components (aluminium, steel) including microsensors and fiber optic-mechanical components, *etc.*
- Moulds (high strength aluminium, copper alloy, steel) for optics lenses and fiber optic element and moulds for reflective foils

After reviewing the state-of-the-art of several commercially available typical ultraprecision machine tools (refer to [10]), initial specifications of the 5-axis milling machine tool as listed in Table 10.2 were worked out.

Table 10.2. Initial specification of the 5-axis milling machine tool

Configuration	5-axis CNC micro-milling machine			
Base	Granite			
Axes	X, Y and Z axis	B axis	C axis	Spindle
Type	Aerostatic slideway	Air bearing	Air bearing	Air bearing
Stroke	X:200mm Y:100mm Z: 50mm	360°	360°	N/A
Stiffness	>400 N/μm	N/A	N/A	50 N/μm
Motion accuracy	Straightness (μm/mm): X, Z<0.01/200, Y<1.0/250	Radial/Axial run out (μm): <1/0.5	Radial/Axial run out (μm): <0.1	≤ 10 nm
Resolution	5nm	0.000001°	0.00001°	N/A
Drive system	Linear motor	Servo motor	Servo motor	DC brushless motor
Maximum speed	N/A	N/A	N/A	300,000 rpm

10.5.2.2 Preliminary Design and Analysis
According to the machine specifications, the machining envelope, the bench-type required dimensions, the types of components/materials to be machined, and the overall accuracy to reach and maintain, the machine tool layout in Figure 10.17 was selected as the preliminary design for analysis. In this stage, some feasible structural configurations available were also reviewed.

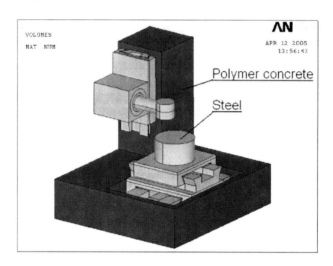

Figure 10.17. Preliminary configuration

This project aims to develop a bench-type machine tool, so the machine tool was initially designed with an open frame as shown in Figure 10.17, which greatly facilitates machining area access for fixturing and part handling.

The structural static and dynamic behaviours of the machine tool were simulated using ANSYS software. Because there are many air bearings in this machine tool configuration (the air bearing slideways and the air bearing spindle), it is difficult to simulate the compressed air in ANSYS; an equivalent method was proposed to simulate the air bearing. In this method spring elements in ANSYS were used to simulate the stiffness in different directions. The stiffness of the spring element is based on the stiffness data provided by experiments. Figure 10.18 shows this equivalent FE model for the air bearing slideway.

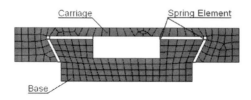

Figure 10.18. An equivalent FE model for the air bearing slideway

Both static and modal analyses were conducted and the first 10 natural frequencies were extracted in modal analysis using the block Lanczos method. It should be noted that all the interfaces in the machine tool are assumed to be rigid; thus, the natural frequencies are estimated higher than the real values.

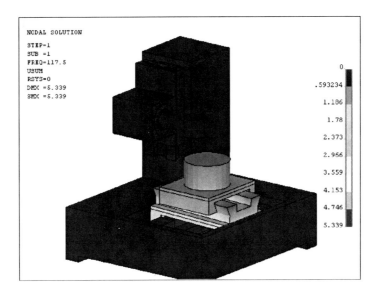

Figure 10.19. First natural frequency and its mode of origin configuration

After running a series of FEA simulations on the structure, we identified the important sensitive component on the machine structure, which can be seen from the mode shape of the first natural frequency (117 Hz). Due to the stack of equipment assembled on top of each other, the X slideway and the Z slideway are subject to the important "tilting" effect, which is likely to affect the machine accuracy, as illustrated in Figure 10.19. Therefore, from the structural modification of the view, improving the stiffness of the slideway is the most efficient method.

10.5.2.3 Redesign and Reanalysis
Following the information gained from the sensitivity analysis based on preliminary FEA results, a gantry type of machine configuration was proposed. This gantry type of machine configuration will enable a much better overall machine stiffness, in static as well as in dynamic mode and overcome the problems identified in the original machine design. In order to improve the overall stiffness of the machine tool, the machine structure material was changed from polymer concrete to granite. Modal analysis was conducted for the new configuration in which the horizontal slideway was neglected to increase computational speed because there is no tilting slideway. Figure 10.20 shows the first natural frequency and its vibration shape. The first natural frequency was increased from 117 Hz in the preliminary design to 134 Hz, and the following natural frequencies were also improved. The final gantry 5-axis machine configuration is shown in Figure 10.21.

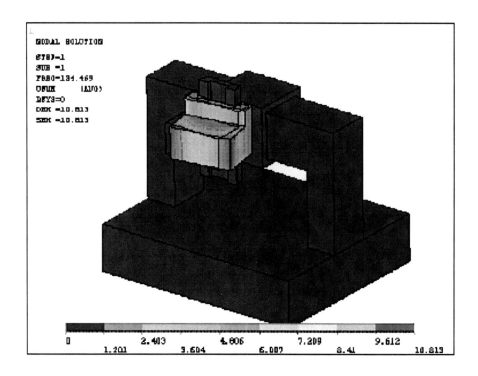

Figure 10.20. First natural frequency and its mode of gantry configuration

Figure 10.21. The gantry type of machine configuration

10.5.3 Design Case Study 3: A Precision Grinding Machine Tool

A grinding machine - the "PicoAce" developed by Loadpoint Ltd, is used in this case study. PicoAce is an ultra precision grinding machine for producing optical quality surface finishes and low levels of sub surface damage on a range of hard and brittle materials. It is suitable for traverse grinding of flat or convex work surfaces up to a maximum of 305 mm diameter or plunge grinding of flat surfaces to a maximum diameter of 200 mm. The principal machine elements of PicoAce include: a grinding spindle with a cup wheel, a rotary work table, and X and Z slideways mounted in a closed loop structure. The structure takes the form of a pyramid shaped frame fixed to a solid rectangular base. The cup wheel spindle has a vertical axis and is arranged to slide up and down in a cylindrical Z slideway centrally positioned over the base. Directly beneath the wheel spindle is the rotary table mounted on the X slideway.

FE analysis was conducted to determine the thermal deformation and stability of the machine and then corresponding structural modification and compensation scheme are carried out based on the FEA results.

A 3D FE model was developed to compute the thermal deformation of the grinding machine structure. The same mesh was created for both the thermal and structural analysis. In the FEA numerical simulation using ANSYS software, three dimensional thermal tetrahedral solid elements with ten nodes (SOLID87) and continuum stress/displacement elements with the same dimension and number of nodes (SOLID92) were used for the thermal and structural analysis, respectively [15]. The model was meshed into the above solid element using a free meshing method and for a typical mesh there were 42551 elements and 11711 nodes.

Table 10.3. Heat input data to the grinding machine tool model

	Power (Watts)	Oil temperature rise (°C)	Flow rate (l/min)
Rotary table	650	3.5	6.0
X Slideway	270	2.3	4.0
X Leadscrew	180	3.8	1.4
X Thrust bearing	80	3.2	1.2
Z Sleeve	112	2.3	1.7
Z Leadscrew	180	3.8	1.4
Z Thrust bearing	80	3.2	1.2

Major heat sources to the model are due to hydrostatic oil based on the assumption that all heat is removed by the oil and no conventional heat is lost through the structure. The heat input data from the experiments are listed in Table 10.3. Also, the temperature distribution obtained by the above thermal FE analysis of the grinding process as a heat source was input to the structural analysis.

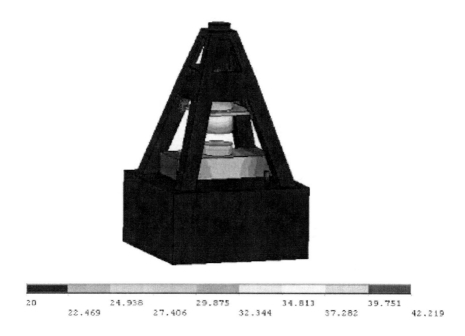

Figure 10.22. Temperature distribution contour of the machine

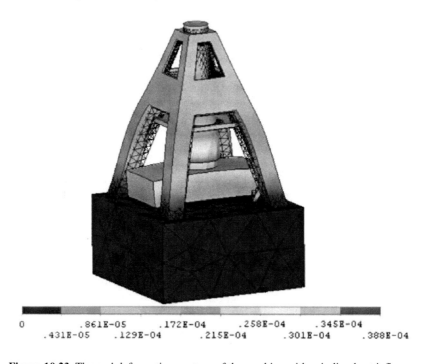

Figure 10.23. Thermal deformation contour of the machine with grinding heat influence

In this model, temperature independent material properties were assumed using spherical graphite cast iron room temperature values (20°C). No latent heat and radiation were considered. The resulting temperature profiles were then applied to a structural analysis model. All material properties were assumed to be temperature independent. This assumption simplifies the model and reduces the computational time. While the simplifications may introduce errors, all calculated results were compared to experimental data so as to further validate the FE simulation.

Based on the above FE model and boundary conditions a transient analysis was conducted. Figure 10.22 presents the temperature distribution on the machine tool structure in the 3.6 minutes snapshot. It is noticed that the temperature of the spindle assembly and rotary table is significantly higher than the rest of the machine tool. The local temperatures obtained by FEA in some machine elements such as the thrust bearings, the slideway, the rotary table and the leadscrew are in better agreement with the experimental data.

Figure 10.23 shows the contour of machine thermal deformation after 3.6 minutes. It can be seen that the maximum displacement occurs at the grinding wheel position, which can be considered as the accumulation of the displacement of the structural legs and grinding spindle itself. The machine base has a slight deformation due to its lower conductivity and coefficient of thermal expansion.

FEA thermal deformation results have been used in the structural modification in an iterative manner. Figure 10.24 shows that the final assembly of PicoAco after the structural modification.

Figure 10.24. PicoAce - a precision grinding machine tool (Courtesy: Loadpoint Ltd)

Acknowledgements

The authors are grateful for the support of the EU 6th Framework NMP Program under the contract number NMP2-CT-2-4-500095. Thanks are due to all partners at the MASMICRO project consortium and within the RTD 5 subgroup in particular.

References

[1] Slocum, A. H. Precision Machine Design, Englewood Cliffs, Prentice Hall, NJ: 1992.
[2] Maeda, O., Cao, Y. and Altintas, Y. Expert spindle design system. International Journal of Machine Tools and Manufacture, 2005, 45(4–5): 537–548.
[3] Park, C. H., Lee, E. S. and Lee, H. A review on research in ultra precision engineering at KIMM. International Journal of Machine Tools and Manufacture, 1999, 39: 1793–1805.
[4] Kim, H. S., Jeong, K. S. and Lee, D. G. Design and manufacture of a three–axis ultra–precision CNC grinding machine. Journal of Materials Processing Technology, 1997, 71: 258–26.
[5] Mekid, S. High precision linear slide. Part I: design and construction. International Journal of Machine Tools and Manufacture, 2000, 40(7): 1039–1050.
[6] Schellekens, P. and Rosielle, N. Design for precision: current status and trends, Annals of the CIRP, 1998, 47(2): 557–584.
[7] Rao, S. B. Metal cutting machine tool design – a review. Proceedings of the Institution of Mechanical Engineers, Part B: Journal of Manufacturing Science and Engineering, 1997, 119: 713–716.
[8] Bryan, J. B. Design and construction of an ultraprecision 84 inch diamond turning machine. Precision Engineering, 1971, 1(1): 55–61.
[9] Stephenson, D. J., Veselovac, D., Manley, S. and Corbett, J. Ultra–precision grinding of hard steels, Precision Engineering, 2000, 15: 336–345.
[10] Luo, X., Cheng, K., Webb, D. and Wardle, F. Design of ultraprecision machine tools with applications to manufacture of miniature and micro components, Journal of Materials Processing Technology, 2005. 167(2–3): 515–528.
[11] Ai, X., Wilmer, M. and Lawrentz, D. Development of friction drive transmission. Journal of Tribology, 2005, 127(4): 857–864.
[12] Deiab, I. M. and Elbestawi, M. A. Effect of workpiece/fixture dynamics on the machining process output. Proceedings of the Institution of Mechanical Engineers, Part B: Journal of Engineering Manufacture, 1994, 218(11): 1541–1553.
[13] Ikawa, N., Donaldson, R. R., Kormanduri, R., König, W., Aachen, T. H., Mckeown, P. A., Moriwaki, T. and Stowers, I. F. Ultraprecision metal cutting–the past, the present and the future. Annals of the CIRP, 1991, 40(2): 587–594.
[14] Benaroya, H. Mechanical Vibration – Analysis, Uncertainties, and Control. Marcel Dekker, New York: 2004.
[15] ANSYS Basic Analysis Procedure Guide, 2002, (ANSYS, Houston).
[16] Lee, S. W., Mayor, R. and J. Ni, Dynamic analysis of a mesoscale machine tool. Transactions of the ASME: Journal of Manufacturing Science and Engineering, 2006, 128: 194–203.
[17] Baker, J. R. and Rouch, K. E. Use of finite element structural models in analyzing machine tool chatter. Finite Elements in Analysis and Design, 2002, 38(11): 1029–1046.

[18] Filiz, I. H., Akpolat, A. and Guzelbey, I. H. Deformations and pressure distribution on machine tool slideways. International Journal of Machine Tools and Manufacture, 1997, 37(3): 309–318.
[19] Mahdavinejad, R. Finite element analysis of machine and workpiece instability in turning. International Journal of Machine Tools and Manufacture, 2005, 45(7–8): 753–760.
[20] Chen, C. Y. and Cheng, C. C. Integrated structure and controller design of machine tools. Automation and Mechatronics, 2005, 2(2): 869–874.
[21] Zeljkovic, M. and Gatalo, R. Experimental and computer aided analysis of high–speed spindle assembly behaviour. Annals of the CIRP, 1999, 48(1): 325–328.
[22] Bais, R. S., Gupta, A. K., Nakra, B. C. and Kundra, T. K. Studies in dynamic design of drilling machine using updated finite element models. Mechanism and Machine Theory, 2004, 39(12): 1307–1320.
[23] Raja, P. V., Pillai, P. R. and Radhakrishnan, P. Thermal analysis of spindle units under high speed machining. Journal of the Institution of Engineers, Mechanical Engineering Division, 2001, 81(4): 155–159.
[24] Anagonye, A. U. and Stephenson, D. A. Modeling cutting temperatures for turning inserts with various tool geometries and materials. Transactions of the ASME: Journal of Manufacturing Science and Engineering, 2002, 124(3): 544–552.
[25] Madhavan, V. and Adibi–Sedeh, A. H. Understanding of finite element analysis results under the framework of Oxley's machining model. Machining Science and Technology, 2005, 9(3): 345–368.
[26] Jeon, S. Y. and Kim, K. H. A fluid film model for finite element analysis of structures with linear hydrostatic bearings. Proceedings of the Institution of Mechanical Engineers, Part C: Journal of Mechanical Engineering Science, 2004, 218(3): 309–316.
[27] Ng, E. and Aspinwall, D. K. Modelling of hard part machining. Journal of Materials Processing Technology, 2002, 127(2): 222–229.
[28] Yan, J. and Strenkowski, J. S. A finite element analysis of orthogonal rubber cutting. Journal of Materials Processing Technology, 2006, 174(1–3): 102–108.
[29] Lee, W. B., Wu, H. Y., Cheung, C. F., To, S. and Chen, Y. P. Computer simulation of single–point diamond turning using finite element method. Journal of Materials Processing Technology, 2005, 167(2–3): 549–554.
[30] Zeljkovic, M. and Gatalo, R. Experimental and computer aided analysis of high–speed spindle assembly behaviour. Annals of the CIRP, 1999, 48(1): 325–328.
[31] Ozel, T. and Altan, T. Process simulation using finite element method – prediction of cutting forces, tool stresses and temperatures in high–speed flat end milling. International Journal of Machine Tools and Manufacture, 2000, 40(5): 713–738.
[32] Weck, M., Fischer, S. and Vos, M. Fabrication of microcomponents using ultraprecision machine tools, Nanotechnology, 1997, 8: 145–148.
[33] http://www.masmicro.net. Accessed on 4th July 2008.

Index

ABAQUS, 300
accelerometer, 208
AC motor, 286
acoustic emission (AE), 146–148
adaptive control, 257
air bearing, 289
ALGOR, 302
aluminium oxide (Al_2O_3), 125
analysis, 294, 298–300
  dynamic, 283, 294, 299–300
  harmonic, 299
  modal, 299
  preliminary, 314
  spectrum, 300
  static, 294, 298
  transient, 300
angular velocity, 188
ANSYS, 300
ARMA, 228, 269
ARMAX, 225, 228, 229, 230
ARX, 225
attritious wear, 245
audio measurement, 97

ball end mill, 122, 186–187
brainstorming, 293
built–up edges (BUEs), 120, 172

CAD model, 157
calibration, 181, 194
capacitance probes, 97
carriage, 314
cast iron, 132, 157, 158, 285
cemented carbide, 124–125

ceramics, 125–126
cermet, 125
chatters, 15, 145, 167
  Arnold–type, 16
  detection, 98–99
  marks, 269
  reduction, 16
  regenerative, 16
  suppression, 79
  turning, 67
  velocity dependent, 16
chemical vapour deposition (CVD), 128, 132
chip, 134, 236
  formation, 236
  geometry, 234
  hammering, 134
  shape ratio, 238
  size, 244
  thickness, 237, 265
chromium nitride (CrN), 128, 129
comb cracks, 134
conceptual design, 294
contact length, 237
control system, 284, 287
COSMOS, 300
crater depth, 135–136
crystallographic orientations, 265, 266, 276–277
cubic boron nitride (CBN), 124, 127
current monitoring, 97
cutter, 178
  run–out, 178
cutting, 244

324 Index

cutting conditions, 197, 205, 213, 312
cutting direction, 267
cutting efficiency ratio, 243
cutting edge density, 258
cutting forces, 22
  axial, 188
  coefficients, 181, 183, 194, 198
  differential, 178, 191
  dynamic, 54, 186, 188, 215
  model, 186, 188
  normal, 174
  proncipal, 172
  tangential, 173, 174
  total, 180, 193
cutting parameters, 163, 197,199, 213
  depth of cut, 163
  feed rate, 163
  specific energy, 172, 175
  speed, 163
cutting plane, 267
cutting process models, 25
cycle time, 257
cylindrical grinding, 248

damping, 9, 290, 291
  coefficient, 12
  joint, 292
  material, 292
  ratio, 12
damper, 292
data dependent systems (DDS), 271
DC brushless motor, 286
deflection, 248, 249, 250
deformations, 30
  machine tool structures, 30
  tools, 32, 33, 34, 36
  workpiece, 32, 33, 37–39
delay, 205
Deform 2D/3D, 302
depth of cut, 163, 188, 250
  axial, 188
  radial, 190
detailed design, 294
diamond, 129

diamond–like carbon (DLC), 128, 129, 132
diamond turning machine, 311
dies, 313
down–hill slope, 198
down milling, 32, 74, 176, 178, 181, 189, 190, 191, 193, 200
dressing, 234
  conditions, 235, 247
  depth, 245
  kinematics, 234, 246
  lead, 245
  tools, 234
drive system, 284, 286
drying machining, 132
ductile regime machining (DRM), 126
dwell stage, 248, 251
dynamic chip thickness, 59
  milling, 62
  turning, 60
dynamic cutting force coefficients (DCFC), 58–59
dynamic cutting system, 269
dynamics, 1, 3, 64, 68, 145, 151, 288
  analysis, 44, 154, 293
  disgnostics, 85
  machine tool, 43, 47–49,
  machining, 2, 3, 145, 151, 312
  milling process, 61
  model, 200, 213
  turning process, 59
dynamometers, 96

edge chipping, 134
elastic deflection, 244
elastic deformation, 244
envelope curve, 206, 207
environmentally friendly machining (EFM), 123–124
excitation, 152
excitation techniques, 93
experimental analysis, 294

fast Fourier transform (FFT), 101, 215, 216, 271

Index   325

fast tool servo (FTS), 303, 308, 311
fibre–optic bragg grating sensors, 98
finite element analysis (FEA), 5,
  157, 159–160, 161, 283, 298–299,
  300–302, 306
fixture system, 284, 286–287
flank wear land width, 134–135
flexure hinge, 305
flute edge, 174
flute geometry, 188
flute number, 188
form errors, 39–40
fracture wear, 245
frequency domain, 99, 100–105
frequency response function (FRF),
  44, 51–54, 294
  experimental testing, 95, 96
  measurement, 92
friction angle, 22
friction coefficient, 241

geometric model, 186, 187
grain fracture, 258
grain shape, 245
granite, 285
grinding, 233
  conditions, 235–236, 246, 247
  control strategies, 253, 255
  cycle, 248, 255
  dwell time, 257
  force, 234, 238, 242
  kinematics, 234, 236, 248
  mechanics, 236
  power, 234, 251, 254, 255, 259
  simulation, 245–246, 255–256
  specific energy, 240
  temperature, 234
  wheel, 234, 247
  vibrations, 234
grinding burn, 253
grit, 235, 236, 238, 241, 243
  density, 245
  shape, 245
gross fracture, 134

hammer test, 44–45, 95, 208
harmonic excitation, 289

helical end mill, 173
helix angle, 174
Helix lag angle, 170
Hertz distribution, 242
high speed machining, 123
high speed steel (HSS), 124
homogeneity, 285

I–DEAS, 302
inertia, 291
infeed rate, 249
infeed stage, 248
information theory, 111
in–process sensors, 96
inspection system, 284, 287
ISO 3685:1993, 135, 136

KDP single crystal, 264
kinematics, 242, 248, 253, 260

Laplace domain, 203, 209
laser Doppler velocometer (LDV),
  94
leadscrew, 317
linear motor, 286
low pollution machining, 123–124
lumped–parameter techniques
  (LPT), 298

machine base, 284
machine column, 284
machine configuration, 294
machine design, 283, 293
machine dynamics, 283, 288
machine layout, 294
machine performance, 284, 287–288
machine specifications, 294
machine structure, 284, 285
machine tool constitutions, 284
machine tool loops, 288, 289
machine tool vibrations, 288
machine–tool–workpiece loop, 7
machining instability, 17, 18, 19
machining processes, 1, 151, 153,
  167, 226
  grinding, 233, 246
  milling, 167

turning, 151, 153
machining systems, 21, 226
machining setup, 213, 214
mass, 290, 291
material anisotropy, 276
material induced vibration, 263, 270
material removal rates (MRR), 41, 76
mean arithmetic roughness (R_a), 277
mechanical components, 313
mechanics, 242, 253, 260
medical components, 313
MEMS, 313
metal cutting, 118
  grinding, 233
  milling, 167
  turning, 151, 153, 263
metal matrix composite (MMC), 132, 137
metrology, 287
micro cutting, 263
micro milling machine, 313
microplasticity model, 265
milling, 167
  ball–end, 169
  dynamics, 213
  face, 169
  model, 176, 200, 213
  peripheral, 169, 171, 213
  plane, 189, 199
  slot, 185, 197
minimum quantity lubricant (MQL), 123–124, 132–133
modal, 87, 294
  analysis, 88–90, 154, 207, 208, 209
  constant, 88
  mass, 88
  parameters, 210, 212
mode, 87, 88
mode coupling, 289
modelling, 295
modularisation, 200
  cutting force, 200
monitoring system, 284
moulds, 313
multi–degree of freedom, 204, 208
multi frequency solution, 68

multiscale modelling, 5–6
mutual information, 111

National Aeronautics and Space Administration (NASA), 128
nanometric cutting, 263
nano–surface, 264
NASTRAN, 302
natural frequency, 9, 155, 159–160, 161
non–interference, 276

oblique cutting, 172
optimisation, 154, 155
optical components, 313
overshoot, 254

PATRAN, 302
physical vapour deposition (PVD), 128, 129
piezo–actuator, 303
  effective stroke, 304
  nominal stroke, 304
piezoelectric accelerometer, 93–94, 208
plastic deformation, 244
ploughing, 236, 243, 244
plunge grinding, 248
poly–crystal diamond (PCD), 126, 137
polycrystalline, 263
polymer concrete, 285, 314
position loop, 289
postprocessor, 302
precision machines, 263, 283
preliminary analysis, 314
preliminary design, 314
preprocessor, 300
process damping, 57
pulsating excitation, 289

radial immersion angle, 170, 190
rake angle, 120, 122, 172
  effective, 170, 175
  normal, 170, 175
  radial, 170, 175
reanalysis, 315

redesign, 315
regenerative displacement, 203, 205
  tool, 203, 289
regenerative effect, 289
removal rate time constant, 250, 252, 253
resonance, 152
rigidity, 43
  dynamic, 43
  fixture, 156
  machine tool, 156
  work material, 156
rotary table, 317
rubbing, 236, 243, 244

scallop hight, 190
scanning electron microscope (SEM), 127, 140, 266
sculptured surfaces, 188
SDRC, 302
sensor sysetm, 287
shear angle, 22, 265
  oscillations, 55–57
signal processing, 92
signal–to–noise ratio (SNR), 45
silicon, 139
  machining, 139–142
silicon nitride (Si_3N_4), 126
simulation, 207, 245–246, 264, 295
single–crystal diamond (SCD), 126
single crystal materials, 264
single–degree–of–freedom (SDOF), 46
single frequency solution, 70
single grain, 234
single mode, 46
single point diamond turning (SPDT), 263
slideways, 284
sliding, 236
slip rings, 98
size effect, 245, 260
size error, 234, 254
soft computing, 110
solution, 302
spark–out, 248, 252

specific energy, 240
spindle frequency (SF), 171, 218, 219, 220, 221, 222, 223, 224
spring element, 314
squeeze film dampers, 292
stability, 64
  limit, 71
  lobes, 72
  milling, 68–74
  turning, 64–67
STAR, 208, 209
steel, 158, 210
step–over, 190
  feedrate, 190
stiffness, 7, 248–249, 290
  dynamic, 9
  dynamic loop, 10, 289
  static loop, 8, 289
stiffness loop, 289
stiffness–to–mass ratio, 291
structural dynamic parameters, 46
structural loop, 297
structural materials, 285
structural modification, 294
sum–squared error (residual), 226, 228, 229
surface generation, 205, 206
surface integrity, 234
surface profile, 206
surface roughness, 118, 234, 252, 253, 273, 312
  peak to valley (Rmax) 119
surface topography, 224, 273, 274, 312
system identification, 224, 225–226

Taylor's model, 142–145
telemetry, 98
temporal stability, 285
thermal loop, 289
three–dimensional (3D) surface analysis, 272, 274
thrust bearing, 317
time domain, 99, 105–109
titanium aluminium nitride (TiAlN), 128, 129, 129–131

titanium carbon (TiC), 128, 129
titanium carbon nitride (TiCN), 128, 129, 129–131
titanium nitride (TiN), 128, 129, 129–131
tool, 117
  coated, 127, 129, 132
  design, 118,
  edge radius, 120, 122
  failure, 133, 134
  geometry, 120, 122, 162
  life, 142
  materials, 162
  rake angle, 120, 122, 172
  relief angle, 120, 122
  round–nosed, 119
  straight–nosed, 119
  wear, 133, 162
tool–work vibration, 275–276
tool–workpiece replication model, 118
tool–workpiece loop, 288
tool interference, 275
tooling system, 284, 286
tooth passing frequency (TPF), 171, 215, 216, 218, 220, 221, 222, 223, 224
tooth passing period (T), 171, 216
topography, 100
transfer function (TF), 44, 45, 46, 203, 205
turning, 151, 153, 263
two–degree–of–freedom, 175

ultra–precision machines, 263
ultraprecision machining, 126, 263
undeformed chip thickness, 122, 172, 175, 238, 244
up milling, 32, 42, 74, 177, 179, 182, 189, 190, 192, 194, 201

variable pitch cutters, 79–82
verification, 186, 198

vibrations, 10
  amplitude, 169
  forced, 13, 167
  free, 10
  frequency, 169
  self–excited, 102, 167, 168–169
virtual surface, 274

wavelets, 99, 109–110
wear, 133, 162
  abrasive, 141
  adhesive, 141
  classification, 134–135
  crater, 134
  diamond tools, 139–142
  diffusion, 141
  effects, 162–163
  evolution, 136
  flank, 134, 137
  material–dependence, 138
  progressive, 134–135
  recoating, 133
  tool, 133, 139
wheel, 234
  conditioning, 235
  diameter, 257
  dressing, 234
  speed, 257
  topography, 234
  wear, 234
  width, 257
workpiece, 162
  materials, 162, 173
  speed, 257
  surface, 242, 243
WYKO interferometric microscope, 277, 278

X–Y plane, 275
XYZ gantry configuration, 315–316

Z–Buffer scanning algorithm, 206
Zygo 3D surface profiler, 312

Printed in the United Kingdom
by Lightning Source UK Ltd.
135940UK00007B/64/P